REGIONAL DEVELOPMENT MODELING:
THEORY AND PRACTICE

Studies in Regional Science and Urban Economics

Series Editors

ÅKE E. ANDERSSON
WALTER ISARD

Volume 8

NORTH-HOLLAND PUBLISHING COMPANY – AMSTERDAM • NEW YORK • OXFORD

Regional Development Modeling: Theory and Practice

Editors

MURAT ALBEGOV
ÅKE E. ANDERSSON
FOLKE SNICKARS

Regional Development Task
International Institute for
Applied Systems Analysis
Laxenburg, Austria

1982

NORTH-HOLLAND PUBLISHING COMPANY – AMSTERDAM • NEW YORK • OXFORD

ISBN: 0 444 86473 3

Publisher:
NORTH-HOLLAND PUBLISHING COMPANY
AMSTERDAM • NEW YORK • OXFORD

Sole distributors for the U.S.A. and Canada:
ELSEVIER SCIENCE PUBLISHING COMPANY, INC.
52 VANDERBILT AVENUE
NEW YORK N.Y. 10017

Library of Congress Cataloging in Publication Data
Main entry under title:

Regional development modeling, theory and
    practice.

    (Studies in regional science and urban
economics ; v. 8)
    "Papers presented at a conference on
theoretical and practical aspects of regional
development modeling, held at the International
Institute for Applied Systems Analysis (IIASA),
Laxenburg, Austria on March 19-21, 1980, ..."
"organized by the Regional Development Task
at IIASA"--Pref.
    Includes indexes.
    1. Regional planning--Congresses. 2. Urban
economics--Congresses. 3. System analysis--
Congresses. 4. Economic development--
Simulation methods--Congresses. I. Albegov,
Murat Mikhaĭlovich. II. Andersson, Åke E.,
1936-   . III. Snickars, Folke, 1944-   .
IV. International Institute for Applied Systems
Analysis. Regional Development Task.
V. Series.
HT391.R2985 1982   338.9'00724   82-10615
ISBN 0-444-86473-3

1-12-85

PRINTED IN THE NETHERLANDS

INTRODUCTION TO THE SERIES

Regional Science and Urban Economics are two interrelated fields of research that have developed very rapidly in the last three decades. The main theoretical foundation of these fields comes from economics but in recent years the interdisciplinary character has become more pronounced. The editors desire to have the interdisciplinary character of regional sciences as well as the development of spatial aspects of theoretical economics fully reflected in this book series. Material presented in this book series will fall in three different groups:

- interdisciplinary textbooks at the advanced level,
- monographs reflecting theoretical or applied work in spatial analysis,
- proceedings reflecting advancement of the frontiers of regional science and urban economics.

In order to ensure homogeneity in this interdisciplinary field, books published in this series will:

- be theoretically oriented, i.e. analyse problems with a large degree of generality,
- employ formal methods from mathematics, econometrics, operations research and related fields, and
- focus on immediate or potential uses for regional and urban forecasting, planning and policy.

Åke E. Andersson
Walter Isard

PREFACE

This volume contains a collection of papers presented at a conference on "Theoretical and Practical Aspects of Regional Development Modeling", held at the International Institute for Applied Systems Analysis (IIASA), Laxenburg, Austria on March 19–21, 1980. The conference was organized by the Regional Development Task at IIASA as part of its work on comparing and synthesizing recent research on applied regional development modeling. Its aim was to bring together theoretical and practical results of applied regional development modeling in both market and planned economies, in an attempt to summarize the state of the subject at the beginning of the 1980s.

Some 50 papers were presented at the conference; of these, 26 were selected for this volume. Although the papers were brought together at the conference, this book should not be seen primarily as a conference proceedings; rather, the contributions have been selected to strike a balance between the theory and the applications of regional systems analysis.

The book has been divided into seven parts. The first is basically an introduction, while the second and third contain overviews of current modeling practice in market and planned economies. In the next two parts the focus shifts to the theoretical problems encountered in structural and multiobjective analysis of regional systems. The final two sections contain examples of regional development models currently ready for use or in operation and analyze the success of these models in clarifying regional planning and policy problems.

Last but not least we must thank the authors contributing both to the conference and to this book for their valuable material and comments, and the staff of the Regional Development Task at IIASA for their skillful and efficient assistance, without which this volume would not have been possible.

The Editors

CONTENTS

PART I

INTRODUCTION

Regional Development Modeling: Theory and Practice
M. Albegov, A.E. Andersson and F. Snickars (editors)
North-Holland Publishing Company
© IIASA, 1982

Chapter 1

REGIONAL DEVELOPMENT MODELING – THEORY AND PRACTICE

Folke Snickars, Å. E. Andersson,* and M. Albegov**
International Institute for Applied Systems Analysis, Laxenburg (Austria)

1.1. Basic philosophy of regional development modeling

The purpose of this book is to present the current theory and practice of regional development modeling in industrialized nations with different resource endowments, economic structures, and political systems. This will be done by giving examples of approaches and applications in various countries, rather than providing a comprehensive overview. The book is primarily intended to give a broad coverage of approaches and is not specifically oriented toward the theoretical aspects. A very important aim is to provide a basis for comparing the regional problems tackled and the quantitative methods employed in market and planned economies.

An attempt will be made to demonstrate how modeling can be and is being used as a tool for solving regional problems in the framework of regional development planning. Regional development planning is defined here as the process of dealing with the long-term overall structural economic problems of regions within a nation. Thus, both the growth strategies for developing regions and the problems of economically well-developed regions facing rapid structural change will be highlighted. The emphasis is on resource distribution between regions rather than on intraregional allocation problems. Although general, these statements imply that regional development modeling is seen as taking a constructive role in regional and national economic development.

It should already be apparent that regional development planning has different connotations in different countries and also at different regional levels. This may be due in part to national variations in the pattern of regional development (balanced regional growth in some countries, interregional imbalances in others). It may also be a

---

* Currently at the Department of Economics, University of Umeå, Umeå, Sweden.
** Currently at the Central Institute for Economics and Mathematics, USSR Academy of Sciences, Moscow, USSR.

reflection of the role of planning and particularly the role of regional development planning in the different countries.

In countries that have pure market economies, i.e., in which planning is viewed primarily as a tool for correcting market imperfections, regional development planning plays a minor role. Regional development is not usually regarded as external to the market and therefore little attention is given to policies with a regional dimension. Such is the case, for instance, in the USA, where regional planning is quite a recent phenomenon; this may also be seen from Chapters 6 and 26 of this volume.

In mixed economies of the western-European type, regional development planning has traditionally been aimed at removing interregional discrepancies in income and employment opportunities. The regional planning framework has been welfare-oriented rather than efficiency-oriented. Even in the analysis of the long-term consequences for Sweden of phasing out nuclear power, the regional effects are evaluated mainly by employment indicators (see Chapter 25).

In planned economies the regional dimension has always been used as a means of increasing overall economic efficiency. Whereas support in the form of increased investment may be given to chronically depressed regions in mixed-type market economies, in planned economies investment is directed primarily to regions with the potential for expansion. Examples of this are provided in Chapters 22 and 24, which relate to the development of territorial production complexes and other industrial expansion projects in Siberia.

Of course, in market economies the efficiency-oriented regional distribution of investment capital is governed by the regional differentials in capital returns. This will, in principle, give rise to an economically efficient regional distribution of labor (although this is questioned, for example, in the so-called center–periphery theory). From this perspective, it is natural that optimal territorial planning should also be used in planned economies to achieve such interregional efficiency. In both types of system, therefore, interregional flows of labor and capital necessary and sufficient to promote growth would tend to emerge. The welfare-oriented role of regional development planning characteristic of mixed economies serves to keep these processes of structural change within socially acceptable limits. Chapter 27 gives a provocative view of the success of Dutch regional policies in this respect over the last decade.

Although this book is not intended to be a critical review of regional planning systems, we believe that it is possible to assess the different approaches to regional development modeling and the models characteristic of different countries using the background outlined above.

Regional development planning undeniably requires a multidimensional approach. Regardless of the geographical level or the economic or political structure of the region under study, the aim is to analyze and influence different components of economic and social processes simultaneously. The complexity of regional development planning stems from the fact that both spatial and temporal interdependencies are present and must be included. Comprehensiveness calls for an explicit analysis of uncertainties and long-term options, since many regional processes are inert and contain temporal indivisibilities. Chapter 11 points to the need to develop new criteria for evaluating such problems and to set new standards for freedom of action and robustness. This type of approach is often more appropriate than attempting to minimize some cost function or maximize some efficiency function.

The above discussion suggests that it is useful to assess the role of systems-analytic approaches in regional development planning. In what directions have the theories and models of regional development planning evolved in market and planned economies? Are current planning problems being tackled with the same type of models used in the last two decades, using methods which were formulated for very different circumstances? Are problems similar to those that occurred in earlier decades now being analyzed with different tools? Are the current quantitative modeling techniques quite general or more specialized, i.e., is the same type of methodology used in countries with different economic and political structures?

In this introductory chapter, these questions will be examined briefly by means of an historical summary and a comparison of some of the approaches mentioned in other chapters. As part of this background, the attempts at the International Institute for Applied Systems Analysis (IIASA) to provide a framework for applied regional development modeling will be described. This will be further illustrated by a summary of the IIASA case studies, which were performed in a variety of economic, social, and institutional settings. This chapter concludes with a summary of the regional issues that this work suggests should be examined using systems analysis in the future.

1.2. Regional development analysis in the last two decades

There are a number of different ways in which the evolution of new theories and models for regional development planning over the last two decades may be examined. Here the framework given in Table 1.1 will be used, though this categorization is in no way unique.

Some important developments in interregional or multiregional explanatory modeling will first be outlined. Following the original work of Leontief (1951) on national input–output modeling, Isard (1951) formulated the general interregional input–output model. This piece of work encouraged the development of techniques to overcome the computational difficulties inherent in these input–output methods; see, for example, Moses (1955) and Leontief and Strout (1963). However, although this model was formulated in the 1950s it has still not been widely applied.

A more direct reformulation of the Leontief model as a regional input–output model has had greater success. It has been applied in various countries at all geographical levels, often complemented by independent econometric estimates of import, export, and consumption functions.

Spatial general equilibrium models are scarce, owing to the lack of theories and the dearth of statistical information to support or reject them. Lefeber (1958) formulated a model of this type, though the brief treatment of the transportation sector has led to problems in implementation. Location theoretical analysis performed by Koopmans

Table 1.1. Classification of regional development modeling research.

Spatial scope	Type of model	
	Explanatory and predictive	Planning and policy

Interregional or multiregional	Input/output
	Spatial general equilibrium
	Central place
	Migration
		Multiregional planning
		Economic growth
		Transport and/or
		    investment cost-
		    minimization

Regional	Input/output Basic/nonbasic Growth pole	Mathematical programming Spatial competition
Intraregional	Urban land equilibrium Transportation Spatial interaction Lowry-inspired	Transportation/land-use optimization Cost–benefit Accessibility

and Beckmann (1957), as well as classical studies by Hotelling (1929), suggest different reasons for the fact that market equilibria may not, even in theory, be sustained in a multiregional system. Koopmans and Beckmann (1957) single out the indivisibility of certain factors as one possible explanation, while Hotelling (1929) stresses the small number of actors. Although both of these results have been questioned in more recent research, applied general equilibrium models of interregional development are still uncommon. However, many recent econometric regional models rely implicitly on equilibrium concepts; see Chapter 6.

There has been a marked development in the use of equilibrium models for the study of intraregional land-use patterns during the last decade. These models, often termed "new urban economics models", stem from the work of Alonso (1964) and Muth (1969) on the functioning of the urban residential market. They have reached the application stage in many urban areas, especially in combination with urban transportation models, which are also quite often of the equilibrium type; see, for example, Florian (1976) and Ben Akiva and Lerman (1977).

At regional and intraregional levels there has also been increased interest in comprehensive land-use and transportation models of the nonequilibrium or partial equilibrium type. The first large-scale models of urban and regional processes fall into this category. Some of them, such as the basic/nonbasic model developed by Lowry (1964), were exceedingly successful and have led to the development of a new class of models, now termed "spatial interaction models". Wilson (1970) is the most prominent exponent of this modeling approach. Other models, for example, some very large-scale transportation models developed in the USA, were not successful, and did not reach the application stage. Complex large-scale models, such as that developed by Forrester (1969), were subjected to heavy criticism, for example, by Lee (1973). The main criticisms were related to the "over-optimistic" view of the usefulness of computer techniques in regional planning (Putman, 1973). Nevertheless, spatial interaction models have provided planners with computer-based tools for quantitative modeling that did not exist in the early 1960s, although the lack of theory at the core of these models may still be questioned.

There has been very slow development in the modeling of intraregional production patterns. The allocation of urban land is generally considered to be a question of dividing land between residential and transport use, presupposing the fact that production always outbids other activities in central locations.

Economic geographers and location theorists have built on the work of Christaller (1966) and Lösch (1954) in trying to validate and apply "central-place" concepts in regional development planning. Rather than working with zonal subdivisions, these approaches presuppose that regional development occurs in a network of interlinked villages, towns, and cities, which form a hierarchy. Chapter 15 outlines a dynamic version of this theory. The concept is to some extent related to the idea of growth poles, which was developed in relation to the French territorial planning system by Perroux (1955), although growth poles generally refer to a nonspatial system. However, central-place concepts, although theoretically pleasing, have not played a very important role in the development of quantitative models. In spite of this, the conceptual importance of central-place and growth-pole models should not be underestimated. These ideas have been used to formulate policies as well as to provide a framework for policy evaluation in several European countries.

There is no clear distinction between "explanatory" regional modeling and "planning" regional modeling. This is particularly true of the mathematical descriptions of different approaches. In many cases the same conditions may be derived from both programming and simulation models.

The primal–dual relationships of linear programming provide a framework within which the results of an optimization approach can be interpreted as a market equilibrium. Alternatively, the primal problem expressed in terms of quantity allocation could be used to optimize resource allocation in a planned economy; see also Chapter 9.

If, then, this similarity in model structure reflected a similarity in actual economic processes, planned and market economies would be truly dual, both converging to the same stationary state. However, in general the same mathematical theories have not been used in different economic systems. Programming models are assumed by western economic planners to be applicable only (or primarily) to planned economies, while eastern economic planners discard the shadow-price system as unrealistic.

Multiregional planning models have been developed by Tinbergen (1967) and also by Soviet workers; see, for example, Aganbegyan et al. (1972). The Tinbergen system was based on a hierarchical framework with sectoral, interregional, and local levels. It was assumed that regional development planning should also take place on these three levels and that planning models appropriate for each level should be used consecutively.

Although the Tinbergen models were set in a comprehensive framework, they were still based heavily on the minimization of cost by linear models. They were followed by a number of other planning models with transport and/or investment cost-minimization as the basic goal. The models developed by Mennes et al. (1969) and Carrillo-Arronte (1970) were both of this type.

In the USSR rapid development of regional modeling began in the early 1960s in an attempt to aid the development of existing and recently established industrial regions. In the first stage, intersectoral relations for the country as a whole and for separate regions were modeled, and national–regional models were developed for the main sectors.

From the beginning, the main emphasis was on national intersectoral analysis (Ephimov and Berry, 1965) but some work was also devoted to investigating regional problems (Kossov, 1973). At the same time experience in modeling sectoral growth and location led to the generalization of existing methods for solving these problems by optimization techniques.

These independent analyses of national, regional, and sectoral problems laid the foundations for a system of national–regional intersectoral models. This work was started in Novosibirsk and Moscow; see Aganbegyan et al. (1972) and Danilov-Danilyan and Zavelsky (1975). Rather elaborate systems were developed to coordinate important features of the national and regional plans (such as final consumption, volume of production, and interregional distribution of capital and labor). These schemes proved to be very difficult to implement on the computer; one example was the general Isard (1951) model. For this reason only one version of the Granberg model was completed. Research work in this field carried out in Moscow is described in Baranov (1969).

The simplified set of interregional–intersectoral models developed by Albegov (1970) was more successful. In this system the interregional distribution of labor is given exogenously and transport is described rather roughly, but at least the model allows the problems to be analyzed practically.

In the early 1970s, interest shifted from sectoral to multisectoral analysis and to the analysis of territorial production complexes (TPCs). In another trend, interest moved away from regional economic analysis using input–output techniques to the development of regional model systems.

In the first of these developments, Albegov and Solodilov (1970) analyzed a multisectoral system, taking into account the nonlinear dependence of resource costs and the effects of agglomeration. Multisectoral analyses on a local level were carried out with the help of multistage models of TPCs. These analyses included not only production, but also population, settlement and service, and environmental systems. This has resulted in well-balanced development of the main sectors of regional economies and auxiliary subsystems; see Bandman (1980).

In later developments of this approach in both East and West, the hierarchical structure has been relaxed and intersectoral relationships have been introduced at the interregional stage (see Paelinck and Nijkamp, 1976, and Granberg, 1978). Current research in market, mixed, and planned economies involves the introduction of multiple objectives into these multiregional models.

There are relatively few linear models of this type at the regional level. When used, they often take the form of cost–benefit models; the cost measures are extended to include all urban gains and losses that can be given monetary values and specified linearly (see Ben Shahar et al., 1969). The housing market models based on the linear programming model of Herbert and Stevens (1960) are good examples of this approach.

Another theoretical advance in regional planning modeling was made by Rahman (1963), in a model which was designed to examine whether there was a conflict between national economic growth and regional equality. The main result was that a technologically less efficient region could compensate for inefficiency by a higher savings ratio, so that for long-run national growth it would be profitable to invest in lagging regions as well as advanced areas. This approach has been extended by several Japanese researchers, for instance Outsuki (1971), to show that the Rahman results are true even in a more general regional economic setting. This class of growth models has recently been analyzed by Fujita (1978), with various production functions.

Although these growth models have not been used in regional development practice, their importance for regional policy-making should not be underestimated. They differ from other models in their explicit treatment of time and show that this may lead to results that are quite different from those of static models.

Models of spatial competition are similar to regional growth models in their treatment of the dynamic aspects of regional development processes. They provide a possible explanation for the formation and development of service centers and are therefore useful in planning and policy-making. Recent analyses by Gannon (1977) and Webber (1977) use techniques that are different from those employed in Hotelling's (1929) original work.

In these more modern approaches the only element that has been changed is the spatial demand elasticity. The fact that there is a distance factor in spatial demand patterns is usually reflected by introducing accessibility indicators into regional models. Such indicators were introduced on an ad hoc basis in classical economic geography but have been adopted over the past decade by research workers such as Hägerstrand (1970). More recent research has pursued axiomatic theories of accessibility measures; see Weibull (1976) and Smith (1978). This research indicates the need to devote more attention to the formulation of goal and interdependency indicators in regional planning models. Normative land-use transportation models using such indicators have been developed in some mixed-economy settings; see, for example, Sharpe et al. (1975) and Lundqvist (1977). Work on the development of econometric intraregional models has also just been initiated in the USSR; a dynamic model of economic growth in the Ukrainian Republic has already been constructed. A survey of multiregional economic models carried out by the Regional Development Task at the International Institute for Applied Systems Analysis (IIASA) during 1980 and 1981 indicates that there has been a considerable increase in the number of econometric multiregional and regional models developed in the USA and western Europe. This new trend first became apparent in the late 1970s.

A number of recent developments in regional modeling have now been considered briefly. The main conclusion is that there have been no major changes in the theoretical bases for this type of modeling. Therefore, with the notable exceptions mentioned above, it must be admitted that current problems are being tackled with the help of old theories. There are no significant differences between the models used in practice in the various countries, whether in East or West.

On the other hand, it must be acknowledged that substantial changes have occurred in the real-world problems facing decision-makers at both the regional and national levels. As a result of economic development, regions have become more vulnerable to external changes beyond regional or even national control. Problems of resource shortage have come to the fore. Pressure groups have made stronger claims for the conservation of resources and the environment, especially in their own areas.

Regional development planning tools, both theories and models, should be adapted to meet this changing and complex situation. The role of computers must be properly assessed; modern theories and models of regional development and its planning should be synchronized with advances in modern computer technology. The drawbacks of earlier theories and models from a systems-analytic perspective should be eliminated in the newer versions; the systems approach may then make it possible to achieve a deeper understanding of current economic, social, and political processes. This view calls for the revival of integrated larger-scale models, a view that is also put forward in Chapter 2. Systems-analytic models of this type should examine the external uncertainties as well as the multiple-objective aspect of regional development planning in the future.

These comments on the historical trends in regional development modeling provide a background against which to view the current state of research in this field. In the next section the various approaches are described using the same general classification by subject that is followed in the rest of this book.

1.3. Current trends in regional development modeling

In the previous section it was pointed out that the theoretical basis of regional development modeling has remained essentially the same during the last two decades, despite a change in the fundamental planning issues that these models are intended to handle. The following are beginning to emerge as important factors in regional planning:

1. Fundamental uncertainties about economic development in the medium and long term have increased.
2. Economic integration has increased at both the local and international levels, implying that regions are increasingly vulnerable to external economic processes.
3. The shortage of fuels and other raw materials has led to a shift in the emphasis of the analyses of critical economic processes toward problems of technological change.

The three points outlined above do not concern regional processes as such, but represent examples of general economic phenomena with a regional impact that may substantially affect future regional development. Shifts in economic development may imply new regional growth patterns at both national and local levels. Economic stagnation or rapid structural change in the industrial sectors, even in highly industrialized countries, are other characteristics of this new order.

1.3.1. Overview of model systems

A substantial part of this book (five chapters) is devoted to overviews of current theory and practice in regional development modeling. Since the contributions come from the USA, western Europe, and the USSR, a balanced picture of trends in a variety of institutional settings should emerge.

It should be noted that there is quite a marked difference in emphasis in regional modeling between market and planned economies. As is evident, for instance, from Chapters 6 and 8, North American research has been concentrated on urban modeling, with a recent shift in interest toward multiregional analyses. The situation is reversed in the USSR. Chapter 4 clearly shows the strong emphasis of Soviet planning on multiregional analyses or regional analyses of nonurbanized regions; the recent shift is toward urban problems. This change in regional level is reflected in an increased interest in econometric analyses and the use of simulation models, which are already in common use for urban modeling in North America.

Chapter 5 contains information about multiregional modeling from a western-European or, rather, a French perspective. It provides an example of another trend in contemporary regional economic modeling, the linkage of real and financial economic models. This is a natural response to the emergence of the problem of economic integration mentioned above. Under conditions of high inflation it is very important to study the interdependence between price systems and technologies.

Another feature of current multiregional modeling as presented in these overviews is the interest in linking models of regional subsystems. The increased interest in larger-scale models appears as a desire for applied multiregional, multisectoral models with a considerable degree of sectoral as well as regional disaggregation. Chapters 3, 5, and 6 illustrate the breakthrough that has been made in the development of applied large-scale economic models of linked subsystems.

Another example of this interest in linked model systems is the growing attention given to the economic aspects of migration processes and labor-supply determinants. Chapter 7 contains some new ideas about the complex problems of modeling the economic factors affecting migration processes. This chapter also illustrates the trend toward a clearer treatment of uncertainty and multiple-objective problems, a topic so important in this field that it will be discussed at greater length in another section.

Chapter 8 draws attention to some important practical problems of transferability between theoretical models, applied models, and the computer software necessary for such models. Applied regional modeling should strike a balance between generality and specificity, not only in theory but also as a matter of computing practice. The compatibility of theoretical and practical versions of regional economic models is crucial for the credibility of modeling exercises and should therefore be taken more seriously by systems analysts.

1.3.2. Multiobjective analyses

Although it has long been argued that regional planning is a field characterized by multiple and conflicting objectives, applications of models using multiobjective techniques are quite rare. This is not to say that there has been no development in the theory — it is simply an acceptance of the fact that the theoretical analyses have not yet been transformed into real-world applications.

The four chapters on multiobjective methods included in this book represent the latest theoretical developments and practical applications in this field. Chapter 9 gives an overview of the multiobjective techniques used in multilevel, multiregional planning. The notion of "compromise solutions" is discussed in a hierarchical setting, using an algorithm proposed earlier to solve decomposable models by direct quantity distribution rather than by price adjustments.

Chapters 10 and 11 give examples of applications-oriented mathematical programming models used to solve problems with conflicting objectives. Chapter 10 considers the problem of providing quantitative specifications for objectives in comprehensive land-use planning, and the question of how indivisible quantities should be modeled is also discussed. Chapter 11 examines the treatment of time in multiobjective analysis, and illustrates various approaches with examples from physical (and regional economic) planning. It is shown that robustness criteria are exceedingly useful in situations of genuine (or static) uncertainty, i.e., in long-term planning.

Chapter 12 shows how modern optimization techniques can be used in applied regional modeling to achieve compromise solutions. The method described builds on a minimax paradigm that reduces the dimensionality of the full optimization problem at the cost of greater mathematical complexity. The most important property of this paradigm is that it actually pays in solution efficiency to carry out this transformation.

The methods used for multiple-objective decision-making are many and various. They range from game theory to multiattribute utility analysis and incorporate elements of both the social and the economic sciences. Chapters 9–12 cover only a limited range of approaches at the forefront of multiobjective methodological research. As already pointed out (for instance, by Rietveld, 1980), the field is characterized not by a lack of alternative approaches but by the need to tackle larger-scale problems. The solutions of multiobjective models are by no means trivial and, in fact, the difficulty of solution is responsible for most of the current limits on the practical usefulness of the models.

1.3.3. Structural analysis of regional development

The changes in real-world problems have led not only to a shift of emphasis in the analysis of objectives in regional development planning but also to a new analysis of the internal structure of the regional economies themselves. Since the interdependence between nations and between regions is increasing, the nature of the linkages should be analyzed in detail. The uncertainty inherent in economic development suggests that analyses of profit distributions, business-cycle variations, and investment behavior should be made. The dynamic behavior of labor markets and other regional economic subsystems also holds more interest than heretofore. Although this book contains no chapter concerned specifically with technological change, the structural analysis of regional economic systems is discussed from various viewpoints in Chapters 13–18.

Chapter 13 describes an original way of analyzing industrial structural change, making good use of existing Swedish data on industrial development. The approach resembles the migration studies so typical of modern demography, viewing the production unit as an entity migrating through a space of gross profits. Previous models of regional industrial-development planning have used too little information of this type to be economically realistic or meaningful.

Superficial treatment of real-world complexity is also a problem in normative labor-market models. Although there have been major advances in theoretical labor-market studies (see, for example, Lippman and McCall, 1976), few models deal with regional problems. Inclusion of the regional aspects is definitely warranted in view of the private and public social costs and benefits associated with commuting, migrating, changing occupation, and unemployment. Chapter 14 uses a variant of linear activity analysis to describe a regional labor market including various friction costs.

Chapter 15 represents a recent development in the economic–geographic field: consideration of the dynamics of central-place systems. This chapter, although concerned mainly with service provision, provides an example of the analysis of regional economic change in a network of distinct nodes. Even if the idea that geographical space is either discrete or continuous has existed as long as regional analysis, there is still no general theory that encompasses both aspects. Chapter 16 gives an example of recent advances in ideas about economic processes in continuous space. This type of work is presented in more detail in Isard and Liossatos (1978) and Beckmann and Puu (1981).

The next two chapters are concerned with examples of functional or structural analysis in the USSR. Chapter 17 has its western counterparts in the work of Isard (1951), Isard et al. (1969), and Stone and Brown (1965), although the last does not deal explicitly with the regional aspects. This chapter makes it possible to compare systems for social and financial accounting in different institutional settings. Chapter 18 (by the late A. M. Alekseyev) represents yet another facet of regional development modeling: the evaluation of the feasibility, adequacy, and timing of practical regional development projects. The management problem is more complex than in western models of the same type because of the more comprehensive geographical and socio-economic scope of the analysis.

1.3.4. Regional development models – some recent examples

The main difference between the chapters reviewed in this section and those summarized above lies in the comprehensiveness of the analysis. The chapters discussed in this section contain some examples of complete regional economic planning and policy models that have recently been developed in IIASA member countries, in some cases in collaboration with IIASA (see also Section 1.4 below).

Chapter 19 describes a normative regional planning model developed for an agricultural region in central Poland; the dynamics of the production system and its service support systems are emphasized. Chapter 20 reports on current work on a decision support system for the development of an agricultural region in Bulgaria. Both models are consistent with the existing national planning systems. Although both systems concern agricultural regions in eastern Europe, they are quite different. Apparently the local economic and social differences are more important than the similarities in organizational setting and political structure.

The Bulgarian model attaches great importance to the analysis of transportation at an intraregional level. However, in Chapter 21 transportation projects in Japan are analyzed at a national level. This chapter has several interesting features: it shows how available theory and methods can be used in a simplified, but workable, model system, and it also illustrates how comprehensive simulation models may be used for policy evaluation in a situation where the number of options is limited.

Chapter 22 summarizes the regional input–output models used in the USSR. It shows how one particular large-scale input–output model can be used to assess the nationwide impacts of alternative patterns of growth in Siberia. It is not intended to provide unique optimal production patterns but rather to illustrate the importance of Siberia in the Soviet economy over the medium term. Thus, the aim is very similar to that of the Japanese transport study: policy evaluation.

1.3.5. Applications of regional systems analysis

In Chapters 23–27, it is the application rather than the method that is of central interest. However, having said this, Chapter 23 is an exception, since it contains both a new theory and its application. It has been included as an example of ex post regional economic policy evaluation in that it attempts to deduce the effects of policies from statistical data. With its discussion of statistical motivations the chapter may be usefully compared with the analysis of migration flows using logit models in Chapter 7.

Chapter 24 contains a summary of methods and, primarily, applications of territorial production complex (TPC) analysis, and provides a good illustration of the range of current Soviet regional economic planning models.

Chapter 25 considers the use of systems analysis in the organization and modeling involved in resolving an important national and regional policy question: should Sweden phase out its nuclear power plants or not? In 1979 a Swedish government commission on the nuclear power question made successful use of systems-analytic methods in both specialist studies and organizational matters to resolve the problem under heavy time constraints.

Chapters 26 and 27 are the most policy-oriented in this volume. They both consider the effects of regional policies on market and mixed economies, but from somewhat different viewpoints. Chapter 26 argues in favor of discarding regional policies in the

USA because they distort the market, whereas Chapter 27 criticizes the inability of Dutch regional policies to influence regional development. The views presented in the two chapters are therefore diametrically opposed. While Chapter 27 stresses the need to coordinate different types of policies to attain certain levels of welfare, Chapter 26 presupposes that policies actually have some effect but that this effect only reduces the welfare levels. Although their conclusions differ, both chapters emphasize the need to assess the effect of regional policies on the welfare of the population. This demand is relevant whatever the type of economy.

In Section 1.2, some important developments in regional modeling over the last two decades were classified as shown in Table 1.1. It is interesting to see where the various chapters in this book fit into the general scheme of Table 1.1, with the proviso that "overview" chapters cannot be included in this classification. Furthermore, since this book is mainly concerned with the regional and multiregional aspects of regional development modeling, the distribution will necessarily be somewhat biased.

The four chapters on multiobjective analysis fall into the general category of programming models. Of the eight chapters having a primarily analytical purpose, two or three also consider policy aspects (Chapters 14, 18, and in part Chapter 7). The rest have an explanatory purpose, but do not correspond with the individual model types mentioned in Table 1.1. One new element here is the use of probabilistic models such as the logit model.

Chapters 21, 23, and 25 all have an explanatory or predictive aim. In these chapters there is also a tendency to include elements of different theories in a way that was not common in the early 1970s.

Thus it may be concluded that regional development modeling is moving toward more comprehensive models, incorporating elements from various different economic, geographic, and demographic theories, and there is also increasing interest in general equilibrium approaches. Both of these trends may owe part of their success in practical applications (and hence their popularity) to rapid developments in computer technology.

1.4. Regional systems analysis — an IIASA approach

The International Institute for Applied Systems Analysis (IIASA) is currently engaged in case studies of regional development in agricultural regions of Bulgaria and Poland and metropolitan regions of Sweden and Italy. This introductory chapter will conclude with a brief description of the ideas behind these studies, as well as some of the methods used.

The purpose of the IIASA research is to create integrated systems for regional and sectoral development and apply them to the regions under study. The basic approach is to analyze the development problems of the four regions within the limits set by national and regional policies. The types of problem analyzed include investments in sectors of production, use of energy and other primary resources, interregional location, population, and transportation policies, planning of new and existing towns, and the construction of industrial and agricultural production complexes.

The broad range of these case studies has made it necessary to find some basic principles of regional systems analysis that can cover different political systems as well as regions at various levels of aggregation; these principles are outlined below.

It is evident from the work on regional development carried out at IIASA that regional policy-makers must be provided with better tools to enable them to cope with the essentially dynamic, uncertain, and interdependent factors regulating long-term economic and technological growth and structural change at the regional level. This can only be done through developments in the basic methodology of regional development analysis.

1.4.1. A definition of applied regional systems analysis

To qualify as applied regional systems analysis, a study must have a clear emphasis on policy. At IIASA this has resulted in detailed case studies of actual regions, the researchers cooperating with the policy-makers and planners of the region and country under study.

Regional systems analysis must also consider the long-term problems of policy-making; this implies more concern with strategic matters than with tactical and operational policy issues. This has also meant that the economic analysis central to these case studies must be long term rather than short term. At the regional level the modeler is then forced to study not only growth problems but also problems of secular decline – a rare phenomenon in national economies but not uncommon at disaggregate levels. It is also obvious that it is necessary to study structural change in regional long-term case studies; the work very often involves generating policies to change the structure of production, real capital, and the labor force.

There are necessarily some well-defined as well as ill-defined uncertainties involved in long-term policy analysis. It is, for instance, difficult to predict the size of a population, its propensity to work, the availability of local resources, and other basic variables. It is even more difficult to predict future demand structures in the region and its surrounding area, and it is almost impossible to make reasonable quantitative predictions about the technology and the political values that will be employed two or three decades into the future. For this reason scenario creation, structural sensitivity analysis, and similar approaches must be used to provide insights into the consequences of the fundamental uncertainties of long-term regional policy-making.

Applied regional systems analysis means that economic, ecological, technological, and demographic systems must be related to each other in an essentially dynamic and spatial systems analysis. This implies that a large number of variables have to be linked, which requires large-scale model-building, and hence a very sensitive tradeoff between realism, simplicity, and ease of parameter estimation.

1.4.2. Purposes of regional development modeling

Applied regional systems analysis cannot be used simply to generate straightforward quantitative results. The emphasis on long-term policy problems makes it impossible to obtain any policy recommendations that are sufficiently definite to warrant immediate action. A more realistic goal would be to use the models to generate qualitative policy recommendations in the form of general guidelines.

In some cases even this aim is unrealistic; however, in this situation applied regional systems analysis can always be used to achieve a better understanding of long-term regional policy problems and their interactions. It is often necessary in regional systems analysis to generate a large number of projections demonstrating the consequences of different courses of action and different patterns of development, and these projections can be very useful to regional planners.

The creation of these scenarios is an important part of the planning process. The planning scenarios can be developed through a purely verbal process, sometimes aided by map sketching, as is often done in physical planning. However, experience has shown that this type of procedure is only viable if the number of planning variables is small. Computer-assisted planning procedures become necessary when the dimension of the problem and the degree of disaggregation increase.

The economic structure of a region can be seen as the first dimension of the planning process; the spatial structure can be viewed as a second dimension, and the temporal sequencing of activities as the third. If it is assumed that the economic structure can be represented by 30 production sectors, the spatial structure by 10 subregions, and the temporal structure by 3 time periods, a model including all interdependencies would have to consider 900 variables. It would be very difficult to construct a plan for such a system without the aid of a formal computer model.

In the early theory of planning, it was often assumed that a large system of this type must, by necessity, have one and only one goal function, which should be maximized, subject to certain technological constraints. Adopting the same planning philosophy, a giant model could be constructed for the whole system, and the goal function could be maximized subject to permissible variations in the variables. These variables may be viewed as the instruments of planning.

As discussed in the review of current research on multiobjective decision analysis, the a priori selection of a single goal function is a difficult and dangerous task. A more realistic approach would be to suggest a number of possible goal functions, and then to study the range of solutions obtained. The problems of aggregating conflicting objectives could thus be avoided, but the approach would still be subject to the difficulties inherent in solving large-scale systems. Unfortunately the numerical capacity of any optimization model containing spatial, sectoral, and temporal dimensions is still very limited.

Yet another difficulty is evident. Most planning models can be given relatively concrete and statistically reasonable technological constraints: this is certainly true for constraints on the use of resources. Most economic planning models contain reasonably accurate constraints on the use of primary resources, labor, and other factors affecting production. The sectoral interdependencies can also be specified with some degree of precision. It is, however, far more difficult to specify the behavioral constraints that regulate the activities of households (consumers) and other decision-makers in the economic system. It is thus probable that any optimization model used for economic planning will not include very accurate descriptions of human behavior. This implies that the behavior of decision-makers and planners and their mutual interactions should be modeled more closely, using optimization models to guide the process.

1.4.3. The choice of approach

Regional systems analysis is an application of systems analysis to regional policy-making, and must include consideration of both time and space. The spatial element can be handled in two ways: the problem can be analyzed in continuous space as proposed by Beckmann (1952), Puu (1977), Isard and Liossatos (1978), Mills (1972), and others, or the total space (for example, the nation) can be subdivided into a set of discrete regions. Work at IIASA has generally been based on the discrete, regional approach to spatial analysis. In most cases, it has also been decided to handle time as a set of discrete periods.

In a purely theoretical general equilibrium analysis of the kind proposed by Debreu (1959), each decision variable, for example, the quantity to be produced, is associated with a particular time, decision-maker, type of commodity, and region. It can be shown that, in a static situation, there will always be a solution to an interregional equilibrium problem of this type, even with an infinitely large number of consumers and producers. However, this approach requires a large number of simplifying and not very realistic assumptions about the convexity of preference and production sets, assumptions that are not normally valid in the real world.

The extension of this approach to realistic transportation–communication technologies and situations of growth and development has never been possible. To cope with policy-making problems in dynamic, interdependent regional production and consumption systems of this type, considerable simplification is necessary. Simplification through decomposition and structuring of regional development problems has been proposed by many analysts for both market and planned economies. Isard, Leontief, Aganbegyan, and Granberg all suggest simplification through linearization of technologies for production, transportation, and consumption of commodities. Using linearization it is possible to solve problems with several hundred regions and a large number of production sectors and different household categories.

Two main criticisms can be made against the linearized approach proposed by Leontief and others. The first concerns economies of scale in the production and transportation of commodities. Economies of scale in production lead to nonlinearities, which, however, for sufficiently large regions are of limited importance. Leontief and other global modelers can thus claim that nonlinearities are not important in their work. However, at a lower level of aggregation this problem is more serious. The regions that make up a nation are normally rather small and economies of scale cannot be disregarded in national–regional models.

The second criticism is that, in general, transportation and communication cannot be linearized but can be nonlinear at any level of aggregation. In fact, it can be argued that the higher the level of aggregation, the more nonlinear these systems become.

For these reasons the IIASA approach to regional systems analysis has not been limited to simplification by linearization, although this technique has been used widely whenever justified.

It was decided to simplify the long-term regional policy-making problems by decomposing the system into a set of interlinked models. The types of submodels chosen (linear, nonlinear, integer, or real-valued) were partly an attempt to achieve the best formulation of the problem and partly a reflection of the systems represented by the submodels.

There are basically three ways of decomposing a regional policy problem (see Chapters 3–8):

1. A procedure based on making policy at an international level, then at the national level, and finally at the regional level (top-down approach).
2. An approach that starts with the planning of an individual region, aggregates these plans up to the national level, and then considers the world markets (bottom-up approach).
3. An approach that analyzes interactions at a multiregional sectoral level, assuming mutual interdependence between the regions.

Although it has often been assumed that each of these approaches is suited to a different institutional framework, this is only partially true. However, international organizations tend to prefer the sectoral approach, and the decentralized Scandinavian countries have shown a marked preference for the bottom-up approach to regional planning.

In actual fact, the scale of the region is more important than the institutional framework in determining the best approach. A relatively large region cannot disregard the impact of its policies on other regions and even on the nation as a whole. In such a case it would be best to use the sectoral approach, in which the large regions are considered simultaneously in their national context. The policies of small regions, however, can have little effect on other regions, or on the nation as a whole. International and national technological and market developments may determine the action possible in the regions, but the action taken in any single region is unlikely to influence national or international development to any significant degree.

The IIASA group has tried to take these factors into consideration in choosing the decomposition technique most appropriate for use in each of the four case studies.

The following section gives an overview of the basic problems, organizational settings, and methods used in the IIASA case studies. It is not intended to summarize the results of the studies, but rather to illustrate the approaches used to overcome the specific problems of each region.

Table 1.2. Summary comparison of four regional development case studies.

Region	Main economic characteristics	Main development problems	Main collaborators	Main methods and models
Noteć, central Poland	Agricultural region	Shortage of water, out-migration	Central- and regional-level planners, water-resource experts	Simplified system of regional models, linked to elaborate agriculture and water-supply models
Silistra, northern Bulgaria	Agricultural region	Slow growth in the basic agriculture sector	Central-level planners, various research institutes in Sofia	Linked models of agriculture, water-supply industry, etc., and breakdown models
South-western Skåne, southern Sweden	Specializes in agriculture, the food industry, and the chemical industry	Rather slow growth, specialization in protected industries, land-use conflicts	Regional Planning Office of South-western Skåne (SSK), University of Lund (water problems), Swedish Council for Building Research	Hierarchical system with emphasis on multi-objective land-use models
Tuscany, central Italy	Specializes in the textile and leather industry, and tourism	Vulnerable position in the world economy, exposed to competition	Regional economic planners in Florence, research workers at the National Research Council, Rome	International and inter-regional trade models of main industrial sectors, and labor-market models

1.4.4. Four case studies

The work carried out in IIASA's Regional Development Task has not been aimed primarily at basic methodological research but at developing systems of models to be applied in a series of case studies. The regions studied have been chosen to provide a number of different environments in which the model packages may be tested. Although a special set of models has been developed for each of the four regions, it is still intended to compare the case studies in terms of both problems tackled and models used. This comparison should include a discussion of the institutional frameworks and problems encountered in the actual collaboration between IIASA and the authorities and planners in the four regions.

Some of the characteristics of the four case studies currently underway are summarized in Table 1.2.

Both eastern-European studies (Noteć and Silistra) concern agricultural regions in which growth has been rather unsatisfactory because of various shortages. Water supply has been a major problem, and the studies looked into the possibility of increasing agricultural productivity by irrigation. Both studies are based on a series of linked models of the regional subsystems in which agriculture has been given the central role and other economic processes are treated less comprehensively. The share of national output provided by these regions has been assessed by partial, local analysis. The Bulgarian case study also includes a model that ties development of the Silistra region to national and interregional economic development. Thus a combination of top-down and bottom-up approaches is used in this study.

It was argued above that local analysis should be used for regions that play a minor role in the national economy. However, since the intraregional approach for small regions leads to problems in handling the indivisible nature of plants and service facilities, local analyses tend to be just as complex, from a methodological viewpoint, as those commonly used for larger regions. The interdependence of the regions and the relationships between them should be treated explicitly when modeling the larger regions, leading to an integrated approach in which it is important to model national–regional ties. This suggests the use of hierarchical systems of models, and this line of attack is therefore pursued in the Skåne and Tuscany case studies.

The case study of southwestern Skåne involves analysis of water-supply and land-use problems in the region. The analysis highlights the conflict between urban and rural use of land in this fertile part of Sweden. Skåne also faces a number of important long-term decisions concerning its energy supply.

As shown in Table 1.2, a hierarchical approach with an emphasis on multiobjective land-use models has been used for the Skåne study. The analysis has been carried out at a local level (in geographical terms), and this calls for a more refined treatment of indivisible factors than is usually attempted in urban land-use models.

Tuscany is a typical open economic region (see Table 1.2); it produces highly specialized industrial products that are traded on the world market under heavy competition, for example, from developing countries. Tuscany's dependence on tourism also makes it vulnerable to external influences. The Tuscany case study therefore deals with alternative scenarios of future industrial specialization under uncertainty in world market development, rather than with the problems of intraregional land use and location of service facilities more typical of the Skåne study.

The Tuscany labor market is unusually free; i.e., it is a market in which intersectoral labor mobility is high, responding to changes in structural demand and business-cycle fluctuations. The volatile nature of this market is said to be one of the factors responsible for the rapid economic development that took place in Tuscany during the 1970s. The IIASA study tried to assess this property in quantitative terms and also to determine its possible negative distributional effects.

The Tuscany case study is based to a large extent on regional and interregional input–output analysis. A stronger emphasis is put on short-term or medium-term problems than in the other studies, and this implies the use of econometric simulation and multiobjective optimization models.

As may be seen from Table 1.2, the four case studies have quite different organizational settings. In the Skåne case study there is direct contact with the planners and politicians of the region, whereas the Tuscany study relies on collaboration with local research workers rather than with the actual decision-makers. The Polish and Bulgarian studies involve contact with research institutes at the national level and planners working with regional problems. The case studies therefore cover a wide range of organizational settings, and it should be possible to draw some conclusions about the application of the systems-analytic approach in different institutional frameworks.

1.5. Some problems for future regional development modeling

This introductory chapter gives a broad description of the evolution of regional development modeling over the last two decades. The theoretical background of this type of analysis has been reviewed, the development of modeling techniques has been summarized, and the practical use of some of the methods and models has been discussed. The rest of the chapters in this book consider these points in more detail.

It seems appropriate to conclude this introduction with a few remarks on the future of regional development modeling, with special emphasis on the work being carried out at IIASA.

A procedure for regional systems analysis, based on case studies, has been proposed in Section 1.4. It is assumed that policy-makers familiar with the region, its historical biases and institutional constraints, should identify the problems to be solved. This link between actual policy-makers and analysis is not always accepted in pure research since it is often argued that problem formulation should be policy-oriented but not necessarily defined by current decision-makers. It might be interesting for systems analysts at institutes such as IIASA to experiment with such a free problem formulation; this would supplement the results of the case studies, which are based on policy-makers' perceptions of policy problems.

The methods discussed in this book are biased in the sense that they all rely on fairly complicated mathematical methods. However, some important problems in decision-making are not very suited to mathematical formulation. In this category belong problems of human relations both at the microlevel and at the political macro-level. This means that systems-analytic approaches employing mathematical models must necessarily be complemented by "softer" approaches to the problems defined by the humanities and some social and behavioral sciences.

There is also reason to investigate new quantitative methods for regional systems analysis. For example, sensitivity analysis can be used to assess the fundamental uncertainties associated with this field. In the very long run, however, it is necessary to use structural stability analysis to examine the possible future changes in the structure of regions.

However, regional development modeling is not the only area in which the fundamental uncertainties should be assessed. The structural design of regions, one of the central problems in regional systems analysis, should also be made to take fundamental uncertainties into account. Development scenarios should be judged not only in terms of their benefits, costs, accessibilities, environmental impacts, and other easily quantifiable consequences, but also with respect to their inherent adaptability, flexibility, and resilience in the face of unforeseen changes in behavior and technology.

It is hoped that this book will stimulate interest in the practical application of these new techniques. Development work of this type will help applied regional systems analysts to keep pace with the rapidly changing problems facing regional planners in countries in all parts of the world.

References

Aganbegyan, A. G., Bagrinovski, K. A., and Granberg, A. G. (1972). Systema Modelei dlya Plani-rovaniya Narodnogo Khozyaistva (A System of National Economic Planning Models). Mysl., Moscow.

Albegov, M. M. (1970). Problemy optimizatsii prostranstvennogo planirovaniya (Spatial planning optimization problems). Ekonomika i Matematicheskie Metody, 6 (6).

Albegov, M. M. and Solodilov, Z. (1970). Voprosy optimizatsii razmescheniya systemy promy-shlennykh komplexov (Issues of locational optimization for a system of industrial complexes). Ekonomika i Matematicheskie Metody, 1.

Alonso, W. (1964). Location and Land Use. Harvard University Press, Cambridge, Massachusetts.

Bandman, M. K. (1980). Territorial production complexes as a form of the spatial organization of production: a subject for planning and management. Paper presented at the Conference on Theoretical and Practical Aspects of Regional Development Modeling, held at the International Institute for Applied Systems Analysis, Laxenburg, Austria on March 19–21, 1980. A summary of this paper is included in this book (Chapter 24).

Baranov, E. F. (1969). Problemy modelirovaniya dinamiki razvitiya khozyaistva ekonomicheskikh rayonov (Problems of modeling the development of economic regions) In Sbornik Optimalnoye Planirovaniye i Sovershenstvovaniye Upravleniya Narodnym Khozyaistvom (Optimal Planning and Improvement of National Economic Management). Nauka, Moscow.

Beckmann, M. (1952). A continuous model of transportation. Econometrica, 20: 643–660.

Beckmann, M. and Puu, T. (1981). A Continuous Transportation Model. International Institute for Applied Systems Analysis, Laxenburg, Austria.

Ben Akiva, M. and Lerman, S. (1977). Disaggregate Travel and Mobility Choice Models and Measures of Accessibility. Paper presented at the Third International Conference on Behavioural Travel Modelling, held at Tenenda, Australia.

Ben Shahar, H., Mazor, A., and Pines, D. (1969). Town planning and welfare maximization: a methodological approach. Regional Studies, 3.

Carillo-Arronte, R. (1970). Empirical Test on Interregional Planning. A Linear Programming Model for Mexico. Rotterdam University Press, Rotterdam.

Christaller, W. (1966). Central Places in Southern Germany. Prentice-Hall, Englewood Cliffs, New Jersey.

Danilov-Danilyan, B. and Zavelsky, M. (1975). Sistema Optimalnogo Perspectivnogo Planirovaniya Narodnogo Khozyaistva (System of Models for Optimal Long-Term Planning of Economies). Nauka, Moscow.

Debreu, G. (1959). Theory of Value. Wiley Publishing Co., New York.

Ephimov, A. and Berry, N. (1965). Metody Planirovaniya Mezhotraslevykh Proportsiy (Methods of Planning Intersectoral Proportions). Ekonomika, Moscow.

Florian, M. (1976). Traffic Equilibrium Methods. Springer Verlag, Berlin.

Forrester, J. H. (1969). Urban Dynamics. MIT Press, Cambridge, Massachusetts.

Fujita, M. (1978). Spatial Development Planning: A Dynamic Convex Programming Approach. North-Holland Publishing Co., Amsterdam.

Gannon, C. A. (1977). Product differentiation and locational competition in spatial markets. International Economic Review, 18: 293–322.

Granberg, A. G. (1978). Matematicheskie Modeli Sotsialisticheskoy Ekonomiki (Mathematical Models of a Socialist Economy). Ekonomika, Moscow.

Hägerstrand, T. (1970). What about people in regional science? Papers of the Regional Science Association, 24: 7–21.

Herbert, J. and Stevens, B.H. (1960). A model for the distribution of residential activities in urban areas. Journal of Regional Science, 2: 21–36.

Hotelling, H. (1929). Stability in competition. Economic Journal, 39: 41–57.

Isard, W. (1951). Interregional and regional input–output analysis: a model of a space economy. Review of Economics and Statistics, 33: 318–328.

Isard, W. and Liossatos, P. (1978). Spatial dynamics: some remarks on the state of the art. In A. Karlqvist, L. Lundqvist, F. Snickars, and J. W. Weibull (Editors), Spatial Interaction Theory and Planning Models. North-Holland Publishing Co., Amsterdam.

Isard, W. et al. (1969). General Theory: Social, Political, Economic, and Regional. MIT Press, Cambridge, Massachusetts.

Koopmans, T. and Beckmann, M. J. (1957). Assignment problems and the location of economic activities. Econometrica, 25: 53–76.

Kossov, B. (1973). Mezhotrasleviye Modeli (Intersectoral Models). Ekonomika, Moscow.

Lee, D. B. (1973). Requiem for large-scale models. Journal of the American Institute of Planners, 40: 163–178.

Lefeber, L. (1958). Allocation in Space. North-Holland Publishing Co., Amsterdam.

Leontief, W. (1951). The Structure of the American Economy, 1919–1939. Oxford University Press, New York.

Leontief, W. and Strout, A. (1963). Multiregional input–output analysis. In T. Barna (Editor), Structural Interdependence and Economic Development. St Martin's Press, London.

Lippman, S. and McCall, J. J. (1976). The economics of job search: a survey. Part I: Optimal job search policies. Economic Inquiry, 16: 155–189.

Lösch, A. (1954). The Economics of Location. Yale University Press, New Haven, Connecticut.

Lowry, I. S. (1964). A Model of Metropolis. RM-4035-RC. Rand Corporation, Santa Monica, California.

Lundqvist, L. (1977). Adaptivity and Freedom of Action in Urban Development Planning. TRITA-MAT-19. Royal Institute of Technology, Stockholm.

Mennes, L. B. M., Tinbergen, J., and Waardenburg, J. G. (1969). The Element of Space in Development Planning. North-Holland Publishing Co., Amsterdam.

Mills, E. D. (1972). Studies in the Structure of the Urban Economy. Johns Hopkins University Press, Baltimore, Maryland.

Moses, L. N. (1955). The stability of intraregional trading patterns and input–output analysis. American Economic Review, 45: 803–832.

Muth, R. F. (1969). Cities and Housing. The University of Chicago Press, Chicago.

Outsuki, Y. (1971). Regional allocation of public investment in an *n*-region economy. Journal of Regional Science, 11.

Paelinck, J. H. P. and Nijkamp, P. (1976). Operational Theory and Method of Regional Economics. Saxon House/Lexington Books, Farnborough, Hampshire.

Perroux, F. (1955). Notes sur la notion de pole de croissance (Notes on the concept of a growth pole). Economie Appliquée, 8:307 ff.

Putman, S. H. (1973). Urban land use and transportation models: a state-of-the-art summary. Paper presented at the Second Intersociety Conference on Transportation, held at Denver, Colorado.

Puu, T. (1977). A proposed definition of traffic flow in continuous transportation models. Environment and Planning A, 9: 559–567.

Rahman, M. A. (1963). Regional allocation of investment. Quarterly Journal of Economics, 77.

Rietveld, P. (1980). Multiple Objective Decision Methods and Regional Planning. North-Holland Publishing Co., Amsterdam.

Sharpe, R., Brotchie, J. F., and Ahern, P. A. (1975). Evaluation of alternative growth patterns for

Melbourne. In A. Karlqvist, L. Lundqvist, and F. Snickars (Editors), Dynamic Allocation of Urban Space. Saxon House/Lexington Books, Farnborough, Hampshire.

Smith, T. E. (1978). A general efficiency principle of spatial interaction. In A. Karlqvist, L. Lundqvist, F. Snickars, and J. W. Weibull (Editors), Spatial Interaction Theory and Planning Models. North-Holland Publishing Co., Amsterdam.

Stone, R. and Brown, A. (1965). Behavioural and technical change in economic models. In A. E. G. Robinson (Editor), Problems in Economic Development. Macmillan Publishing Co., New York.

Tinbergen, J. (1967). Development Planning. Weidenfeld and Nicolson, London.

Webber, M. J. (1978). Spatial interaction and the form of the city. In A. Karlqvist, L. Lundqvist, F. Snickars, and J. W. Weibull (Editors), Spatial Interaction Theory and Planning Models. North-Holland Publishing Co., Amsterdam.

Weibull, J. W. (1976). An axiomatic approach to the measurement of accessibility. Regional Science and Urban Economics, 6: 357–379.

Wilson, A. G. (1970). Entropy in Urban and Regional Modelling. Pion Press, London.

PART II

SYSTEMS ANALYSIS FOR REGIONAL PLANNING

Regional Development Modeling: Theory and Practice
M. Albegov, A.E. Andersson and F. Snickars (editors)
North-Holland Publishing Company
© IIASA, 1982

Chapter 2

REGIONAL THEORY IN A MODELING ENVIRONMENT

Britton Harris
School of Public and Urban Policy, University of Pennsylvania, Philadelphia, Pennsylvania (USA)

2.1. Philosophical background

In a paper presented at the European Regional Science Association Conference in August, 1979, I proposed three main directions for future research in regional science. These were: the need to model large and complex social systems; the need to come to grips with the computer as a tool for research and instruction; and the need for careful attention to policy questions. These ideas will now be developed in the context of regional theory, with the main emphasis being placed on the first two research directions.

I should first outline the philosophical basis on which I approach these issues; my philosophy of science is largely Cartesian in that I believe that building theories can lead only to a deductive process. Building theories involves finding axioms and including them in the axiomatic system, finding and proving theorems, and using them to formulate testable hypotheses. A testable hypothesis may be derived from an axiom, but is more often derived from a theorem. The empirical part of a science lies in the testing of hypotheses.

This approach raises interesting speculations about the role of fantasy in generating hypotheses (Boulding, 1980), speculations which are, however, largely outside the scope of the present discussion. Fantasized statements may be regarded either as hypotheses to be tested or as theorems to be proved. One might naturally question whether a theorem that cannot be proved should be tested. Some economists might argue not only that testing unproved hypotheses should be avoided, but also that a theorem, once proved, does not *need* to be tested. (This assumes that the axioms can also be proved.) I reject both of these positions, and also the idea that finding theorems is an orderly or logical process.

Theorists in fields that are preparadigmatic or that have newly established and untested paradigms are actually trying to create or design completely new deductive systems. Pursuing this course takes one between the Scylla of logic and the Charybdis of utility. Logic requires that a system should be incapable of generating contradictory hypotheses; when such contradictions appear, the remedy requires the elimination or weakening of one or more axioms. This is not a trivial design issue. On the other hand, Gödel's theorem establishes that any axiomatic system has undecidable propositions, so that a utilitarian approach to theory suggests a strong axiomatic system to settle the truth of a large number of operational questions. The remedy for a weak theory is of course the opposite of the cure for a contradictory one – namely, the use of more or more-powerful axioms. The possibility that a theorem may be undecidable or may have a long or complex undiscovered proof justifies the testing of unproven theorems.

Thus the principal properties of deductive systems are those of consistency and completeness; completeness is sometimes aesthetically described as richness, and this perhaps tends to emphasize a subjective and teleological aspect of theory-building. Testing systems for consistency is a kind of "empiricism of the mind". This implies that the conformity of theorems with rules of transformation which permit them to be derived from axioms, and with other rules of noncontradiction, is in a sense empirically observable. Unlike other mental activities such as creative writing, where such rules, if they exist, may be deeply hidden, theoretical deduction excludes fantasy. Similar remarks apply to the testing of completeness, but the definition of this term may have some empirical content. In number theory, it has been argued that the richness of an axiomatic system does not depend on any real-world references, but only on mental images and perhaps on aesthetic sensations; in contrast, the theories of natural and social sciences have a correspondence with the real world by which their completeness or incompleteness may be tested. Both tests contain elements of subjectivity and personal preference.

It is clear that if scientific systems are to be defined as a subset of deductive systems, they must have, in addition to consistency and completeness, something that may be defined vaguely as realism. That is, a scientific theory produces not only theorems, but also hypotheses that may be tested for at least some correspondence with reality. If a set of axioms can generate a theorem that in some application reproduces reality, then these axioms may be regarded as sufficient conditions for a theory of this reality. Such a set of axioms can never be proved necessary – even passing over all sorts of epistemological problems, the identification of a set of axioms necessary to reproduce reality would imply truth rather than realism in an axiomatic system, thus violating the well-established precept that scientific hypotheses are falsifiable, but not verifiable. This proposition impeaches at least the superficial meaning of efforts to "validate" social science theory and models.

There is an important sense in which a theory can be tested by empiricism of the mind rather than by systematic observation and statistical analysis. The mind of a mature person is a storehouse of empirical knowledge acquired and processed either by direct experience or vicariously through social communication. Of course, much of this empirical information will have been misapprehended or misinterpreted, but it still provides a built-in approximation of a reality against which hypotheses can be tested. (Some social scientists regard introspection as the strongest form of this type of empirical observation.) The existence of this substratum of experience is especially important for the nonexperimental sciences. In astronomy, geology, and the social sciences, theory-building aims at reproducing observed and, frequently, aggregated system behavior. The foundation for this behavior usually exists at a more disaggregated level and may be deduced from experiments and theory in other disciplines such as physics, chemistry, and psychology.

2.2. The importance of the modeling environment

We can now approach the problem of the modeling environment, which I believe to be very important for regional theory-building. On the one hand, models provide a framework in which to organize empirical observations and perform statistical tests of the realism of hypotheses. In this context, they are, as I have said elsewhere (Harris, 1966), a form of experimental design based upon a theory. On the other hand, models can be seen as a form of mental experiment, a test by empiricism of the mind. Most narrowly, a model is a theorem or a hypothesis, possibly, together with its derivation, in an axiomatic system. In a more general sense, a model can be regarded as an extended form of a theorem or a hypothesis intended to permit occasional tests for systematic contradictions, more frequent explorations of completeness (or ability to sustain a number of generalizations), and very frequent tests of its ability to reproduce salient features of reality. These features are largely in the mind of the experimenter and have been suggested by previous observation; frequently the extraction of "salient features" is based on a process which records previous successes and failures in hypothesis formulation.

Viewed in this way, models have varying levels of complexity. Static models may require some principle of equilibrium or optimality for their solution. Dynamic models generate development paths and sequential states and do not necessarily contain such principles. Nevertheless, the two types of models can be combined. It is now very well understood that only the simplest models of social systems can be mathematically represented and solved in closed form. If this were the end of the story, the power of axiomatic systems in social affairs would be very limited, but, paradoxically, this limitation is accepted in many disciplines, with pride by some. In the transition to a much richer form of modeling, some relatively simple systems can be solved by numerical analysis, now greatly aided by the advent of the electronic computer.

Complex systems easily exceed the ability of the theoretician either to formulate and analyze them in closed form or to imagine the results on the basis of intuition and heuristic analysis. Consequently, large and complex systems are natural objects for computer modeling. Such modeling is undertaken for three purposes: to perform the mental experiments necessary for theory-building just discussed; to engage in empirical testing; and to provide an operating tool for practicing analysts and decision-makers.

There is another issue concerned with the scale or closure of models. A complete model of a large closed social system could be very complex, but would be subject to a limited number of external influences. At the other end of the spectrum, a model of a class of behavioral units could be intrinsically relatively simple, but would be almost completely open and therefore subject to a wide variety of external influences. An intermediate case, a model containing more than one type of behavioral unit but representing an incomplete system, will have some degree of internal complexity and be open to a certain number of external influences. The apparent complexity of social science models and their almost baroque design may be a consequence of this double-edged nature. On the one hand, there is no standard format for the interactions between different subsystems and behavioral units, so that the provision for these interactions is often idiosyncratic. On the other hand, the provision of external connections is conducted on a very ad hoc basis and can greatly increase the apparent complexity. Finally, a (partially misplaced) emphasis on model-fitting and data analysis can lead to bizarre distortions of theoretical models in order to cope with the unavailability of data; these modifications include noncomparable geographic subdivisions, proxy variables, inappropriate classifications of actors, and so on. The addition of these features adds another layer of apparent complexity to an already unwieldy structure.

This last difficulty is almost entirely avoidable at the theory-building level. Data may be generated or simulated from a common-sense interpretation of experience, and computer techniques used in a consistent and realistic way to test the realism of models. At this stage of theoretical investigation the structure of the models should be kept as clear and as simple as possible. More complex structures can be used later in deriving parameters, in empirical testing, and in operational applications. The Forrester model of urban dynamics (Forrester, 1969) is an extreme and arguably useful example of the first part of this procedure. While many regard the structure of Forrester's behavioral functions as highly arbitrary, there is no doubt that the simulated data that he extracted from experience have provided a feasible basis for a model of some complexity and interest; its realism and applicability, however, are more questionable.

2.3. Theory and modeling of large-scale social systems

The outlines of a suitable procedure for modeling and developing the theory of large-scale social systems are now beginning to emerge. The essential structure of such large-scale models should be based on two simple but very important principles: (i) the behavior of actors and subsystems at the most elementary level should be represented in an extremely simple fashion; (ii) the interactions between these components should also be simple, but very large in number. When modeling is used as an aid to theory-building, attention should be focused on the consequences of these interactions. Given multiple paths of causation and feedbacks in the operation of a large-scale model with several components and many interactions, the propagation of impacts through the entire system can be followed over space and time. In policy-making, it is by now a truism that these expanding ripples of causation are responsible for the unintended consequences of single-purpose actions and suboptimization. In systems theory, it is equally clear that these chains of causation and feedback explain why the behavior of the system as a whole is not simply the sum of the behavior of its components.

It should come as no surprise to note that many of the most productive urban and regional models are indeed very simple, in accordance with the first of the principles given above; the gravity model of the attenuation of action over distance is an excellent example. Classical input–output analysis is based on a very simple and almost tauto-logical set of accounting definitions. The behavioral assumptions behind most housing market models are based on simplified versions of classical utility theory, which in general lead to procedures from which the willingness to pay or a set of bid prices can be estimated. A very useful model of retail trade location, which was employed in the Penn Jersey Transportation Study and later analyzed from a theoretical point of view (Harris and Wilson, 1978), contains two simple behavioral assumptions. Consumers are expected to behave according to a gravity model of interaction and producers are expected to behave so that the supply of retail services is adjusted to the demand.

Most of these models at the simplest behavioral level lack interest and effectiveness until they are placed in a larger context. However, these larger contexts usually involve (either implicitly or explicitly) a number of additional assumptions. A behavioral model of bidding in the housing market can be expanded into a locational model by a variety of mechanisms. The Herbert–Stevens model (Herbert and Stevens, 1960) and the NBER model (Ingram et al., 1972) use different linear programming techniques to accomplish this. Anas (1973) proposes a version of entropy-maximization together with the local equilibration of supply and demand, which in the limit could lead to a linear programming solution. Modeling the responses of the suppliers to residential demand requires a different set of models and a new level of linkage. Gravity or entropy-maximizing models of travel behavior appear relatively naïve in a cross-sectional transportation analysis, but begin to acquire much more interest and signi-ficance in a wide variety of market-type situations, some of which are sketched above.

All of the interconnections that have been discussed so far are simple one-stage linkages. Larger models with apparently greater complexity can be constructed very simply using such linkages. The Lowry (1964) model of the metropolis links residence with employment, and service industry with residence, with virtually no feedback. The multiplier model of input–output analysis (which is of much greater interest than the basic accounting model) assumes multiple iterations of the elementary model, but does not provide constraints and feedbacks. It is therefore intrinsically of this same elementary one-stage nature.

2.4. Economies of scale and externalities

An important set of characteristics of most regional models has to do with economies of scale and with externalities. In a simple behavioral or descriptive model, economies and diseconomies of scale introduce nonlinearities. In larger models, both features create feedback loops, positive and negative, and can produce unusual outcomes. The impacts of simple externalities have been shown by Schelling (1978) to have surprising long-term consequences and the prevalence of these grossly nonlinear features have many important theoretical consequences. In dynamic models, they lead to divergences and bifurcations. The definition of bifurcations in catastrophe theory is based upon the assumption of a manifold of equilibrium states, and wherever bifurcations occur, there are potentially multiple local equilibria. This fact dashes any hope of a unique general equilibrium and also challenges the existence of a unique optimum associated with general equilibrium. This fact is very important in policy-making and planning, since it undermines the assumption that economic behavior in production, exchange, and consumption necessarily leads to an optimum optimorum in the allocation of resources. Finding and reaching such an optimum may thus require planning and controls.

Frequently, of course, the effects of externalities, economies, and diseconomies are felt across several models. Many such effects are spatial and are propagated through spatial interactions such as transportation costs or pollution. Not all effects, however, are entirely spatial. Certain aspects of health, education, and employment interact quite strongly, but in a conventional model of any of these areas the interactions would be sacrificed or would at best appear as exogenous variables. The interactions that propagate all of these externalities through a larger system will generate complex feedback loops and lead to the appearance of strong nonlinearities and the introduction of economies and diseconomies. In a situation in which any of these subsystems is being analyzed and then optimized in the design sense, the fact of suboptimization is obvious; the extent of this suboptimization and the difficulties caused by it depend on the degree of the interactions and their cumulative impact within the model.

Self-contained models including interactions of this kind are generally not suitable for closed-form solutions. Close examination shows that some disaggregated and superficially complex models which can be solved in closed form do not have very realistic feedback properties. Examples include the Lowry model, which can be solved by matrix inversion in a different formulation developed by Garin (1966), the extended input–output model, and some of the new urban economics models dealing with urban form. It is interesting to see how rapidly this class of models is exhausted. Very simple modifications of a monocentric housing model lead economists like Muth (1969) to numerical analysis. The retail trade model discussed in Harris and Wilson (1978) has a closed-form solution in certain very standard cases, but the introduction of economies of scale in retailing destroys even this limited property. Clearly, mixed models containing submodels with different properties and different modes of solution are almost certain not to have a closed-form solution.

2.5. Sensitivity analysis

The lack of methods for direct solution suggests the need for a more extended consideration of computer modeling; such a consideration has arisen repeatedly as models have been developed and applied in a variety of circumstances. A situation is thus evolving in which computers are being asked to fulfill a number of different roles in modeling; these roles must be identified so that the computer system can be made more responsive to the needs of users.

Perhaps a good way to approach this problem is by way of sensitivity analysis, which, as will be shown later, can also be used to evaluate models and theories. Conventional sensitivity analysis in its simplest form investigates the sensitivity of models to changes in parameters or "independent" variables. In a simple model the sensitivity can be gauged directly from the parameter values, but where chains of causation and feedback are involved this method does not work. A second conventional approach tests the sensitivity of the models to constraints such as budgets, land availability, and so on. Still other tests examine the sensitivity of the model to initial conditions (for dynamic models) and major inputs such as levels of income, industrial composition, and public policies. Somewhat less conventional sensitivity analysis could extend to the functional forms used within models and to the response of a large system to the replacement of one set of models by another. The replacement of models could implicitly reorganize the interactions between the models.

It should be noted that sensitivity testing depends largely on the kind of sensitivity being measured. Simple correlation coefficients are not an adequate measure of the goodness of fit of a model, and simple directions and magnitudes of change are not sufficient for sensitivity testing. A number of examples in which special aspects of model performance and consequently of sensitivity need to be examined are outlined below.

Several simple cases involve problems concerned with the housing market. For example, does a model of housing market choice reproduce the patterns of income, occupation, and ethnicity observed in the buyers, and what are the implications of the model's performance in this area? If the abandoning of dwellings is being studied, it would be useful to know more about the various ways in which different geographical patterns of abandoned dwellings can arise. In an unpublished study of housing in the New York region, one model was found to give very good predictions of the rents bid by various classes of the population for various classes of housing in various locations but suffered from one serious structural difficulty: well-to-do households were outbidding the poor for dwellings in poor neighborhoods. This did not affect the R-squares greatly because very few cases were involved, but it would lead to outrageous results in any practical application. The problem was finally overcome by adding a variable or variables reflecting neighborhood poverty and ethnicity. Examples may also be found in retail trade modeling. A retail trade model that will automatically generate clusters of concentrated activity is quite unusual — most retail trade models start from assumptions about the existence of centers. The model discussed in Harris and Wilson (1978) is an exception in that it creates centers and does not make this assumption. More generally there should be regional models that generate centers of activity, cities and metropolises, in appropriate locations — but such models are extremely rare, if they exist at all.

The examples given above are intended to support the idea that models used in regional science should be able to reproduce distinctive features of the phenomena under study and should satisfy consistency conditions that cannot always be imposed in the statistical analysis. This is in itself a form of sensitivity testing, since the ability to meet these criteria may require the model to fulfill the same conditions as the most rigorous sensitivity test. The distinctive system characteristics which the models should reproduce may be selected on the basis of experience, which is generally sufficiently empirical for this exercise.

There is a more rigorous form of model validation or testing that is sometimes not independent of empirical data collection and statistical testing. This is the sensitivity of models to drastically changed conditions, often referred to as transferability or "portability". It is useful to note that, in an empirical context, portability implies that a model can be used outside the range of the observed variables, and further that this is necessary for policy-testing. Policies that may be described by combinations of observed variables are not novel and do not need to be examined by the use of models. The study of new policies requires concepts and possibly models that may be transferred into unknown territory, and the more unusual the policies under study, the more severely the portability of the models is tested. Portability is a reflection of the sensitivity of models to wide variations in income, culture, institutions, economic activity, and governmental policies, and reflects the ability of the models to produce new results as some or all of these factors are varied.

2.6. Computerized systems of large-scale models

This overview suggests a number of important features that a computerized system should have if it is to permit the research worker to engage in a partially speculative exploration of the characteristics of large-scale systems.

The experimenter would in principle prefer to use simplified systems with relatively few actors and relatively few geographical subdivisions to keep the computational load to a minimum. Detail in modeling is by no means equivalent to complexity, but the desire for compactness may be countered by the fact that complex system behavior may disappear if the model is excessively simple.

The experimenter will want to be able to substitute one model for another in any particular system, and to substitute one functional form of computation for another within models. This implies a well-linked system of modular subroutines that can be inserted into a large model at will. More generally, the experimenter will want to be able to link these models in a variety of ways, not only by substitution, but also by rearranging the structure, the order of computation, and the transmission of information from one model to another. This implies a flexible and powerful system containing models which can be altered without massive rewriting and reorganization by computer specialists.

The research worker will also require a general-purpose data base covering the complete set of phenomena being modeled. A data base can either be built up from experience, as discussed above, or be adapted from observed experimental results.

In many applications these approaches are equally useful, but if the statistical properties of the models are being investigated an observed data base is almost certainly necessary. Comparative studies involving portability require either a second data base or the systematic variation of characteristics in the initial data base. The difficulty in synthesizing or systematically changing data bases lies in the need to retain consistency and to avoid altering important internal characteristics of the data; these characteristics may simply be correlations between variables, or, at a higher level, may involve large-scale structural features.

If a computational system of this type could be put into operation, it is conceivable that many theoretical and practical aspects of regional analysis would be substantially altered. A great deal of regional analysis deals with partial models and subsystems whose performance in a "richer" context is not easy to predict. Models that appear perfectly reasonable in one setting might prove to be counterproductive or even perverse in their implications in another setting. On the other hand, models that appear to be rather unimaginative, or that do not seem to fit the data, or that are unreasonable at first sight, might prove to be accurate, productive, and useful in a different context. These hypothesized changes in submodel behavior are based on the assumption that feedback loops and interactions between models will generate large-scale system behavior that cannot be predicted from the performance of the individual models. While this assumption has a firm foundation both in systems theory and common experience, it could be tested in many different ways in the computational environment proposed above, thus adding yet another dimension to the concept of empiricism of the mind.

2.7. Questions of experimental design

A wide variety of experimental design questions arise in enterprises of this kind. The main difficulty faced in the experimental design of model systems is very similar to the chief problem encountered in designing optimal policies. In both cases, the decisions to be made are generally discrete or lumpy, and, in any event, the interaction between decisions creates a series of outcomes that are only locally optimal — that is, in the case of policy-making, small perturbations of the system cannot lead to any improvement in its overall performance. If it is accepted that there is some way of measuring the performance of a system of models then the analogous statement would be that small perturbations in parameters, functions, models, and interconnections could not improve the overall performance of the whole system.

In the continuous case, small perturbations in policies or models are made by incremental changes in continuous quantities. If the model is fixed, these quantities are variables; if the conditions are fixed, these quantities could be parameters. It is also possible to make some changes in functions continuous, as in the Cox–Box transformation in statistics. Generally speaking, however, the substitution of one function for another or one model for another is a discrete step of the smallest possible kind. In this case a system of models with, say, twenty binary decisions could be formulated in over a million possible ways.

There are thus two main problems to be faced in the experimental design of model systems in this environment. First, in the absence of signposts such as partial derivatives (which exist only in the continuous case), the experimenter must use either a systematic approach or his own judgment in the search for new general model designs. Second, because of the expected abundance of local optima, an improvement process often does not tell the experimenter where to start the design process or how to choose starting points in order to improve his chances of success. The seriousness of these difficulties has yet to be tested in an operational environment of the type described above.

2.8. Problems of policy synthesis

These problems have been approached largely from the point of view of developing theories in regional science, but their study has partly been motivated by a belief that sound theory can help policy-makers and politicians to make better decisions and, all other things being equal, lead to an improvement in the human condition. This belief motivates many people working in the field of regional science and regional analysis, and requires a brief consideration of the problems of policy synthesis that have not been discussed already.

The problems caused by the unintended consequences of policy actions and the dangers of suboptimization are among the most fundamental difficulties of policy-making. These difficulties are obscured whenever policy-making becomes reactive, short-term, or placatory. But if these modes of action are indeed dangerous, they will prove to have an inferior survival value for the parties that adopt them. Consequently, regional analysts can usefully devote themselves to a consideration of the kinds of models that will avoid shortsighted policy-making, even if this involves generating counterintuitive results.

It is of course entirely possible that our current theories and our ability to extend them are still not sufficiently advanced to produce accurate long-term projections for large-scale interactive systems. However, I feel that we must assume the opposite and make every effort to reproduce the behavior of such systems both in theory and in models. The large number of small-scale models in existence testifies to the individualism of much research in the social sciences, to the lack of resources in the field, and to the lack of a suitable computational environment. Unfortunately, it also suggests a preoccupation with detail and with a potentially spurious statistical accuracy which can probably never lead to success in reproducing the behavior of large-scale systems.

2.9. Concluding remarks

The main conclusion of this survey is that some researchers working with regional theory should redirect their attention and possibly change their style of work to deal with large-scale systems and their overall interactive behavior. This would extend and build upon the work of Isard and many others who have had a broad vision of the interconnected nature of the world in which regional planning and development played an important role. It would also recognize the connection between economic, social, moral, and political affairs, and the interconnection between many types of subsystems, including not only the firm and the individual consumer, but also the family, the neighborhood, the social group, and the nation. At the same time, however, it seems necessary to abandon the idea that complex systems can be described by models which can be solved in closed form and which are based on traditional model-building concepts. It is also necessary to abandon the idea that models, when used to examine various policy options, will lead to a single optimum or equilibrium position and therefore make the creative and inventive aspects of policy-making unnecessary.

Unfortunately, this last conclusion, if transferred back into the realm of model design, leads to the possibility (already recognized at other levels of model-building) that models with completely different structures can lead to very similar (satisfactory) levels of performance and that new criteria will have to be devised to choose between them. This result is not unique in the history of science, and gives some hope that regional analysts will not become unemployed in the near future as a result of their own success.

References

Anas, A. (1973). A dynamic disequilibrium model of residential location. Environment and Planning A, 5: 633–647.

Boulding, K. E. (1980). Science: our common heritage. Science, 207: 831.

Forrester, J. W. (1969). Urban Dynamics. MIT Press, Cambridge, Massachusetts.

Garin, R. (1966). A matrix formulation of the Lowry model of intra-metropolitan activity allocations. Journal of the American Institute of Planners, 32: 361–364.

Harris, B. (1966). The uses of theory in the simulation of urban phenomena. Journal of the American Institute of Planners, 32: 258–273.

Harris, B. and Wilson, A. G. (1978). Equilibrium values and dynamics of attractiveness terms in production-constrained spatial-interaction models. Environment and Planning A, 10: 371–388.

Herbert, J. and Stevens, B. H. (1960). A model for the distribution of residential activities in urban areas. Journal of Regional Science, 2: 21–36.

Ingram, G., Ginn, J. R., and Kain, J. F. (1972). The Detroit Prototype of the NBER Urban Simulation Model. Columbia University Press, New York.

Lowry, I. S. (1964). A Model of Metropolis. Memorandum RM-4035-RC. Rand Corporation, Santa Monica, California.

Muth, R. (1969). Cities and Housing. Chicago University Press, Chicago, Illinois.

Schelling, T. C. (1978). Altruism, meanness, and other potentially strategic behaviors. American Economic Review, 68: 229–230.

Regional Development Modeling: Theory and Practice
M. Albegov, A.E. Andersson and F. Snickars (editors)
North-Holland Publishing Company
© IIASA, 1982

Chapter 3

A SIMPLIFIED SYSTEM OF REGIONAL MODELS

Murat Albegov[*]
International Institute for Applied Systems Analysis, Laxenburg (Austria)

3.1. Introduction

Although many different regional models have been constructed, there is still no generally applicable system of models which covers the main sectors of the regional economy and which provides practical solutions to regional problems. Any system of models designed to tackle these problems should fulfill five major requirements:

1. The development of the region should be compatible with national development plans.
2. All of the important sectors of the regional economy and their interrelationships should be included in the analysis.
3. Economic, social, environmental, and institutional problems should all be examined.
4. The multidimensional nature of regional problems and the factors of uncertainty should be considered in the analysis.
5. Effective measures for controlling regional development should be investigated.

The last point is particularly important because the aim of regional development depends largely on regional conditions – it should be possible to control regional development using the methods most suited to conditions in each region.

This chapter will concentrate on two regions which have similar problems and similar aspirations: the Silistra region in Bulgaria and the Noteć region in Poland. Case studies of these regions are currently being carried out as part of the work of the Regional Development Task at the International Institute for Applied Systems Analysis (IIASA).

---

[*] Currently at the Central Institute for Economics and Mathematics, USSR Academy of Sciences, Moscow, USSR. Leader of the Regional Development Task at IIASA from November 1976 to December 1980.

Both Noteć and Silistra are largely agricultural areas in which irrigation plays an important role in regional development; both regions also require rapid expansion of the existing system of settlements and services. The main aims of regional development in the Silistra region were initially assumed to be:

(i)   to maximize the rate of regional growth (in general terms)
(ii) to ensure that there is no decrease in the region's already significant share in national agricultural production
(iii) to attain a level of industrial development at least equal to the national average
(iv) to decrease rural–urban migration
(v)   to maximize the growth in average wages for the region as a whole

However, discussions with the Bulgarians taking part in the Silistra case study suggested some additional (or slightly modified) goals:

(vi)   to maximize agricultural production
(vii) to develop an irrigation system that would make it possible for farmers to achieve optimal production efficiency
(viii) to develop local agriculture and industry such that no serious environmental problems are created

Since the local conditions and development aims of the two regions are virtually identical (Albegov and Kulikowski, 1978) (the main differences are the need to equalize the subregions in the Noteć region, and the fact that agriculture is largely in the hands of private farmers in the Noteć area while it is state-run in the Silistra region) the same system of models is used to analyze the problems of both areas. The most important sectors are seen to be agriculture, industry (as a sector complementing agriculture), water supply (including irrigation), and population growth and migration.

Models of these sectors form the basis of the system. The number of submodels has deliberately been kept to a minimum to ensure that the system is simple and practical at the first stage of analysis. At later stages it is planned to include additional submodels dealing with factors such as regional settlements and services and environmental quality.

3.2. Generalized regional agriculture model (GRAM)

The Generalized Regional Agriculture Model (GRAM) has been designed principally to aid policy-makers in decisions concerning agricultural specialization in the regions. The most suitable combination of crops and/or livestock for a given region depends on a large number of factors, which include the type of land, the amount of labor available, and the supply of animal feedstuff; the model takes all of these factors into account in its analysis.

GRAM is strictly limited to the problems of agricultural regions, though it may include significant feedbacks and results from other subsystems dealing with water, industry, and labor. The model treats the following areas in detail: regional agricultural specialization; production of various crops and different types of livestock; problems connected with land use, with special reference to irrigation, drainage, and the use of pastures; the composition of animal feed (relative proportions of protein, roughage, and vegetable material); crop rotation; the possibility of producing a second crop; and the regional availability of resources (labor, capital investment, fertilizers, water, and so on).

The structure of the model is described in detail in Albegov (1979), and therefore only a summary of the variables and constraints will be given here. The main variables in the model are the size and type of crops, the number and type of livestock, the amount of food required for human consumption, the amount of food required for the livestock, and the levels of purchases and sales. The most important constraints limit land use, consumption, production, the availability of resources, and the levels of purchases and sales. There are also a number of financial constraints.

Some problems are ignored by the model, while others are included but not examined in detail. For example, the transportation of final products is included only indirectly through a system of transportation costs, and the relationship between production and processing is omitted. It would therefore be necessary to use different models to analyze these problems in detail.

3.3. Regional industry model

It is necessary to introduce a general model to analyze industrial growth in the regions. The model adopted here was developed at the Central Institute for Economics and Mathematics (CIEM) in Moscow (Mednitsky, 1978); it contains a general description of the production process and it is therefore possible to consider a large number of resources, final products, and nonlinear cost dependencies.

An outline of this model is given below; it is first necessary to explain the notation.

l represents the possible locations of production units within the region

s represents the points where demand is concentrated (within the region and on the boundary)

r represents the type of plant or production unit (variant)

ϵ represents the rate of return on capital investment

R represents the set of plant variants; this is independent of l

Z^0 represents the final demand (including demand both within and outside the region)

Z_l represents the local demand for nontransportable resources at point l

a^l represents the fixed demand for transportable commodities at point l

c_{lr} represents the unit cost of producing commodities using plant r at point l

K_{lr} represents the capital investment per unit of output from plant r at point l

f_{lr} represents the local resources available to plant r at point l

T_{ls} represents the cost of transportation (of individual commodities) from point l to point s

A_{lr} represents the standard level of transportable resources and commodities input to plant r at point l

F_{lr} represents the standard level of nontransportable resources and commodities input to plant r at point l

B_{lr} represents the standard level of output from plant r at point l (transportable commodities)

E_{lr} represents the standard level of output from plant r at point l (nontransportable commodities)

L_{lr} represents the level of use of plant variant r at point l

U_{ls} represents the volume of commodities transported between points l and s

σ_{lr} represents the integer variables that indicate whether plant r should be located at point l

The model contains the following constraints:

1. The demand for transportable resources within and outside the region under analysis must be satisfied.

$$\sum_l B_{lr} L_{lr} \geqslant Z^0$$

2. Local demand for nontransportable resources must also be satisfied.

$$E_{lr} L_{lr} \geqslant Z_l$$

3. The volume of commodities transported from a given point corresponds to the volume of transportable commodities produced at that point.

$$B_{lr} L_{lr} = \sum_s U_{ls}$$

4. The demand for transportable commodities (both fixed and from new enterprises) at each point is met by the volume of goods transported to that point.

$$a^l + A_{lr} L_{lr} = \sum_m U_{ml}$$

5. Local consumption of nontransportable resources is confined to the available supply.

$$F_{lr} L_{lr} \leqslant f_{lr} \sigma_{lr}$$

All of the variables are nonnegative and some are also integers:

$$L_{lr} \geqslant 0; \quad U_{ls} \geqslant 0; \quad \sigma_{lr} = \{0 \text{ or } 1\}$$

The most common objective function is minimization of production and transportation costs, namely

$$\min \left\{ \sum_{lr} [(c_{lr} \cdot L_{lr}) + \epsilon K_{lr} \cdot \sigma_{lr}] + \sum_{ls} (T_{ls} \cdot U_{ls}) \right\}$$

although it is possible to use other objective functions in this model.

The model has many quite general features and can be used to represent a multiproduct system. In this case the products are aggregated and all inputs and outputs are calculated for this aggregate. The appropriate capacity and technology of the production unit can be selected from a number of capacity/technology variants for each location. The model may include transportable and nontransportable products, local and external demand (the latter often concentrated at points near the boundary of the region under analysis), as well as substitutable elements of production and consumption. Finally, the transportation matrix can easily be modified to incorporate, for example, the costs of transporting a variety of products.

A detailed package of associated programs makes it possible to use this model even when the problems to be solved are multisectoral.

3.4. Water-supply model

A water-supply model was required to describe regional water supply and demand and to produce a set of water costs; for a formal description of the model used, see Chernyatin (1980). In the Silistra case study the model considers both the seasonal and spatial problems of water supply (Albegov and Chernyatin, 1978), but since regulation over a period greater than a year is not included, the problem is basically that of water distribution. The following assumptions are made:

1. The demand for water at each point in each period of time is known.
2. The volume of water available is almost unlimited.
3. All users consume water irreversibly.
4. The water resources are regulated over a period no longer than one year.
5. The time taken in water distribution may be ignored.

The main goal is to meet the demand for water over a given period with minimum cost. This period is considered to be the last year of the time period under analysis. Problems of water quality are not considered.

The following notation is used in the model:

α represents the water inflows ($\alpha = 1, 2, \ldots, s$)

j represents the nodes:

reservoir nodes ($j = 1, 2, \ldots, r$)

pumping nodes ($j = r + 1, r + 2, \ldots, r + s$) (exogenous)

pumping nodes* ($j = r + s + 1, r + s + 2, \ldots, r + s + m$) (endogenous)

distribution nodes* ($j = r + s + 1, r + s + 2, \ldots, r + s + m + 1$)

k represents the period of time ($k = 1, 2, \ldots, N$)

i represents the arcs ($i = 1, 2, \ldots, n$)

$I = \{1, 2, \ldots, n\}$ represents the set of all arcs

I_j^+ represents the subset of arcs entering node j

I_j^- represents the subset of arcs leaving node j

a, b, γ, e represent the costs of exploitation

W_j^k represents the outflow through j in period k

t^k represents the duration of period k

V_j represents the capacity of reservoir j

Z_i represents the discharge capacity of canal/arc i

X_j represents the capacity of pumping station/node j

q_α^k represents the inflow α in period k

y_i^k represents the flow in arc i in period k

S_j^k represents the active water storage in reservoir j at the beginning of period k

The constraints used in the model are outlined below.

Flow balances at pumping and distribution nodes:

$$q_\alpha^k + \sum_{i \in I_j^+} y_i^k - \sum_{i \in I_j^-} y_i^k = 0 \quad \text{(pumping)}$$

where $\alpha = 1, 2, \ldots, s$; $j = r + \alpha$; and $k = 1, 2, \ldots, N$.

$$\sum_{i \in I_j^+} y_i^k - \sum_{i \in I_j^-} y_i^k = W_j^k \quad \text{(distribution)}$$

where $j = r + s + 1, r + s + 2, \ldots, r + s + m + 1$, and $k = 1, 2, \ldots, N$.

Mass balances for reservoirs:

$$S_j^1 = t^N \left(\sum_{i \in I_j^+} y_i^N - \sum_{i \in I_j^-} y_i^N \right) + S_j^N \quad \text{(stationarity condition)}$$

where $j = 1, 2, \ldots, r$.

---

* Outflows ($j = r + s + 1, r + s + 2, \ldots, r + s + m + 1$) have the same sequence of index numbers as the endogenous pumping nodes and distribution nodes.

$$S_j^{k+1} = t^k \left(\sum_{i \in I_j^+} y_i^k - \sum_{i \in I_j^-} y_i^k \right) + S_j^k$$

where $j = 1, 2, \ldots, r$, and $k = 1, 2, \ldots, N-1$.

Upper bounds:

$$t^k \sum_{i \in I_j^-} y_i^k - S_j^k \leqslant 0 \quad \text{(for reservoir nodes)}$$

where $j = 1, 2, \ldots, r$, and $k = 1, 2, \ldots, N$.

$$S_j^k - V_j \leqslant 0 \quad \text{(for reservoir nodes)}$$

where $j = 1, 2, \ldots, r$, and $k = 1, 2, \ldots, N$.

$$\sum_{i \in I_j^+} y_i^k - X_j \leqslant 0 \quad \text{(for pumping stations)}$$

where $j = r + s + 1, r + s + 2, \ldots, r + s + m$, and $k = 1, 2, \ldots, N$.

$$y_i^k - Z_i \leqslant 0 \quad \text{(for canals/arcs)}$$

where $i = 1, 2, \ldots, n$, and $k = 1, 2, \ldots, N$.

Objective function:

$$E = \sum_{j=1}^{r} a_j V_j + \sum_{j=r+1}^{r+s+m} b_j X_j + \sum_{i=1}^{n} \gamma_i Z_i + \sum_{\alpha=1}^{s} e_{r+\alpha} \sum_{k=1}^{N} t^k q_\alpha^k + \sum_{j=r+s+1}^{r+s+m} e_j \sum_{k=1}^{N} t^k \sum_{i \in I_j^+} y_i^k$$

Decision variables:

$$q_\alpha^k, y_i^k, S_j^k, V_j, X_j, Z_i$$

The model has four general features. All the elements of the water-supply system (with the exception of regulating dams) are included and any configuration of the system can be considered. Regional space can be represented in the model in terms of point subregions. The model takes into account seasonal variations in water consumption, and it is possible to analyze a number of water-supply systems simultaneously. However, the model could be improved by considering the effects of water regulation over a year, calculating the optimal development trajectory for the construction of water systems over a number of years, and analyzing the problems of water quality (including the disposal of waste water).

3.5. Population and migration models

The population and migration models are used to forecast the labor force expected in each subregion and in the region as a whole. This requires a thorough evaluation of previous trends and may suggest the need for possible human intervention to alter these trends.

Three elements of regional population change are considered: fertility, mortality, and migration. All are influenced by demographic, economic, and social factors. However, in developed economies, fertility and mortality do not change as rapidly as migration and therefore fertility and mortality rates are assumed to be constant. An econometric approach is used to analyze migration.

It should be noted that the model is used to forecast, not to plan, population growth. Planning models consider the means for regulating regional population growth; forecasting models merely project the observed population into the future, making certain assumptions about rates of birth, death, and migration. In this study the population and migration models were used to forecast the number of people moving into and away from the region, the future population of the region, the future populations of the subregions, and the future labor force in the subregions and in the region as a whole.

Different types of model can be used to forecast migration in different regions. The following logit model has been adopted in the case study of the Silistra region (Andersson and Philipov, 1979):

$$P_{ij} = \frac{\exp v_j}{\exp v_i + \exp v_j} = \frac{\exp(v_j - v_i)}{1 + \exp(v_j - v_i)} \tag{3.1}$$

where P_{ij} is the probability of moving from region i to region j and v_j, v_i are measures of the attractiveness (to migrants) of regions j and i, respectively.

The form of the function v is

$$v_i = \sum_{k=1}^{n} \alpha_{ik} \cdot X_{ik} + \beta_i$$

where the X_{ik} represent the characteristics k of region i and α_{ik}, β_i are coefficients to be estimated by an econometric approach.

Model (3.1) can then be rewritten as

$$P_{ij} = \frac{\exp\left[\sum_{k=1}^{n} (\alpha_{jk} \cdot X_{jk} - \alpha_{ik} \cdot X_{ik}) + (\beta_j - \beta_i)\right]}{1 + \exp\left[\sum_{k=1}^{n} (\alpha_{jk} \cdot X_{jk} - \alpha_{ik} \cdot X_{ik}) + (\beta_j - \beta_i)\right]}$$

In the Bulgarian case study, the X_{ik} denote: the percentage of the total national population in region i; the percentage of the total regional population working in the service sector; the mean wage in region i (per employee); the average dwelling space (square meters per employee); and the distance (in kilometers) between the regional city centers.

When the net interregional migration is known, the population growth in the region under analysis can be studied easily.

The interregional migration rates can change from year to year, depending on the results of migration model runs. The age and sex structure of migrants may be assumed to remain constant, as a first approximation. The results obtained in this analysis are then used in the analysis of intraregional problems.

Given the data on regional population growth, intraregional population growth can be analyzed using the model developed by Willekens and Rogers (1978):

$$\{K^{t+1}\} = G\{K^t\}$$

where

$\{K^t\}$ is the age and subregional distribution of the population at time t

    G is the multiregional (in this case, multisubregional) matrix growth operator or generalized Leslie matrix

$t + 1$ is the time period following t (5-year periods are usually analyzed).

Using this model it is possible to deduce the size and age and sex structure of the subregional populations during each period of time. These regional and subregional populations and their age/sex structure are used to assess the size of the labor force. The results obtained from the population and migration models are then used as constraints in modeling other components of regional growth.

3.6. Structure of the system of models

This section discusses the overall structure of the system of models, which is outlined in Figure 3.1. The general idea is to coordinate the development of the main sectors of the regional system using the optimum amount of local and migrant labor, under various external restrictions. These external restrictions are derived from: a system of prices for raw materials and final products, which are used to assess the efficiency of local industry and agriculture; a system of averaged data for the country as a whole, including information which can be used for migration forecasts (wages, quality of services, dwelling space per person, etc.); and information on total external investment in the regional economy – the distribution of this investment can be used to regulate future regional development.

Production in the various sectors is coordinated in three stages. First, the resources of a given region are evaluated and those important for the region's growth are identified. The most important resources are usually labor and capital, but additional resources (for example, water) can be included. These additional resources should be limited to no more than three, because otherwise the coordination procedure becomes too complicated. Second, optimal and near-optimal solutions are calculated for each production sector and a special functional relationship is derived. This function indicates how efficiency in each sector depends on the amount of resources allocated to that sector. Third, the following problem should be solved:

$$\sum_i E_i(C_i, L_i, W_i) \to \max \tag{3.2}$$

subject to

$$\sum_i C_i \leqslant C_{\max} \tag{3.3}$$

$$\sum_i L_i \leqslant L_{\max} \tag{3.4}$$

$$\sum_i W_i \leqslant W_{\max} \tag{3.5}$$

$$0 \leqslant C_{\min} \leqslant C_i \leqslant C_{\max} \tag{3.6}$$

$$0 \leqslant L_{\min} \leqslant L_i \leqslant L_{\max} \tag{3.7}$$

$$0 \leqslant W_{\min} \leqslant W_i \leqslant W_{\max} \tag{3.8}$$

where

i represents the sector or group of sectors
E_i represents the efficiency function of sector i
L_i represents the number of employees in sector i
C_i represents the capital investment in sector i
W_i represents the consumption of any other resource by sector i

The solutions to problems (3.2)–(3.8) give the intersectoral distribution of the most important resources, and could be used as input to find more precise sectoral solutions.

The main and auxiliary sectors are coordinated in accordance with the importance of the intersectoral links. For example, it seems evident that development in the agricultural sector depends significantly on the water supply, whereas the dependence of the industrial sector on the water supply is relatively low. Therefore, it may only be necessary to coordinate the water-supply model to the agricultural and not to the industrial model (see Figure 3.1).

The scheme considered here links the growth of the main sectors and the size of the labor force. Given that local population growth is largely independent of the development of regional industry and agriculture, it is clear that control is only possible through migration.

The Silistra case study has shown that the most important factors affecting migration are: the average regional wage, the amount of dwelling space per employee, and the quality of services (which can be measured indirectly by the number of employees in the service sector). These three factors are all included in the analysis. The mean level of wages in the region may be achieved by weighting the wages paid in the various sectors according to the optimal weight of each sector in the regional economy. The wage level in each sector i may be taken to be the national average, and these aggregated

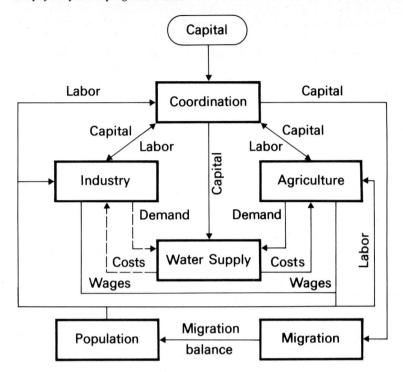

Figure 3.1. A simplified system of models.

data introduced into model (3.1). If the quality of services is measured in terms of the number of employees in this sector, the method of control is rather simple. The labor resources in the coordination block can be allocated in several ways. By changing the number of employees in the service sector, one can (to some extent) influence the regional migration process. The same is true for the distribution of capital investment.

3.7. Generalization of the approach

The above discussion makes it clear that the proposed system of models is still incomplete. Many more interactions occur in the real world than can be described by this system, and therefore the structure of the system should be made more flexible. One possible approach is outlined below.

1. The system of models should have a structure that can be modified depending on the problems to be solved.
2. There should be a variety of frameworks depending on the institutional structure of the region under investigation (for example, one framework for planned, and one for market economies).

3. The models describing each subsystem should be carefully classified and the most appropriate model used to analyze each problem.
4. The software developed for the system of models should include all potentially important links and feedbacks between the subsystems.

Going through these points in more detail, the first implies that the structure of the system must be easy to modify. A region generally suffers from only a few of the problems analyzed by the system and therefore, in principle, it should be possible to make the system of models more compact in each practical case.

The second point is necessary because of the institutional differences between nations; two basic models are needed for regional development planning and management, one for market economies and one for planned economies. In many cases, however, the sectoral models will be the same for both planned and market economies (for example, models for water supply and population growth).

Point 3 above states that a thorough classification of models is necessary for each sector, to ensure that the most appropriate model is used for a particular problem. For instance, regional water-supply models may be classified according to: their treatment of time (static or dynamic models); their treatment of water-quality problems; their treatment of water-regulation problems; and the size of the water-supply system considered (large irrigation system, system with mainstream regulation, etc.). The problem is then to find the optimal balance between comprehensive and partial models: both have advantages and disadvantages. Specialized models are usually very compact, but they need to be linked to the rest of the system and this is not always easy. The overall system structure should be such as to minimize the modifications needed if additional subsystems are to be included.

Point 4 above is self-explanatory; the software should include all potentially important links even if only a small proportion of them are used in any one application.

References

Albegov, M. (1979). Generalized Regional Agriculture Model (GRAM): Basic Version. WP-79-93. International Institute for Applied Systems Analysis, Laxenburg, Austria.

Albegov, M. and Chernyatin, V. (1978). An Approach to the Construction of the Regional Water Resource Model. RM-78-59. International Institute for Applied Systems Analysis, Laxenburg, Austria.

Albegov, M. and Kulikowski, R. (Editors) (1978). Noteć Regional Development – Proceedings of Task Force Meeting I on Noteć Regional Development. RM-78-40. International Institute for Applied Systems Analysis, Laxenburg, Austria.

Andersson, Å. E. and Philipov, D. (Editors) (1979). Proceedings of Task Force Meeting I on Regional Development Planning for the Silistra Region (Bulgaria). CP-79-7. International Institute for Applied Systems Analysis, Laxenburg, Austria.

Chernyatin, V. (1981). Modeling Regional Water Supply: The Silistra Case Study. WP-81-93. International Institute for Applied Systems Analysis, Laxenburg, Austria.

Mednitsky, V. (1978). Special Optimization Methods. Nauka, Moskow.

Willekens, F. and Rogers, A. (1978). Spatial Population Analysis: Methods and Computer Programs. RR-78-18. International Institute for Applied Systems Analysis, Laxenburg, Austria.

PART III

OVERVIEWS OF MODEL SYSTEMS FOR
REGIONAL ECONOMIC ANALYSIS

Regional Development Modeling: Theory and Practice
M. Albegov, A.E. Andersson and F. Snickars (editors)
North-Holland Publishing Company
© IIASA, 1982

Chapter 4

REGIONAL ECONOMIC MODELING IN THE USSR

Abel G. Aganbegyan
Institute for Economics and Industrial Engineering, Siberian Branch of the USSR Academy of Sciences, Novosibirsk (USSR)

4.1. Introduction

In the USSR considerable advances in national economic management techniques have been made in the last few years, in particular with respect to the reorganization of regional development management and the improvement of territorial planning methods. The measures adopted are aimed at satisfying the demands of the population, improving social relations and production efficiency, balancing regional economic development, rationalizing the use of natural resources, and ensuring environmental protection.

Regional economic policy in the Soviet Union is based upon three interrelated concepts of regional development: the direction of capital to those centers in which natural resources are located, the institutional development of regions, and the gradual equalization of levels of regional development. These regional policies are implemented in a hierarchically organized territory ranging from large zones to local industrial centers.

At present, regional policy has three main objectives. The first is to determine the principal directions of national plans and resource allocation for the integrated development of certain areas (Non-chernozem region, Tumen region, Krasnoyarsk territory,[*] Moscow, Leningrad, Siberian and Far Eastern territorial production complexes,[**] etc.).

The second goal is to promote wider intraregional cooperation, and the assignment of labor to areas of rapid economic development. This may be achieved by establishing higher wage coefficients, accelerating the growth of the social infrastructure, or by offering certain privileges and payments depending on period of service. Such increased intraregional cooperation may also be brought about by a wider use of investment

---

[*] In general, the term "territory" is used for an area whose boundaries are not so clearly defined as those for a region.
[**] For a description of a "territorial production complex," see Chapter 24.

funds for infrastructural development, greater material incentives, and greater investment in research and development to improve production technology.

The third aim is to use regional policy to stimulate the development of labor-saving technologies and to encourage the implementation of technological systems specifically adapted to northern and Siberian climatic conditions.

4.2. Brief survey of regional models used for regional planning in the USSR

The models at present in use in the USSR may be classified according to geographical scale: national (national economy) models with regional elements (share of regions in the national economy, interregional flows of goods, migration flows, etc.); regional models, which in turn may be divided into two subgroups, those associated with the independent and relatively autonomous Soviet republics and those related to large economic regions; subregional models, most frequently encountered as the models of territorial production complexes (TPC models); and local models, i.e., urban agglomeration models and models of human settlement systems. These four principal levels can be classified according to the combination of models they include. The models can also be classified according to the extent to which they include regional economic activities. According to such a subdivision, one may distinguish comprehensive (national economy) models, with differing degrees of aggregation of all the elements of the regional economy, from partial models which include only the most important production units of a given region.

4.3. Examples of Soviet national–regional models

The economy of a region is characterized by a certain degree of autonomy but at the same time it is linked to those of other regions and the whole national economy by a series of interdependences. The importance of these interdependences increases with the growth of the national economy, the extension of cooperation and specialization, and the rate of technological change. Furthermore, most resources are distributed between the regions from the central level, giving rise to the problem of how best to coordinate resource distribution in the interests of the national economy. The solution of this problem requires an integrated systems approach, in which the interregional and intersectoral dependences are dealt with explicitly.

4.3.1. Coordination models

For optimal long-term planning of the national economy, a model system that coordinates sectoral and territorial plans is now being developed at the Central Institute for Economics and Mathematics (CIEM) of the USSR Academy of Sciences (see Baranov and Matlin, 1976). The core of the system consists of models of the different national economic sectors, models of the Soviet republics and economic regions of the Russian Federation, and a sectoral–territorial coordination model on the upper (national) level. The regional model tests various possibilities for fulfilling sectoral plans, given the available regional supply of labor and natural resources and the need to match regional interests with national economic development. This system operates with an interactive information exchange between the models.

A series of calculations for the system of models has shown that it may be used within the framework of a computer-based system of planning calculations (ASPC)* to coordinate planning decisions at the upper level of the national economy.

Current research is concentrating on improving some blocks of these models and also their coordination scheme.

4.3.2. Models for regional development planning

Coordination of all the regional development plans of the country is also possible using the optimization input–output interregional model being developed at the Institute for Economics and Industrial Engineering (IEIE) of the Siberian Branch of the USSR Academy of Sciences (Aganbegyan et al., 1972; Rayatskas, 1975, 1978); see also Chapter 22. The structure of the model is based on regional relations between inputs and outputs and on relations between the production and distribution of goods, represented by intersectoral and interregional commodity flows. The objective of the model is to maximize nonproductive consumption for the national economy as a whole.

This model is mainly used for studying the share of large regions (including Siberia) in the national division of labor. The investigation of interregional relations has led to a number of important conclusions about sectoral specialization in some regions of the country. The model includes between two and 11 territorial zones and between 16 and 50 sectors of production.

4.3.3. Industrial growth models

A system of models for optimal industrial growth and location has been developed by the Production Forces Allocation Council of the State Planning Committee. It is used to determine the most advantageous location of various enterprises, taking into account the nonlinear dependence of resource costs on the scale of their use and the agglomeration effects (Albegov, 1970). This system of models has been tested using data from several regions of the USSR.

---

*Various activities are underway in the USSR to develop and implement this computer-based system.

4.3.4. Regional modeling at the republic level

Modeling of socioeconomic processes at the level of economic regions or republics is aimed at solving the most urgent economic planning problems, while meeting the requirements of the planning bodies. Standard economic and computational methods are used for this purpose. The most successful among them are the planning input–output models (with various modifications), demographic forecasting models, population income and balance-of-payments models, production organization models, economic growth models, and so on.

Overall approaches to the modeling of development in certain economic regions and Soviet republics are based on the traditional input–output scheme with additional constraints on labor resources and with some elements of optimization. The development forecasts for the Byelorussian Republic, for example, were obtained using this method. However, dual estimates for the main indicators of economic development, including the volume of the gross national product, capital investment, imports and exports, and labor supply, were also derived.

4.3.5. New types of model

As a result of the generalization of experience gained in regional planning using the standard models described above, it is now possible to concentrate on applying the systems approach to modeling the regional economy. This kind of research is being carried out in the Baltic Republics, Azerbaijan, Kazakhstan, and other Soviet republics (Baizakov, 1970; Rayatskas, 1976). Certain theoretical and practical results have already been obtained. For example, the upper level of the regional model dealing with development of the Latvian Republic is represented by a new type of macro-model describing the national economy. Detailed blocks are also used for a more comprehensive description of labor-supply development.

A variant of the regional model system has been designed for the economy of Kazakhstan (Baizakov, 1970). Optimal development is determined using sectoral and territorial regulatory parameters. The economic system is represented by a group of sectoral and intersectoral input–output models, which provide information for further specification of the regulatory parameters. The system of regional models described above has been tested and a gaming experiment reflecting the interrelations of economic projects within a given republic has been undertaken on the basis of the results.

A second variant of this regional model system has been built for the Lithuanian Republic and implemented in the Republic State Planning Committee (Rayatskas, 1976). The approach taken involves integration of the planning system with the economic calculations. The system is used to obtain several variants of regional plans. Similar systems of models are now being developed in all the Soviet republics.

The next stage of these activities consists in classification of the systems, followed by development of a common conceptual framework on which simulation models of territorial growth with wider coverage of socioeconomic processes can be based.

4.3.6. Establishment of new regions

To reflect the conditions of integrated economic development of new regions, the IEIE has created a multilevel system of models, including a network model for implementation of regional development programs or projects, and optimization models that specify the conditions for the supply of the necessary production resources (see Chapter 18).

The methods available for fulfilling the given long-term objectives are described in a network model (for example, the creation of a set of enterprises with a given capacity, which determines the degree and nature of specialization in a particular region). Moreover, these methods can also be used to solve infrastructural problems (production and social), environmental protection problems, etc. Within the given system, the calculation of costs is carried out using dual estimates from lower-level models. Calculations carried out using the network model make it possible to obtain the optimal scheme for achieving given objectives. In addition, the demand for the resources required for this optimal scheme is determined. The lower-level models define the optimal paths of development for all sectors supplying these resources. After a number of iterations and by means of parameter exchange, an approximate version of the scheme is found for which the aims of the regional program are achieved with the minimum production and resource costs. This methodology is being implemented, in particular, to construct a long-term economic development program for the Baikal–Amur Railway zone.

4.3.7. Analysis of territorial production complexes

Program implementation for the territorial production complexes (TPCs) is being used and improved at present for solving major national economic problems (see Chapter 26). The approach is being used for a number of major projects, including the development of the oil and gas resources of western Siberia, the hydroelectric-power, mineral, and forest resources of the Angara–Enisei region, and the Kansk–Achinsk coalfield.

For each TPC, population, services, and the natural environment are analyzed jointly to achieve the aims defined at the national level. This territorial organization of production ensures balanced development of all the sectors of the territorial economy, including social services.

The optimization models dealing with the establishment and development of TPCs are at different stages of development. The most advanced are the long-term optimization models of production and spatial structure. The dynamic models of TPC formation and development have been successfully tested and work on medium-term planning and simulation models of TPC formation has begun. The next step is to include the TPC models in the national computer-based system of planning calculations. In addition, the construction of optimization models for the spatial production structure of subregions within the established unit has begun.

The available models of TPC development are used for solving problems at five levels of the territorial system: large regions, administrative territorial units (e.g., region, territory), single TPCs, industrial centers, and isolated urban areas.

4.4. Urban analysis

Development of the regions is closely associated with urban growth. At present, a policy of equalization of urban socioeconomic growth requires the elaboration of special types of plan. Some of the major aims with respect to large and medium-sized cities (Faerman and Oleinik-Ovod, 1977) may be summarized as follows: to restrain urban growth that does not lead to rational development of sectoral specialization (primarily, by amalgamation of scientific institutes and production enterprises); to promote population growth in medium-sized cities; to provide investment for protection of the environment and for production and social infrastructure; and to achieve balanced development of the main sectors and the infrastructure.

The major problems faced by small cities are, on the other hand: to provide multidimensional development; to increase the professional and skilled-manual potential; to develop a social infrastructure similar to that of large cities; and to define an appropriate role for each small city within the local settlement system.

To determine and optimize the more important contributions made by each type of center to the national economy it is necessary to analyze the advantages of agglomeration for industry and for commerce. The scheme outlined below has been developed at the CIEM and is used to determine the impact and effectiveness of scientific activities (themselves dependent on the type of city) on the development of a given city or region. This is a fairly typical example of the modeling of agglomeration effects.

Regional differentials in the effectiveness of scientific activities are represented by a special formula, determined for cities of different types with different population and dwelling-unit densities. The parameters of this formula represent the number of staff employed and the capital of scientific institutes, design bureaus, etc., as well as various indexes of urban infrastructure.

At present, the integrated forecasting and planning models representing intersectoral links are among the most advanced urban development models.

A method of systems forecasting is also used to consider the role of a given city in the national division of labor. It consists in the elaboration of certain correction mechanisms for independent forecasts of regional intersectoral links. These mechanisms make the links consistent with those forecast at the upper level of the territorial hierarchy. The same procedures, often based on information-theory arguments, are used for various levels of analysis.

At an interurban level (agglomeration, settlement system), the models developed at the IEIE for the demoeconomic development of cities in settlement systems should be mentioned. These models assess the economic growth of a city and its demographic development in relation to the investment required in the urban sector (the data for these models have been gathered from the towns of the Irkutsk territory).

4.4.1. Intraregional models

Simulation models are used to represent intraurban conditions and their effect on industrial development. The sectors are described using production functions. The distribution of labor and investment among the sectors is iteratively improved in such a way that the multisectoral system is internally balanced under the constraints of limited urban resources. A set of indicators estimating the imbalance caused by the failure of the sectors to achieve normative levels of development has been designed. These indicators may be employed to determine the optimal rate and scale of the development of services. The parameters of these indicators are derived on the assumption that the optimal trajectories deviate only little from previous development trends.

During recent years, forecasting and planning models describing a city in terms of socioeconomic relationships at an urban level have been used in planning urban development. One example is the integrated forecasting model for a large urban region elaborated at the Institute for Socioeconomic Problems (ISP) of the USSR Academy of Sciences. In this model, forecasting is carried out in several iterations involving coordination of a number of factors in six main blocks: economic (specialization), demographic (population), social (social structure and way of life), urban services (infrastructure), ecological (environment), and managerial (coordination of territorial and sectoral aspects of management). The model has been implemented in an integrated forecast of the development of Leningrad as well as for a number of towns in Tataria and the northern part of the Kolsky Peninsula. Similar models are now being developed at the Institute for Economics of the Ukrainian Academy of Sciences (Timchuk, 1974).

The aim of the spatial models is to find a distribution of economic sectors, housing, and services that will meet the spatial limitations of the total forecast levels of urban infrastructural development. The social aspect is taken into account by assigning a quantitative value to the level of services provided and by estimating the loss of time and effort incurred by, for example, commuting. Among such models are a comprehensive location model of long-term urban construction that has been built at CIEM, and another model including location variables for housing and services that has been built at the Institute for Systems Studies (ISS) and at the Central Institute for Urban Construction Research (CIUCR).

4.4.2. Technical subsystem models

The sectoral analyses make use of models of transportation systems (including estimates of transportation flows, selection of modes of transport, etc.), models for allocating urban services, optimization models of housing structure, and optimization models of financial flows (Kovshov, 1977) (see also Chapter 17).

In allocating services to each sector of the urban economy, an assessment of the urban resources is necessary. In particular, an economic appraisal of the use of time is necessary. Appraisal indicators of urban land use which take into account the actual cost of construction are also included.

According to the constitution of the USSR and the Union Republics, the people's deputies should examine and coordinate all the plans for the development and operation of all nonproductive sectors within their territory. The resolutions adopted by the CPSU Central Committee and the infrastructural units should be controlled by municipal authorities. At present, nearly half of such units are in the hands of different ministries.

Conflicts between certain types of production, social, and ecological interests, as well as varying urban and regional conditions are considered in gaming models. These models allow one to study how the socioeconomic optimum for the national economy can be determined through a bargaining procedure.

4.4.3. Uses of regional models in the planning process

Economic models of some regions and of TPCs are widely used for long-term planning purposes in the eastern part of the USSR and for planning the development of natural resources in the Siberian and Far Eastern regions (see, for instance, Chapter 24).

The results of optimization models were used to make proposals about production development in Siberia for the ninth and tenth five-year plans and for the period up to 2000, and for the Krasnoyarsk territory up to 1980 and 1990. They were also used to make proposals regarding the oil and gas industries in western Siberia and the coal industry in southern Yakutia.

The regional projects for Krasnoyarsk territory, Irkutsk territory, Sharypovsky industrial center, and Sayansky TPC were developed on the basis of these results. Thus, the work of Siberian scientists on regional modeling was directly used by the State Planning Committee and the Russian Federation, by the State Construction Committee, and by local planning bodies.

The development and location model for the agro-industrial complex of the Non-chernozem zone constitutes one example of the application of regional modeling by the State Planning Committee. Optimization calculations concerning the location of the food, fertilizer, and light-engineering industries, and the construction-material, agriculture, and energy sectors were used for long-term planning in Kazakhstan. The results were obtained from an intersectoral regional model.

In the Estonian Republic, modeling the development of the fuel, energy, and chemical sectors formed the basis for policy decisions on the scale and location of major enterprises in Estonia (see also Chapter 17).

In the field of urban growth modeling, much work has been done for large urban agglomerations such as Moscow and Leningrad, as well as for a number of large cities in the Ukraine and Byelorussia, for towns in Irkutsk territory, and for the northern part of the Kolsky Peninsula.

4.5. Some directions for future research

Through the use of regional and urban development models, it is possible to consider many development scenarios and to avoid the disadvantages of simply extrapolating from existing planning trends. Models may also help to determine turning points in the development of the planning system (Borschevsky et al., 1975).

The following problems are considered the most important for future regional research: the classification of forecasting models and their coordination in an interregional setting; the quantification of regional development criteria; multiobjective optimization of regional plans; the elaboration of interregional coordination methods and formalized approaches for equalizing the economic development of different regions; and the improvement of coordination algorithms in the sectoral and regional models. The problem of developing tools at the regional level for coordinating national, regional, and ministerial economic interests is also of great importance. This latter problem may perhaps be solved, for example, through a system of deductions from the profits of enterprises and a subsequent amalgamation of rental payments with local budgets.

There should be further research on development project techniques and their coordination with the available planning and forecasting models of the national economy, of the republics, and of the economic regions. The system of models of regional economies should be integrated and coordinated with the newly developed computer-based systems of planning calculations (ASPCs) for each republic.

References

Aganbegyan, A. G., Bagrinovski, K. A., and Granberg, A. G. (1972). Systema Modelei dlya Planirovaniya Narodnogo Khozyaistva (A System of National Economic Planning Models). Mysl., Moscow.

Albegov, M. M. (1970). Problemy optimizatsii prostranstvennogo planirovaniya (Spatial planning optimization problems). Ekonomika i Matematicheskie Metody, 6 (6): 864–871.

Baizakov, S. B. (1970). Systema Modelei Razvitiya Narodnogo Khozyaistva Kazakhstana: Mezhdypromyshlenniye Voprosy Optimalnogo Planirovaniya (A System of Models for Kazakhstan Economic Development: Interindustrial Issues of Optimal Planning). Nauka, Alma-Ata.

Baranov, E. and Matlin, N. S. (1976). Ob eksperimentalnoi realizatsii systemy modelei optimalnogo perspektivnogo planirovaniya (The testing of a system of optimal long-term planning models). Ekonomika i Matematicheskie Metody, 12 (4): 627–649.

Borschevsky, M. V., Uspensky, S. V., and Shkorotan, O. I. (1975). Gorod: Metodologicheskie Problemy Integrirovannogo Sotsialnoekonomicheskogo Planirovaniya (The City: Methodological Problems of Integrated Socioeconomic Planning). Nauka, Moscow.

Faerman, E. Yu. and Oleinik-Ovod, Yu. A. (1977). Voprosy Planirovaniya Rosta Gorodov (Urban Growth Planning Issues). Central Institute for Economics and Mathematics, USSR Academy of Sciences, Moscow.

Kovshov, G. N. (1977). Transport v sisteme modelei perspektivnogo planirovaniya narodnogo khozyaistva (Modeling the transportation system for long-term economic planning). Ekonomika i Matematicheskie Metody, 13 (5): 1033–1054.

Rayatskas, R. L. (1975). Vkhodniye–Vykhodniye Tablitsy Pri Analize Prostranstvennykh Proportsii v SSSR (Input–Output Tables for the Analysis of the Spatial Proportions of the USSR). Nauka, Novosibirsk.

Rayatskas, R. L. (1976). Systema Modelei Planirovaniya i Predskazaniya (A Planning and Forecasting Model System). Ekonomika, Moscow.

Rayatskas, R. L. (1978). Opyt Razrabotky Modelei Planirovaniya Vkhoda-Vykhoda dlya Ekonomicheskikh Regionov (Experience in Elaborating Input–Output Models for the Planning of an Economic Region). Nauka, Moscow.

Timchuk, X. R. (1974). Methods for Evaluating Urban and Regional Development. Budivelnik, Kiev.

Regional Development Modeling: Theory and Practice
M. Albegov, A.E. Andersson and F. Snickars (editors)
North-Holland Publishing Company
© IIASA, 1982

Chapter 5

MULTIREGIONAL MODELING: A GENERAL APPRAISAL

Raymond Courbis
Group for Applied Macroeconomic Analysis, University of Paris-X-Nanterre, Nanterre
(France)

5.1. Introduction

Since the beginning of the 1970s, the problem of interactions between regional and national development has become of central concern in regional and multiregional modeling: at a regional level, for analyzing the regional impact of national activity and national policies; at a national level, because it is recognized that space is not neutral but has feedback effects on national development.

For multiregional models (and also for their regional counterparts), these reciprocal effects lead to the problem of modeling the interaction between regional and national development. Several approaches to solving this problem have been proposed (see Section 5.2). However, to analyze regional and national data together in an integrated way, a "regional–national" model characterized by interdependent regional and national variables is required.

When constructing such regional–national models, a new problem arises: what is the most appropriate spatial level (national, regional, or subregional) for determining each variable? In this chapter, the conditions under which it is necessary to adopt a regional approach (Section 5.3), a national approach (Section 5.4), or an interregional approach (Section 5.5) are examined. The chapter ends with some concluding remarks about the importance of analyzing each variable at the appropriate level (Section 5.6).

5.2. Interaction between regional and national development and classification of multiregional models

The problem of regional–national relationships is posed in various ways in the different types of model which consider the national economy as divided into a number of regions.* To examine these different methods more closely it is useful to classify multiregional models into four types (Figure 5.1).**

5.2.1. Top-down models

In top-down models, the values of the regional variables are connected to those of the corresponding national variables. Such models follow the lines proposed by Klein (1969) for regional (single-region) models: they assume that the regional economy is dependent upon the national economy and that the size of the region is sufficiently small to have no significant impact on national development.

Very often such models are only "regionalization" models that simply allocate, among the regions, the total national estimates (determined elsewhere, for example by a national model). This regionalization procedure can be performed in two ways: by a pure allocation, whereby the national figures are decomposed by means of structural share coefficients (extrapolated or endogenous) whose sum is equal to 1; or by a constrained linkage, whereby regional variables are determined by linkage to the corresponding national variables. In this latter case, if the system of regional equations does not represent a perfect allocation system, reaggregation of the regional figures will not necessarily reproduce the initial national aggregate totals, and so the regional values need to be adjusted to the national totals.

The top-down approach is quite simple. Its practical interest lies in the fact that a large number of regions (and industries) can be considered, thus allowing multiregional predictions that are consistent with national forecasts to be made and the multiregional impact of national development or policy decisions to be simulated.

Several models of this type have been built. For the United States they include the Harris model (1970, 1973, 1978, 1980), MULTIREGION (Olsen, 1976), IDIOM (income determination input–output model) (Dresch and Goldberg, 1973; Dresch and Updegrove, 1978; Dresch, 1980), and the Milne–Adams–Glickman model (Milne et al., 1980). For Canada, the Candide-R model (see below) is partly of this type (for industrial production and investment, and for wage rates), but labor supply and demand and housing investments, in contrast, are obtained using a direct regional approach. Similarly, the Funck and Rembold model (1975) for the Federal Republic of Germany and the Balamo model for Japan (Kawashima, 1977) can be cited. For

---

* There are also some multiregional models that do not consider entire countries but only several regions, for example the models built by Crow (1973) for the northeast corridor of the United States, by Ballard and Glickman (1977) for the Delaware Valley, by Treyz et al. (1977) for Massachusetts, by Carter and Ireri (1970) for California and Arizona, and by Riefler and Tiebout (1970) for California and Washington.

** This classification emphasizes the structure of multiregional models, but other classifications, based on the mode of operation or purpose of the model, such as simulation, optimization, projection, forecasting, policy, or planning can also be used. All the models referred to in this chapter are simulation models (for forecasting and impact analysis), but optimization models have also been constructed, often using a top-down approach. For a description of work in the Soviet Union, see Albegov (1977).

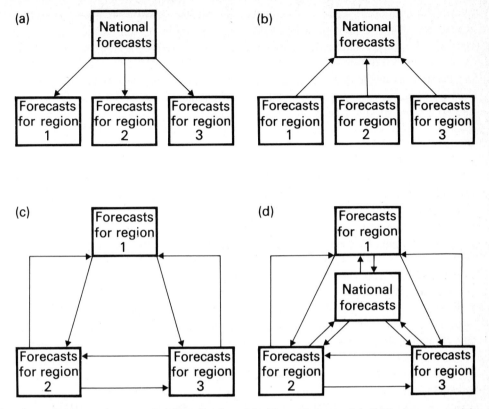

Figure 5.1. The four types of multiregional model: (a) top-down model; (b) bottom-up model; (c) interregional model; (d) regional—national model.

France, the detailed regional projections made for planning purposes by the National Institute for Statistics and Economic Studies (INSEE) also result from a region-alization approach.

Since top-down multiregional models do not introduce any feedback from the regional to the national level, they assume a dichotomy between regional and national analysis. Such models are useful only for analyzing the impact of the national economy upon the region in question without considering the reverse effects, or for making a multiregional forecast consistent with national forecasts which one does not wish to reestimate (for example, because the national model is assumed to be more reliable).

5.2.2. Bottom-up models

When the bottom-up approach is used the regional figures are completely determined at the regional level and the national variables, in contrast to the top-down procedure, result entirely from the aggregation of regional variables. However, only a few multiregional models exist that completely determine national development by aggregation. The Fukuchi model (1978) for Japan more or less corresponds to this type. By adopting a neoclassical approach, this model determines regional production (analyzed for three industries) as a function of the regional labor supply and of the capital stock available in each region. The regional submodels (for the nine regions of Japan) are almost independent, except for regional labor supply, which depends on migration between the regions.

5.2.3. Interregional input–output models

Interregional input–output models emphasize the interrelations existing between regions through interregional flows of products (determined by regional propensities to import or to export, or by a gravity-type method).*

This approach has been developed in great detail for the United States (mainly with respect to analyzing interregional transportation) by Polenske (1970a, 1972, 1980) in the MRIO (multiregional input–output) model and has frequently been applied.**

In such models, regional final demand is most often exogenous and is calculated outside the model by means of regionalizing national final demand. However, it can also be endogenous for household consumption, in which case it is calculated simply on the basis of budget coefficients and the assumption of a given share of household disposable income in the total value added created by each industry (Hill, 1975; Zuker, 1976).

In the Mitsubishi multiregional model for Japan (Suzuki et al., 1973/1978), however, there is a detailed econometric analysis of regional final demand (household consumption, government investment, housing investment, and productive investment). This model demonstrates the possibility of combining an interregional input–output approach and an econometric approach. The determination of interregional flows is also interesting and takes into account both the demand of importing regions and the available capital stock of exporting regions.

---

* It should be noted that a multiregional input–output model is not necessarily the same thing as an interregional model that is used to describe interregional flows. IDIOM for the United States and the Funck–Rembold model for the Federal Republic of Germany correspond to an unlinked system of regional input–output models linked to a national model by the top-down approach. In contrast, in the model proposed by Leontief (1953), while there are no interregional interrelations, there is regional–national interaction: the regional production of "national" industries is tied to total national demand and has a feedback on the latter through its impact on regional demand.

** Other models have been developed: for the United States by Isard (1951), Moses (1955), Leontief and Strout (1966), Greytak (1970), and Evans and Baxter (1980); for Canada by Zuker (1976); for the United Kingdom by Gordon (1974, 1977); for Italy by Chenery (1953); for the Federal Republic of Germany by Carlberg (1979); for Japan by Suzuki et al. (1973/1978) and Polenske (1970b). See also the surveys by Tiebout (1957), Meyer (1963), Richardson (1972, 1978), Miernyk (1973), and Riefler (1973) that relate to these models.

5.2.4. Regional–national models

Regional–national models simultaneously analyze regional and national development in an integrated manner, national and regional elements being of equal importance. These models are characterized by the existence of feedback relations, regional→ national and national→regional.

The determination of national figures is an integral part of these models, which combine the top-down and bottom-up approaches. They also take into account the fact that decisions are not all made or carried out at the same level (see below). Using such models, it is possible to analyze the impact of regional factors, behavior, and policy on national development; conversely, the impact of national development and policies on regional development may be studied.

Since regional–national models also take account of the interrelationships between the regions (but often more fully than interregional input–output models), they can be considered as a synthesis of the three model types presented above.

The construction of this type of model began in the early 1970s. The REGINA (regional–national) model was proposed for France by Courbis and Prager (1971) (see also Courbis, 1972, 1975a, 1978, 1979a, 1979b; Courbis et al., 1980); a regional–national model for Italy was proposed by Brown et al. (1972, 1978); and the RENA (regional–national) model was built for Belgium at the same time by Thys-Clement et al. (1973, 1979). These three models, which were constructed simultaneously but independently, together constituted the first real attempt at regional–national modeling.* This was followed closely by the construction of the RM (multiregional economic) model for The Netherlands (van Hamel et al., 1975, 1979); the Candide-R model for Canada (d'Amours et al., 1975, 1979); and the Macedoine model for Belgium (Glejser et al., 1973; Glejser, 1975), which was only experimental. Recently, the SERENA (sectoral–regional–national) model was also constructed for Belgium (d'Alcantara et al., 1980).

In the United States regional–national modeling began only recently. The NRIES (national–regional impact evaluation system) model of the Bureau of Economic Analysis, in which the United States is disaggregated into 51 regions, was the first to be constructed (Ballard et al., 1980). Another model is being constructed by Wharton EFA, in which the United States is divided into 16 regions (Adams et al., 1977; Fromm and McCarthy, 1978; Fromm et al., 1980).

5.3. Aggregation of regional behavior and the bottom-up approach

When assuming that national figures are independent of the regional values taken for the corresponding variables, the top-down approach makes the further implicit

---

* The purpose of the REGINA and RENA models, constructed for use by the planning bureaus of France and Belgium, respectively, is to analyze explicitly the interaction between regional and national development. In using a regional–national framework for the Italian model, the aim was to improve the consistency of the regional analyses.

assumption that either the national variables are determined by a national decision-making process (this case is analyzed in Section 5.4), or that it is possible to aggregate regional behavior in a perfect way. Since the quality of national data is generally better than that of regional data, national figures are determined first and regional variables are then calculated through a regionalization method. However, evidence shows that the perfect aggregation assumption is unacceptable for a number of particular variables because (i) regional relationships are nonlinear, thus rendering perfect aggregation at the national level impossible (except, perhaps, in the very short term and under conditions of structural rigidity) and (ii) regional mechanisms do not necessarily follow the same behavioral laws in all regions. This leads to direct determination of the regional figures and the adoption of a regional approach, the national figures being calculated by aggregation.

A good example of this problem is given by the determination of wage increases. In order to explain the increase in nominal wage rates, the Phillips–Lipsey scheme of analysis can be used. However, it should be taken into account that there is no perfect and unique labor market, but rather an entire set of micro markets. As Lipsey (1960) has demonstrated, the nonlinearity of the relationship between unemployment and the wage-rate increase implies that the average national increase in wage rates, which is the aggregation of a number of micro relations, depends on the distribution of unemployment among the labor micro markets, and in particular among the regions.

Thus, the average national increase in wage rates depends not only on the national rate of unemployment (and on price increases) but also on the degree of dispersion of regional rates of unemployment. It appears that the rate of change in wage rates at the national level becomes greater if the degree of interregional dispersion of unemployment increases. This effect of unemployment dispersion has been econometrically verified by several authors when directly analyzing the rate of increase in national wage rates. For accounts of studies of the United States see Archibald (1969), Brechling (1973), and Azevedo and O'Connell (1980); for the United Kingdom see Archibald (1969) and Thomas and Stoney (1971). For France, Lecaillon (1976) pointed out that the national increase in wage rates depends on the tightness of both national and some regional labor markets.

Such results imply that it is necessary to take into account the dispersion of regional unemployment or to analyze the wage increase *directly* at the regional level. The latter approach is in fact the better one because, from the studies made on the dynamics of regional wages, it appears that the coefficients of wage relationships are not necessarily the same, and that these relationships can be specified differently from one region to another; for a survey of these studies see Courbis and Cornilleau (1978b) and Courbis et al. (1980). In short, they imply that the average national increase in wage rates depends explicitly on *all* regional unemployment rates and not only on their dispersion.

In particular, it is important to notice that the labor markets of some regions can be "leading" markets within a given nation. By a process of diffusion, any increase in wages in these regions has an impact on wage increases in the other regions. London and southeast England appear to act as such leading markets for the United Kingdom

(Cowling and Metcalf, 1967; Thomas and Stoney, 1971; Hart and Mackay, 1977*);
the Paris region does the same for France (Deruelle, 1974; Courbis, 1975a; Courbis
and Cornilleau, 1978a); Ontario plays this role for Canada (d'Amours, 1972); and
several regions with high wage levels in the United States have a leading influence
(Brechling, 1973).**

The existence of leading regions has very important consequences for national
development. Thus, for example, the effects of job creation on national wage increases
(and thus on inflation and national competitiveness) differ greatly according to the
regions in which the new jobs are located. Let us assume that there is a single leading
region, namely region 1. In this case, we have

$$\dot{w}_1 = \alpha_1 - \beta_1 u_1 + \gamma_1 \dot{p}_1$$

$$\dot{w}_r = \alpha_r - \beta_r u_r + \lambda_r \dot{w}_1 \qquad (\text{for } r \neq 1)$$

where

$\dot{w}_r$ is the rate of increase in nominal wage rates in region r
$\dot{p}_1$ is the rate of increase in prices in region 1
u_r is the unemployment rate in region r
λ_r is the coefficient of wage-increase diffusion from the leading region to the non-
    leading region r

If x_r is the weight of region r in the national index of wage rates, and assuming
for the sake of simplicity that the λ_r values are equal ($\lambda_r = \lambda$), then

$$\dot{w}_n = \sum_{r=1}^{R} x_r \dot{w}_r = \alpha_n + \mu \gamma_1 \dot{p}_1 - \mu \beta_1 u_1 - \sum_{r=2}^{R} x_r \beta_r u_r$$

where

$$\mu = x_1 + \lambda(1 - x_1)$$

and

$$\alpha_n = \mu \alpha_1 + \sum_{r=2}^{R} x_r \alpha_r$$

The value of $\dot{w}_n$ depends on all the regional unemployment rates. But, since the
value of the coefficient of diffusion in practice is close to 1, it appears that the impact
on $\dot{w}_n$ of the rate of unemployment u_1 in the leading region is much greater than

---

* In addition to the general (national) leading character of the labor market of the London region,
Hart and Mackay point out that some other regions have markets which are also "leading" but
in a more local sense.
** For another study of the United States see also Reed and Hutchinson (1976). In the work of
King and Forster (1973), there is a reciprocal and symmetrical interaction instead of leadership
linkages.

that of the rate of unemployment u_r in a "follower" region ($r \neq 1$). The simulations made for France with the REGINA model have demonstrated that such a mechanism has important consequences; the distribution of jobs among the regions (and, in particular, between the leading region and other regions) has a significant impact on national development (Courbis, 1978, 1979a).

Some authors, however, view diffusion not as a process that occurs between leading and following regions* but rather as one that takes place on a national level by the linkage of regional wages and the average national wage; for an application of this approach to the United Kingdom see Thirlwall (1970), and for a study of Italy see Brown et al. (1972, 1978).

Following this interpretation of diffusion, if we consider only a single region and if this region is sufficiently small to have no feedback effect on national development, we can take average national wages to be exogenous and thus adopt a top-down approach (regional wages are determined as a function of national wages and regional conditions). However, if we wish to construct a multiregional model, such a simplification is no longer possible. In the multiregional case

$$\dot{w}_r = \alpha_r - \beta_r u_r + \gamma_r Z_r + \lambda_r \dot{w}_n \qquad (r = 1, \ldots, R)$$

$$\dot{w}_n = \sum_{r=1}^{R} x_r \dot{w}_r \qquad \left(\sum_{r=1}^{R} x_r = 1 \right)$$

where Z_r are specific regional explanatory variables. Therefore

$$\dot{w}_n = \left[\sum_{r=1}^{R} x_r \alpha_r + \sum_{r=1}^{R} x_r (\gamma_r Z_r - \beta_r u_r) \right] \bigg/ \left[1 - \sum_{r=1}^{R} x_r \lambda_r \right]$$

The result is that, even in this case, $\dot{w}_n$ depends on all the u_r (and Z_r) values, and we have to determine wage rates using a regional approach, but taking account of the links between the regions and the national level.**

This example of the determination of wage rates shows the importance of starting with regional analysis and obtaining national values by aggregation. This has been done for wages in the REGINA model for France, the RENA and Macedoine models for Belgium,*** and the regional–national Italian model. It has also been done in the NRIES model recently constructed for the United States, and has been proposed for the multiregional model of Wharton EFA (Fromm and McCarthy, 1978; Fromm

---

* King and Forster (1973), in their study of the United States, use interactions between all the regions and thus a symmetrical linkage; consequently, $\dot{w}_n$ depends on all the regional conditions.
** It is preferable to speak of a regional rather than a bottom-up approach because of the national linkage $\dot{w}_r \rightarrow \dot{w}_n \rightarrow \dot{w}_r$.
*** However, in the SERENA model, which has recently been built at the Belgian Bureau of Planning, wages are determined only at the national level.

et al., 1980).* Bearing in mind the importance of using a regional approach for wage determination, the Candide-R model for Canada and the RM model for The Netherlands, which both use a purely top-down approach, appear to be of only limited usefulness.**

More generally, a regional approach should be adopted for all variables that are determined on a regional market*** basis or that result from the decisions of regional agents (such as households, local authorities, and regional firms). This is particularly important for the following variables:

- labor supply (a regional approach is used for the nine regional–national models mentioned above);
- production processes; it should be noted here that, even if the individual regional production functions[†] were the same, the same combination of labor and capital would not necessarily be adopted because factor prices, in particular the wage. level, are not the same for all regions;
- household consumption (only explicitly analyzed at the regional level in REGINA, Macedoine, and the Italian model, although there are significant differences between the regions);[‡]
- residential investment (explicitly analyzed with regional relations in REGINA, Candide-R, and the Italian model);
- investment of local authorities (only in the REGINA and RM models is this both endogenous and determined at the regional level).

5.4. National versus regional approaches: the case of production and investment

As we have seen, a regional approach appears to be more suitable for several particular variables, but this does not mean that such an approach should be used for all variables. Since a regional economy is largely an open economy, a number of decisions affecting it are taken within a much broader framework than that of the region.

Let us consider the example of regional production, which is important for regional equilibrium determination. Traditionally, regional theory emphasizes the distinction between national activities, which have a national market, and regional (or local) activities, which have only a regional (or local) market.

---

* In the Wharton EFA model, there is however no explicit relationship for regional wage rates. They are determined indirectly from the equalization of the total supply of labor and the sum of total demand for labor and unemployment (all these variables being a function of the wage rate and other specific explanatory variables).
** Both of these models determine the average increase in wage rates directly at the national level; in the Canadian model, these results are then used to determine regional wage increases. A top-down approach is also used for wages in the Milne–Adams–Glickman model for the United States, but this is only a regionalization model.
*** Thus, for wages, if there is a national market for some groups of workers, a "national approach" should be followed. Therefore, the choice of the appropriate level of analysis is not arbitrary.
† For an interregional comparison of production functions, see Lande (1978).
‡ See Courbis et al. (1980) for France, Gillen and Guccione (1970) for Canada, and Lee (1971) for the United States.

For regional (or local) activities, production depends only on regional (or local) demand and regional (or local) conditions, and it can be determined directly at a regional (or local) level. So, for these activities, a regional approach can be adopted, and national production may be calculated by aggregation (the bottom-up approach).

However, for national industries, national demand (or the aggregated demand of all the regions) should be considered. In this case, emphasis should be placed on the national (or multiregional) market rather than on activities directed toward the regional market. The proposal of Klein (1969) for a prototype regional model gives a leading role to national industries and thus to regional exports, which are themselves determined by national production:

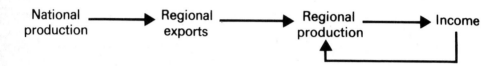

Consequently, for these national activities a top-down approach should be adopted. The consequences are important for those regions whose development depends markedly on national conditions. As the income created by the exporting industries itself creates supplementary regional demand, there is a feedback effect that reinforces the impact of exporting (national) industries on the region in which they are located:

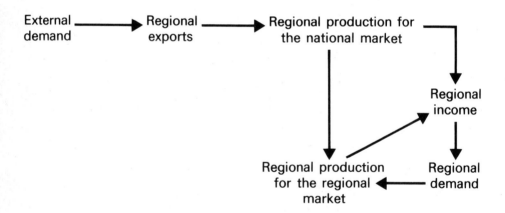

Such an approach is adopted quite frequently in regional and multiregional models. For these models, the regional production of national industries is determined by a top-down method: regional production is related directly to national demand* by

---

* For some models that do not explicitly consider regional production, this type of linkage is made directly for employment in national industries. This is often the case for multiregional models, for example, Olsen's MULTIREGION model (1976) for the United States.

means of regional shares. These shares can be exogenous, as in Leontief's multiregional input–output model (1953) and IDIOM, both constructed for the United States, and in the RENA model for Belgium. Alternatively, they can be endogenous, as in the Milne–Adams–Glickman model for the United States, the Brown–di Palma–Ferrara regional–national model for Italy, the Candide-R model for Canada, and the SERENA model for Belgium. In the Milne–Adams–Glickman regionalization model and the Italian regional–national model, shares by industry and by region are expressed as a function of comparative regional costs* (regional costs relative to national costs or costs of the other remaining regions). For the Candide-R and SERENA models, regional output shares are dependent on regional investment shares.

Analysis of the main distinguishing features of national and regional activities is interesting since it shows that, depending on the sector considered, it is necessary to begin either from a regional and bottom-up analysis (activity in the regional market), or from a national and top-down approach starting with national demand.

The dualist theory of the "economic base" makes it possible to find the appropriate level of analysis for determining the regional and national production of each industry. However, by placing emphasis on external demand and market outlets, the "economic base approach" assumes (Richardson, 1978, p. 12) that there is sufficient production capacity in the region(s) considered. If capacity is insufficient, production will be determined not by demand, but by supply.

From this point of view, it is significant to note that regional investment is usually considered only as a simple component of regional demand: its impact on supply as a factor of production is thus totally neglected. In practice, if regional investments are insufficient, there is a progressive limitation of regional production, and bottlenecks appear at the regional level.

Since they ignore the impact of regional production capacity on regional production, the economic base model and the "demand approach" have limited validity. They can be used principally for the short term. But, in the medium term, one obviously cannot neglect the impact of regional investment behavior on production capacity and regional supply. As proposed by Courbis and Prager (1971), three types of activity can be considered:

- activities whose location is determined by geographical factors (agriculture and extraction industries);
- activities whose location is determined by demand; for these activities (including most of the tertiary sector), regional production is determined by regional demand;
- activities that can be located anywhere; for these activities, which operate in a market larger than the single regional market, location depends not on regional demand, but on investment opportunities (this is generally the case for manufacturing industries).

---

* Comparative regional costs for labor and energy in the Milne–Adams–Glickman model; comparative regional production prices in the Italian model.

For the third type of activity, regional production depends on investment location choices. For given investment decisions at the national level, the locational behavior of multiregional firms determines regional investments, and thus regional accumulation of capital. The latter, according to the value of the capital coefficient, in turn determines regional production in the medium term (assuming that the level of capital utilization is "normal"). Therefore the following scheme exists:

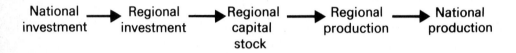

National ⟶ Regional ⟶ Regional ⟶ Regional ⟶ National
investment   investment   capital   production   production
                                  stock

In this case, equilibrium between demand and supply is assured by the external trade of the region (because of the total substitution possibilities that exist in the medium term between regional production and external trade).

Such a determination process emphasizes the impact of the regional capital stock (and regional production capacity) on regional production. It was introduced in the French REGINA model (Courbis, 1972) in the early 1970s. A mechanism of this type can also be found in the RM model for The Netherlands (van Hamel et al., 1975), in Fukuchi's (1978) Japanese model, and in the Macedoine model for Belgium (Glejser, 1975).*

It is interesting to note that in the Canadian Candide-R model (d'Amours et al., 1975) and the new Belgian SERENA model (d'Alcantara et al., 1980), the regional breakdown of total national production depends on the share of investment in each region.** Recently, the builders of the multiregional Wharton EFA model for the United States (Fromm et al., 1980) proposed that regional production be determined as a function of regional capital stock, the demand of the given region and other regions, and comparative costs.

Note that the introduction of capital stock and supply factors, such as those included in the REGINA model for determining regional production, is considered more frequently nowadays.***

In addition, it appears necessary to take into account not only the fact that the market is national (or even multinational) but also that the spatial framework within

---

* There are, however, differences between these models. In the REGINA model, regional capital stock and the capital coefficient (a function of comparative factor costs) directly determine the regional production of "free-located" industries. In the RM, Macedoine, and Fukuchi models, regional production is tied to both regional capital stock and regional employment. Other variables, for example business-cycle variables and external factors, are also introduced in certain cases. In the Macedoine model, regional capital stock (and regional employment) and national demand determine regional output.

** And, more precisely for Candide-R, on the sum of regional investment over four years, which can be considered as a proxy for regional capital stock.

*** The importance of the role of regional capital stock and supply factors in determining regional production has recently been rediscovered by Crow (1979).

which enterprises make decisions is national. Thus, multiregional firms can be seen to follow a two-stage pattern of investment (Courbis, 1972, 1975a, 1979b):* first, they set the national total of their investments (based on total demand and financial possibilities); they then distribute this national total among the regions, taking into account the investment opportunities in the different regions.**

Such a two-stage approach appears essential. It leads to an emphasis on supply and means that national development has a significant influence on regional figures. But since national costs are the result of regional costs, there is therefore a feedback effect on national figures. It also appears that the choice between regional (bottom-up) and national (top-down) approaches is not one of arbitrary selection, but depends on the level of the particular decision considered and on the framework within which the decision-maker operates.***

5.5. An interregional approach and interregional links

The choice to be made is not only between a regional and a national approach: it may also be necessary to adopt an interregional approach, i.e., to consider all the regions. This is obviously necessary for variables such as interregional flows of products. Generally, in multiregional input–output models, a direct determination of interregional product flows allows the regional surplus and thus the regional production of exporting industries to be calculated. Regional production therefore depends on the demand of each region. However, this implicitly assumes that there is regional specialization and hence a set of regional markets in which exporting industries can sell. In the national top-down approach, in contrast, it is assumed that there is a unique, but national, market in which the producers of all regions are competing.

In the NRIES (national–regional impact evaluation system) model (Ballard and Wendling, 1980; Ballard et al., 1980a, 1980b), an approach similar to that of the gravity-type interregional input–output models is adopted for determining regional output. For each region i, the determinants of production (gross output, personal income, retail sales, and population) in all the other regions j are scaled by the distance from region i to region j, and then summed. This gives aggregated indicators (calculated for each region), which are then incorporated in the NRIES production equations for output of durable[†] and nondurable goods, retail sales, transportation, communication,

---

* The REGINA, RM, and Candide-R models consider such a two-stage process. But in Macedoine, as in the Fukuchi model and the Wharton EFA project (Fromm et al., 1980), regional investment is determined directly at the regional level.

** An interesting model that adopts this two-stage determination process for an interregional allocation of capital stock has recently been proposed by van Rompuy and de Bruyne (1979).

*** Although thus far we have considered only two levels (national and regional), more levels can be distinguished. The REGINA model (Courbis, 1975b) introduces three levels: national, regional, and zonal (rural and urban zones); the labor force is analyzed at the zonal level.

† For durable-goods output equations, the NRIES model also introduces the comparative manufacturing-sector wage rate in a particular region relative to the national rate as a relative cost variable. This is similar to the method used in the Milne–Adams–Glickman model for calculating output shares by industry and by region (see above).

and public utilities. Fromm et al. (1980) propose a similar approach for the Wharton EFA multiregional model. However, in this latter case demand originating from outside the regions is considered globally, and the capital stock of the producing region is also introduced.

More generally, an interregional approach should be followed whenever inter-regional variables are considered. This is particularly important for migration (of workers and population), which depends on the comparative opportunities available in each region.

However, such interregional links are not introduced only for regional production (to take into account interregional flows of goods*) and regional labor markets (because of migration**). As we have seen in Section 5.3, the leading character of some labor markets (or a more general process of interaction between the regions) also introduces an interregional link for wage determination. The consequences of this are extremely important not only at the regional level but also (as pointed out in Section 5.3) for national development.

In the Macedoine model for Belgium (Glejser, 1975), an interregional link, which, for each region considered, depends on demand in the other regions and in other countries (and of course on conditions in the region considered), is also introduced for regional investment.***

These examples show the importance of correctly analyzing interregional links and of considering not only regional–national but also regional–regional interactions.

5.6. Concluding remarks

The main conclusion of this chapter is that the choice of a bottom-up, top-down, or interregional approach is not (or at least should not be) arbitrary.

Although a pure top-down approach is the simplest and can be used to calculate the regional impacts of national policies or to make regional forecasts that are consistent with national figures, national feedback effects are ignored. This approach is therefore unsuitable for analyzing the national impacts of regional disequilibria and regional policies. At the regional level, it can be used only if exogenous modifications of national figures are considered. Using the top-down approach, in which

---

* The "interregional flows" approach puts the emphasis on demand, while the "capital stock" approach emphasizes supply. However, it would be possible to unify these two approaches by assuming that trade flows depend on the pressure of demand and, thus, on the rate of capital utilization and supply.

** This refers not only to definitive migration but also to commuting, as in the RENA model for Belgium and the REGINA model for eastern and northern France with the adjacent foreign border regions.

*** Such a formulation can be interpreted as the reduced form of a two-stage (national and regional) process of investment determination. In the Macedoine model, unemployment in a region depends on employment both in that region and in the neighboring regions. It can also be considered as a reduced form introducing implicitly the impact of migration (and the effects of job-creation) on decisions to migrate.

it is assumed that total national figures are not modified, if modifications were made for one region, this would lead automatically to modifications for all other regions. In this case, any results obtained would be meaningless. Thus, although a top-down approach can be used for regional impact analysis or regional forecasting, it has certain limitations.

The bottom-up approach, on the other hand, allows national impacts to be analyzed. Nevertheless, a purely bottom-up approach ignores national linkages and the influence of other regions on the region under consideration.

The interregional approach is useful if interregional links are to be emphasized. However, several variables are determined only at the level of a single region or at the national level.

Thus, it appears necessary to consider in parallel the three types of interaction described above: regional→national, national→regional, and regional→regional. Over the last few years, several models based on a combination of these three approaches have been developed. However, it is important to note that, for each variable, the choice of approach to be used cannot be an arbitrary decision. It is necessary to determine directly at the regional level (bottom-up) the variables that are determined by a regional market or that result from the decisions of regional agents (production processes, employment and labor supply, wage determination, household consumption, residential investment, investment by local public authorities, etc.).* A top-down approach is more appropriate for variables determined by a national market or by national agents (prices, investment of multiregional firms, interest rates, government demand, etc.).** An interregional approach can be used for interregional variables (interregional flows, migration and commuting, etc.) and also for variables determined in terms of comparative regional opportunities or influenced by an interregional diffusion process (such as wage leadership).

An integrated regional–national–interregional approach is therefore necessary in order to perform a meaningful analysis. As demonstrated by the simulations made for the French economy using the REGINA model, such an integrated approach is quite important in analyzing the impacts of regional disequilibria and regional policies on national development.*** Simulations made with this model (Courbis, 1979a) have also demonstrated the influence of regional policies on national development. If we consider, for example, a policy aiming to relocate 40,000 jobs in the manufacturing sector from the Paris region to the provinces over a 10-year period, REGINA simulations have shown that, at the end of the period, the overall effect is an important increase in total national employment (about 100,000 additional jobs) and improvements

---

* If possible, an intraregional level should also be considered, to deal with intraregional markets or behavior. It was for this reason that a "zonal" (rural–urban) level, in particular for labor supply, was introduced into the REGINA model (Courbis, 1975b).
** To take into account the behavior of multinational firms, it would be interesting to introduce a world or multicountry level into the model.
*** The REGINA model was constructed by the Group for Applied Macroeconomic Analysis (GAMA) for the French Planning Bureau. A simplified version, the REGIS (Regionalized Simulation) model, has also been developed (Courbis and Cornilleau, 1978b).

in the external balance of payments and public finances. Such results can be explained by the fact that the Paris region is a wage-leading region and by the inflation-reducing effects of relocating manufacturing jobs to the provinces. Thus, the importance of analyzing wage determination at the regional level, taking into account the wage-leadership phenomenon, is evident. The effect of this interaction is that the national optimum in terms of economic efficiency does not correspond to an equalization of unemployment rates by region.

It is evident that there is widespread interest in considering regional effects and in using integrated regional–national models, and it is clear that regional policy can be used not only to reduce regional inequalities but also to help national development.

From an economic point of view, it is important to take into account regional investment location behavior and the impact of regional capital stock on the regional distribution of free-located activities. This leads to the reintroduction into regional analysis of supply considerations, so that national investment has a direct effect on regional development.

References

Adams, F. G., Hill, J., and McCarthy, M. D. (1977). A National–Multiregional Economic Model for Forecasting Electricity Consumption and Peak Load: A Proposal. Report for the Electric Power Research Institute. Wharton EFA, Philadelphia, Pennsylvania.

Albegov, M. (1977). Modeling for inter-regional and intra-regional allocation of development. In A. Straszak and B. V. Wagle (Editors), Models for Regional Planning and Policy-Making. Proceedings of a joint IBM/IIASA Conference, held in Laxenburg, September 1977. IBM UK Scientific Centre, Peterlee, Durham, pp. 142–151.

d'Alcantara, G., Floridor, J., and Pollefliet, E. (1980). Major Features of the SERENA Model for the Belgian Plan. Working Paper 2279. Bureau du Plan, Brussels.

d'Amours, A. (1972). Salaires, prix et chômage: une approche régionale (Wages, prices, and unemployment: a regional approach). L'Actualité Economique, 47(4):587–620.

d'Amours, A., Simard, G., and Chabot-Plante, F. (1975). Candide-R. L'Actualité Economique, 51(4):603–633.

d'Amours, A., Fortin, G., and Simard, G. (1979). Candide-R, un modèle national régionalisé de l'économie canadienne (Candide-R, a regionalized national model of the Canadian economy.) In R. Courbis (Editor), Modèles Régionaux et Modèles Régionaux–Nationaux (Regional and Regional–National Models). Cujas and Editions du CNRS, Paris, pp. 175–184.

Archibald, G. C. (1969). The Phillips curve and the distribution of unemployment. American Economic Review, 59(2):124–134.

Azevedo, R. E. and O'Connell, G. E. (1980). Migration, unemployment dispersion and the Phillips curve. Regional Science and Urban Economics, 10(2): 287–296.

Ballard, K. P. and Glickman, N. J. (1977). A multiregional econometric forecasting system: a model for the Delaware Valley. Journal of Regional Science, 17(2):161–177.

Ballard, K. P. and Wendling, R. M. (1980). The national–regional impact evaluation system: a spatial model of U.S. economic and demographic activity. Journal of Regional Science, 20(2): 143–158.

Ballard, K. P., Glickman, N. J., and Gustely, R. D. (1980a). A bottom-up approach to multiregional modeling: NRIES. In F. G. Adams and N. J. Glickman (Editors), Modeling the Multiregion Economic System: Theory, Data, and Policy. Heath/Lexington Books, Lexington, Massachusetts, pp. 147–160.

Ballard, K. P., Glickman, N. J., and Wendling, R. M. (1980b). Using a multiregional econometric model to measure the spatial impacts of federal policies. In N. J. Glickman (Editor), The Urban Impacts of Federal Policies. Johns Hopkins University Press, Baltimore, Maryland, pp. 192–216.

Brechling, F. (1973). Wage inflation and the structure of regional unemployment. Journal of Money, Credit and Banking, 5(1) Part 2: 355–379.

Brown, M., di Palma, M., and Ferrara, B. (1972). A regional–national econometric model of Italy. Papers of the Regional Science Association, 29: 25–44.

Brown, M., di Palma, M., and Ferrara, B. (1978). Regional–National Econometric Modeling with an Application to the Italian Economy. Pion Press, London.

Carlberg, M. (1979). Ein interregionales, multisektorales Wachstumsmodell dargestellt für die Bundesrepublik Deutschland (An Interregional Multisectoral Growth Model for the Federal Republic of Germany). Vandenhoeck und Ruprecht, Göttingen, FRG.

Carter, H. O. and Ireri, D. (1970). Linkages of California–Arizona input–output models to analyze water transfer patterns. In A. P. Carter and A. Brody (Editors), Applications of Input–Output Analysis. North-Holland Publishing Co., Amsterdam, pp. 139–167.

Chenery, H. (1953). Regional analysis. In H. Chenery and P. Clark (Editors), The Structure and Growth of the Italian Economy. United States Mutual Security Agency, Rome, pp. 96–115.

Courbis, R. (1972). The REGINA model: a regional–national model of the French economy. Economics of Planning, 12(3): 133–152.

Courbis, R. (1975a). Le modèle REGINA, modèle du développement national, régional et urbain de l'économie française (The REGINA model: a model of the national, regional, and urban development of the French economy). Economie Appliquée, 28(2–3): 569–600.

Courbis, R. (1975b). Urban analysis in the regional–national model REGINA of the French economy. Environment and Planning, 7(7): 863–878.

Courbis, R. (1978). The REGINA model: presentation and first contributions to economic policy. In R. Stone and W. Peterson (Editors), Econometric Contributions to Public Policy. Macmillan Publishing Co., London, pp. 291–311.

Courbis, R. (1979a). Le modèle REGINA, un modèle régionalisé pour la planification française (The REGINA model, a regionalized model for French planning). In G. Gaudard (Editor), Modèles et Politiques de l'Espace Economique (Models and Policies of Economic Space). Volume II. Editions Universitaires, Fribourg, Switzerland, pp. 225–251.

Courbis, R. (1979b). The REGINA model: a regional–national model for French planning. Regional Science and Urban Economics, 9(2–3): 117–139.

Courbis, R. and Cornilleau, G. (1978a). Etude Econométrique de la Dynamique des Taux de Salaires Régionaux (An Econometric Study of the Dynamics of Regional Wages). Working Paper 205. GAMA, Paris.

Courbis, R. and Cornilleau, G. (1978b). The REGIS Model: A Simplified Version of the Regional–National REGINA Model. Paper presented at the XVIIIth European Congress of the Regional Science Association, held at Fribourg (Switzerland), 29 August–1 September 1978.

Courbis, R. and Prager, J. C. (1971). Analyse régionale et planification nationale: le projet de modèle "REGINA" d'analyse interdépendante (Regional analysis and national planning: the REGINA model project for interdependent analysis). Paper presented at the French–Soviet Conference on Using Models for Planning, held in Paris, 11–15 October 1971. Published in 1973 in Collections de l'INSEE R, (12): 5–32.

Courbis, R., Bourdon, J., and Cornilleau, G. (1980). Le Modèle REGINA (The REGINA Model). GAMA Report for the French Planning Office. Edition Economica, Paris. (Forthcoming.)

Cowling, K. and Metcalf, D. (1967). Wage–unemployment relationships: a regional analysis for the U.K., 1960–65. Bulletin of the Oxford University Institute of Economics and Statistics, 29(1): 31–39.

Crow, R. T. (1973). A nationally linked regional econometric model. Journal of Regional Science, 13(2): 187–204.

Crow, R. T. (1979). Output determination and investment specifications in macroeconometric models of open economies. Regional Science and Urban Economics, 9(2–3):141–158.

Deruelle, D. (1974). Détermination à court terme des hausses de salaires: études sectorielles et régionales (Short-term determination of wage increases: sectoral and regional studies). Annales de l'INSEE, 16–17:97–153.

Dresch, S. P. (1980). IDIOM: a sectoral model of the national and regional economies. In F. G. Adams and N. J. Glickman (Editors), Modeling the Multiregion Economic System: Theory, Data, and Policy. Heath/Lexington Books, Lexington, Massachusetts, pp. 161–165.

Dresch, S. P. and Goldberg, R. D. (1973). IDIOM: an inter-industry, national–regional policy evaluation model. Annals of Economic and Social Measurement, 2:323–356.

Dresch, S. P. and Updegrove, D. A. (1978). IDIOM: A Disaggregated Policy Impact Model of the U.S. Economy. Working Paper, Institute for Demographic and Economic Studies, New Haven, Connecticut.

Evans, M. and Baxter, J. (1980). Regionalizing national projections with a multi-regional input–output model linked to a demographic model. Annals of Regional Science, 14(1):43–56.

Fromm, D. and McCarthy, M. D. (1978). An Econometric Policy Model with Spatial Dimensions. Wharton EFA, Philadelphia, Pennsylvania.

Fromm, D., Loxley, C., and McCarthy, M. D. (1980). The Wharton EFA multiregional, econometric model: a bottom-up, top-down approach to constructing a regionalized model of a national economy. In F. G. Adams and N. J. Glickman (Editors), Modeling the Multiregion Economic System: Theory, Data, and Policy. Heath/Lexington Books, Lexington, Massachusetts, pp. 89–106.

Fukuchi, T. (1978). Analyse économico-politique d'un développement régional harmonisé (Economic–political analysis of integrated regional development). In La Planification en France et au Japon, Collections de l'INSEE C, (61):227–253.

Funck, R. and Rembold, G. (1975). A multiregion, multisector forecasting model for the Federal Republic of Germany. Papers of the Regional Science Association, 34:69–82.

Gillen, W. J. and Guccione, A. (1970). The estimation of postwar regional consumption functions in Canada. Canadian Journal of Economics, 3(2):276–290.

Glejser, H. (1975). Macedoine, un Modèle Régional de l'Economie Belge (Macedoine, a Regional Model of the Belgian Economy). Bureau du Plan, Brussels.

Glejser, H., van Daele, G., and Lambrecht, M. (1973). First experiments with an econometric regional model of the Belgian economy. Regional and Urban Economics, 3(3):301–314.

Gordon, I. R. (1974). A gravity flows approach to an interregional input–output model of the U.K. In E. L. Cripps (Editor), Space–Time Concepts in Urban and Regional Models. Pion Press, London, pp. 56–73.

Gordon, I. R. (1977). Regional interdependence in the United Kingdom economy. In W. Leontief (Editor), Structure, System and Economic Policy. Cambridge University Press, Cambridge, pp. 111–122.

Greytak, D. (1970). Regional impact of interregional trade in input–output analysis. Papers of the Regional Science Association, 25:203–217.

van Hamel, B. A., Hetsen, H., and Kok, J. H. M. (1975). Un modèle économique multirégional pour les Pays-Bas (A multiregional economic model for The Netherlands). In Utilisation des Systèmes de Modèles dans la Planification (Use of Model Systems in the Planning Process). United Nations Organization, Geneva, pp. 212–267.

van Hamel, B. A., Hetsen, H., and Kok, J. H. M. (1979). Un modèle économique multirégional pour les Pays-Bas (A multiregional economic model for The Netherlands). In R. Courbis (Editor), Modèles Régionaux et Modèles Régionaux–Nationaux (Regional and Regional–National Models). Cujas and Editions du CNRS, Paris, pp. 147–173.

Harris, C. C., Jr. (1970). A multiregional, multi-industry forecasting model. Papers of the Regional Science Association, 25:169–180.

Harris, C. C., Jr. (1973). The Urban Economies, 1985. Heath/Lexington Books, Lexington, Massachusetts.

Harris, C. C., Jr. (1978). A multiregional econometric forecasting and impact model. Paper presented to the US/USSR Joint Working Group on the Application of Computers to Management, held at Shenandoah National Park (Virginia), 16–19 May 1978.

Harris, C. C., Jr. (1980). New developments and extensions of the multiregional multiindustry forecasting model. Journal of Regional Science, 20(2):159–171.

Harris, C. C., Jr. and Nadji, M. (1980). The framework of the multiregional multiindustry forecasting model. In F. G. Adams and N. J. Glickman (Editors), Modeling the Multiregion Economic System: Theory, Data, and Policy. Heath/Lexington Books, Lexington, Massachusetts, pp. 167–176.

Hart, R. A. and Mackay, D. J. (1977). Wage inflation, regional policy and the regional earnings structure. Economica, 44 (175):267–281.

Hill, E. (1975). Calculation of Trade Flows and Income Multipliers Using the Multiregional Input–Output Model. MRIO Working Paper 3. MIT, Cambridge, Massachusetts.

Isard, W. (1951). Interregional and regional input–output analysis: a model of a space economy. Review of Economics and Statistics, 33(4):318–328.

Kawashima, T. (1977). Regional impact simulation model BALAMO for government budget allocation policy in Japan. In A. Straszak and B. V. Wagle (Editors), Models for Regional Planning and Policy-Making. Proceedings of a joint IBM/IIASA Conference, held in Laxenburg, September 1977. IBM UK Scientific Centre, Peterlee, Durham, pp. 152–180.

King, L. J. and Forster, J. J. H. (1973). Wage-rate change in urban labor markets and intermarket linkages. Papers of the Regional Science Association, 30:183–196.

Klein, L. R. (1969). The specification of regional econometric models. Papers of the Regional Science Association, 23:105–115.

Lande, P. S. (1978). The interregional comparison of production functions. Regional Science and Urban Economics, 8(4):339–353.

Lecaillon, J. (1976). La propagation des hausses de salaires (The diffusion of wage rate increases). Revue d'Economie Politique, 86(5): 679–701.

Lee, F. Y. (1971). Regional variations in expenditure patterns in the US. Journal of Regional Science, 11(3):359–367.

Leontief, W. (1953). Interregional theory. In W. Leontief, H. B. Hollis, P. G. Clark, J. S. Duesenberry, A. R. Fergusen, A. P. Grosse, R. N. Grosse, M. Holzman, W. Isard, and H. Kistin (Editors), Studies in the Structure of the American Economy. Oxford University Press, New York, pp. 93–115.

Leontief, W. and Strout, A. (1966). Multiregional input–output analysis. In W. Leontief (Editor), Input–Output Economics. Oxford University Press, New York, pp. 223–257.

Lipsey, R. G. (1960). The relation between unemployment and the rate of change of money wage rates in the United Kingdom, 1862–1957: a further analysis. Economica, 27(105): 1–31.

Meyer, J. R. (1963). Regional economics: a survey. American Economic Review, 53(1):19–54.

Miernyk, W. H. (1973). Regional and interregional input–output models: a reappraisal. In M. Perlman, C. J. Leven, and B. Chinitz (Editors), Spatial, Regional, and Population Economics. Essays in Honor of Edgar M. Hoover. Gordon and Breach, New York, pp. 263–292.

Milne, W. J., Glickman, N. J., and Adams, F. G. (1980a). A framework for analyzing regional growth and decline: a multiregional econometric model of the United States. Journal of Regional Science, 20(2):173–189.

Milne, W. J., Adams, F. G., and Glickman, N. J. (1980b). A top-down multiregional model of the U.S. economy. In F. G. Adams and N. J. Glickman (Editors), Modeling the Multiregion Economic System: Theory, Data, and Policy. Heath/Lexington Books, Lexington, Massachusetts, pp. 133–145.

Moses, L. N. (1955). The stability of interregional trading patterns and input–output analysis. American Economic Review, 45(5):803–832.

Olsen, R. J. (1976). MULTIREGION: A Socioeconomic Computer Model for Labor Market Forecasting. Paper presented at the Conference on the Dynamics of Human Settlement Patterns, held at the International Institute for Applied Systems Analysis, Laxenburg, Austria, 13–16 December 1976.

Polenske, K. R. (1970a). Empirical implementation of a multiregional input–output gravity trade model. In A. P. Carter and A. Brody (Editors), Contributions to Input–Output Analysis. Volume I. North-Holland Publishing Co., Amsterdam, pp. 143–163.

Polenske, K. R. (1970b). An empirical test of interregional input–output models: estimation of 1963 Japanese production. American Economic Review, 60(2): 76–82.

Polenske, K. R. (1972). The implementation of a multiregional input–output model for the United States. In A. Brody and A. P. Carter (Editors), Input–Output Techniques. North-Holland Publishing Co., Amsterdam, pp. 171–189.

Polenske, K. R. (1980). The U.S. Multiregional Input–Output Accounts and Model. Heath/Lexington Books, Lexington, Massachusetts.

Reed, J. D. and Hutchinson, P. M. (1976). An empirical test of a regional Phillips curve and wage rate transmission mechanism in an urban hierarchy. Annals of Regional Science, 10(3): 19–30.

Richardson, H. W. (1972). Input–Output and Regional Economics. Wiley Publishing Co., New York.

Richardson, H. W. (1978). The state of regional economics: a survey article. International Regional Science Review, 3(1): 1–48.

Riefler, R. (1973). Interregional input–output: a state of the art survey. In G. G. Judge and T. Takayama (Editors), Studies in Economic Planning Over Space and Time. North-Holland Publishing Co., Amsterdam, pp. 133–162.

Riefler, R. and Tiebout, C. M. (1970). Interregional input–output: an empirical California–Washington model. Journal of Regional Science, 10(2): 135–152.

van Rompuy, P. and de Bruyne, G. (1979). Specification and Estimation of an Interregional Allocation Model for Capital Stock. Regional Science Research Paper, 25. C.E.S., University of Leuven, Belgium.

Suzuki, N., Kimura, F., and Yoshida, Y. (1973/1978). Regional Dispersion Policies and Their Effects on Industries – Calculations Based on an Interregional Input–Output Model. August 1973 and March 1978. Mitsubishi Research Institute, Tokyo.

Thirlwall, A. P. (1970). Regional Phillips curves. Bulletin of the Oxford University Institute of Economics and Statistics, 32(1): 19–32.

Thomas, R. L. and Stoney, P. J. M. (1971). Unemployment dispersion as a determinant of wage inflation in the U.K., 1925–66. Bulletin of the Manchester School of Economics and Social Studies, 39(2): 83–116.

Thys-Clement, F., van Rompuy, P., and de Corel, L. (1973). RENA, un Modèle Economique pour l'Elaboration du Plan 1976–1980 (RENA, an Economic Model for Elaborating the 1976–1980 Plan). Bureau du Plan, Brussels.

Thys-Clement, F., van Rompuy, P., and de Corel, L. (1979). RENA, a regional–national model for Belgium. In R. Courbis (Editor), Modèles Regionaux et Modèles Regionaux–Nationaux (Regional and Regional–National Models). Cujas and Editions du CNRS, Paris, pp. 103–122.

Tiebout, C. M. (1957). Regional and interregional input–output models: an appraisal. Southern Economic Journal, 24: 140–147.

Treyz, G., Friedlaender, A. F., McNertney, E. M., Stevens, B. H., and Williams, R. E. (1977). The Massachusetts Economic Policy Analysis Model. Department of Economics, University of Massachusetts, Amherst, Massachusetts.

Zuker, R. (1976). An Interprovincial Input–Output Model. Version III. Ministry of Regional Economic Expansion, Ottawa.

Regional Development Modeling: Theory and Practice
M. Albegov, A.E. Andersson and F. Snickars (editors)
North-Holland Publishing Company
© IIASA, 1982

Chapter 6

USING EMPIRICAL MODELS FOR REGIONAL POLICY ANALYSIS

Norman J. Glickman*

Department of Regional Science, University of Pennsylvania, Philadelphia, Pennsylvania (USA)

6.1. Introduction

The effects of national-level policies and economic activity on a country's regional system have become important public issues in many countries. For instance, questions concerning the differential regional (and urban) effects of public investment, tax and transfer policies, energy costs, and the business cycle have become topics of public debate as well as of scholarly inquiry. In the United States, the discussion of the distribution of federal funds to states and localities has erupted into what has become known as the "Second War between the States",** as the North and South are sometimes seen to compete for federal aid. Of course, this "Frostbelt" versus "Sunbelt" debate has also been replicated in other nations, as subnational areas compete for the benefits of national-level policies.

The concern over regional aspects of national policies has been translated recently into formal requirements within the U.S. federal government to take account of the spatial effects of some of its programs. The March 1978 National Urban Policy statement (U.S. Department of Housing and Urban Development, 1978) called for a wide variety of urban-related legislative and administrative initiatives. Among the latter was a requirement that federal executive branch agencies prepare an "urban impact analysis" (UIA) for each major new policy change (whether it be of an expenditure, taxation, or regulatory nature). The UIA process, codified in Office of Management and Budget (OMB) Circular A-116 (see Salamon and Helmer, 1980), calls for an estimation of the effects on central cities, suburbs, and nonmetropolitan areas of

---

* Financial assistance from the U.S. National Academy of Sciences (NAS) is gratefully acknowledged. I thank Komei Sasaki for helpful comments on an earlier draft. The views presented here are my own and do not necessarily represent those of the NAS or the University of Pennsylvania.
** See, for example, Havemann et al. (1976), Congressional Budget Office (1977), Markusen and Fastrup (1978), and Perry and Watkins (1977).

these federal programs. Variables for which impacts are to be calculated include population, employment, income, and local government fiscal condition.

The Office of Policy Development and Research of the U.S. Department of Housing and Urban Development (HUD) and others have been developing appropriate techniques for estimating the urban impacts of federal policies; for example, see Glickman and Jacobs (1979a, 1979b). This has been in an effort both to advance the state of policy analysis — by extending it to a more deliberately "spatial" focus — and to help understand the spatial implications of federal policies. Too often, the unintentional consequences of a particular policy have been as important as the stated goals of such programs.* Another objective of HUD's efforts has been to make the UIA process work better and, in so doing, to help implement the National Urban Policy.

In undertaking work on UIAs, the role of empirical models in the analytic process has been discussed. Previously, I have argued (Glickman, 1980a) that the role of models in the governmental UIA process is limited because of the nature of most existing models and the short period often allowed for analysis. A number of particular problems may be identified:

1. Few models have significant spatial components.
2. Many urban or regional models have been built for purposes other than policy study (for example, forecasting) and do not have the proper "policy levers" to aid in such spatial analysis.
3. Existing models sometimes do not have the spatial detail necessary for UIAs — that is, they may be built for states or large regions and do not provide the central city–suburban–nonmetropolitan area breakdowns important to urban impact studies.
4. Often program or urban data sources are inconsistent or unavailable for some kinds of studies, making the use of models difficult.
5. The government analysts who are to carry out UIAs may not have the models at hand, the ability to execute the models, or the time to undertake extensive model-development or fine tuning.**

Despite these problems, research has been undertaken using empirical models in urban and regional policy analysis. In a recent compendium of the methodological

---

* For instance, the interstate highway system probably did far more in the promotion of urban and regional deconcentration and the destruction of many viable urban neighborhoods than was generally expected in the mid-1950s. Also, the Congressional Budget Office (1979a, 1979b) has shown some possible differential spatial effects of the "Tokyo Round" trade agreements, which is another "nonurban" policy with potentially important urban and regional consequences. Chapter 25 shows the effects on regions of the closing of some nuclear power plants in Sweden. Sakashita (1974) indicates the impacts of rail policy for regional development in Japan. For a discussion of the role of UIA in federal decision-making, see Glickman (1979b).

** Notable exceptions are the work at the Department of Health and Human Services' Office of the Assistant Secretary for Policy and Evaluation (ASPE) and the Department of Commerce's Bureau of Economic Analysis (BEA). The work of both ASPE and BEA will be discussed below.

bases for UIAs (Glickman, 1980b), several models were used. One of the major purposes of this chapter is to outline the uses of models in urban impact analysis and to make suggestions for their improvement. However, it must be emphasized that urban impact analysis is only one type of urban or regional policy study for which models could be employed. Thus, a broader look at the uses of models in urban and regional analysis is in order. The second purpose of this chapter, therefore, is to review some possible roles for empirical models in more general forms of regional policy analysis.

I write this chapter from the standpoint of one who has built models and who, while serving at the U.S. Department of Housing and Urban Development during 1978 and 1979, was in the position of using models for policy purposes, especially with regard to the development of UIA methodologies. I thus reflect the "producer" as well as the "consumer" segments of the "model market".

In order to keep the exposition reasonably brief, I have restricted the analysis as follows. First, I will discuss models which are primarily multiregional — that is, they model the system of regions within a nation. Second, I will analyze only three types of models (econometric, microsimulation, and input–output) which are usually employed in market economies (although, of course, planned economies also use some of these model types). Third, I will illustrate the use of models with models developed in the USA. To partially compensate for these limitations, an extensive reference list is provided.

The chapter consists of three main sections. In Section 6.2, some of the principal conceptual issues in modeling for regional analysis are discussed. Following this discussion, in Section 6.3 a number of major models are described along with some of their policy applications. Conclusions and suggestions for future work are contained in Section 6.4.

6.2. Using models in regional analysis: some conceptual issues

There are several important conceptual and methodological issues involved in the use of empirical models for understanding the relationship between national and urban/regional economies.* First, single-region models should be contrasted with those which are multiregional in nature. Where urban and regional data bases are adequate, there has been a progression from using single-region models** to multiregion models. There are several reasons for this phenomenon.

---

* This section draws in part on Ballard et al. (1980a) and Adams and Glickman (1980b).
** See, for instance, Bell (1967), l'Esperance et al. (1968), Isard and Langford (1969), Glickman (1971, 1977a), Crow (1973), Hall and Licari (1974), Adams et al. (1975), and Saltzman and Chi (1977) for discussions of single-region input–output and econometric models. For collections of studies of multiregion models, see Glickman (1977b, 1979a), Adams and Glickman (1980a), and Stevens (1980).

6.2.1. Single-region models

First, single-region models built a decade or more ago abstracted from the real world the integration of economic activity with neighboring regions and treated all other regions as the "rest of the world". Although the relationship to other regions had been conceptually identified, the empirical work did not reflect trade flows within a nation's regional system. As a result, interregional feedbacks from changes in one region could not be captured within this framework. Second, all regions were regarded as homogeneous, without due recognition being given to intraregional differences in wages, factor endowments, and other aspects.

Third, individual equations in single-region models were sometimes incorrectly specified econometrically in that they did not reflect existing regional economic theory (see Engle, 1974). Fourth, the national models often incorrectly forecasted national economic activity, so that local forecasts derived from them tended to have compounded errors. This resulted from the fact that single-region models treated variables from the national model as exogenous. Fifth, the usefulness of single-region models for policy purposes was reduced because of their geographical limitations: often policies (especially at the national level) have a wide range of spatial effects which, like private market activity, can easily transcend metropolitan-area or state boundaries. Sixth, even if individual models were available for each of the regions, there would be no way of ensuring consistent forecasts of national activity. Because these individual models are not linked (except to a national model), summing gross regional output across the regions would not guarantee that the total would add up to the gross national product.

6.2.2. Multiregion models

Given these and other limitations, attention has more recently been focused on modeling multiregion systems. For instance, the multiregion input–output (MRIO) model (Polenske, 1975) was an attempt to view interregional linkages in an input–output framework. Harris (1973) used cross-sectional econometric relationships, employing a national input–output table for national control totals, to allocate economic activity among 3111 "county-type" areas. Olsen et al. (1977) have produced a multiregion econometric model based on small economic areas, using pooled cross-sectional data with transportation time indices linking the regions. Ballard and Glickman (1977) and Milne et al. (1978, 1980) have constructed pure time-series multiregion models; the first of these was for the Delaware Valley and the second for the United States. Finally, Golladay and Haveman (1977) have constructed a large-scale microsimulation model with Polenske's multiregion input–output table embedded within it.

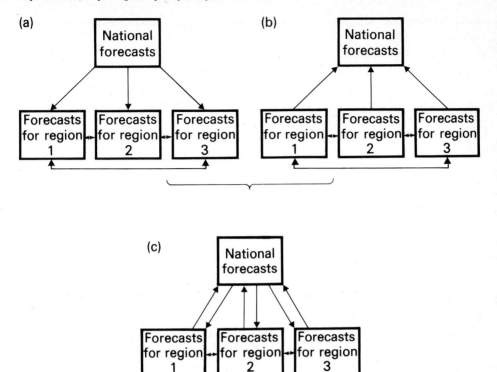

Figure 6.1. The two main types of multiregional model: (a) top down (TD) and (b) bottom up (BU); these may be combined to produce a model (c) which contains both TD and BU elements.

The top-down approach. When we are analyzing the multiregion approach,* it is useful to distinguish between two types of model: "top down" (TD) and "bottom up" (BU) (see Figure 6.1). The first type, the more common of the two, takes national control totals as forecast by a national model and distributes economic activity to its component regions. As shown in Figure 6.1(a), national variables (the "top") are used to determine regional activity (the "bottom"). Estimates of regional shares of national economic activity are typically based upon measures of relative regional attractiveness, such as accessibility, input costs, and industry mix. These models have been employed to analyze the effects on the various regions of both changes in national policy and region-specific events.** In the more sophisticated version of the models, there are

---

* This discussion refers particularly to econometric models but is also relevant to other kinds of models such as input–output (see the discussion of IDIOM and MRIO in Section 6.3.3).
** For instance, Harris (1973) analyzed the regional distribution of impacts of national military expenditure cutbacks and Olsen et al. (1977) considered the regional distribution of projected growth in the national economy between 1975 and 2020. Milne et al. (1978) examined both the regional impacts of increased federal expenditures in the Northeast and decreased expenditures in the remainder of the United States. This research will be discussed more fully in Section 6.3.1.

links between the regions within the systems. This type of model, then, is in the same intellectual spirit as its single-region predecessors. Examples of the TD approach include the work of Milne et al. (1978).

There are two characteristics of TD models that can produce misleading results at both the national and regional levels. First, by modeling a control total at the national level, TD models do not allow the effects of changing regional conditions on the national economy to be considered. For example, if national policy produced spatial shifts in national government spending (that is, among regions), but left the level of national government spending unchanged, the control totals of a TD model would be unaffected. Such models could not be used to determine the extent to which this spatial reallocation of expenditures affected the level of national economic activity. This is a particular problem when the focus of the analysis is on short-run fluctuations in the economy.

A second characteristic of TD models can lead to spurious forecasts at the regional level. Specifically, even if the national control totals produced are correct, regional totals may be in error because the regional share equations are constrained. For example, if the sum of regional shares predicted by a TD model does not equal 100% of the national activity, these shares must somehow be adjusted in order to account for all of the national economic activity. Given the existence of regional differentials in the speed of adjustment to the business cycle, it is probable that the shares of some regions should be increased more than proportionately and those of other regions less than proportionately to the total national change. This issue is particularly important in view of the fact that regional cycles have been found to differ substantially; see Bretzfelder (1973) and Vernez et al. (1977) for regional business-cycle analysis.

The bottom-up approach. As demonstrated in Figure 6.1(b), the alternative BU approach aggregates regional activity (the "bottom") to a systemwide total (the "top"). By formulating a BU model of the economy with single-region models as a basis, some of the difficulties inherent in the TD approach can be mitigated. First, because economic activity is determined initially at the regional level and then summed to produce national aggregates, BU models can be used to estimate the national economic impacts of changes in the regional distribution of economic activity. Second, by modeling specific changes in each region's level of economic activity throughout the business cycle and incorporating such unconstrained relationships into a model, the BU approach can be used to analyze regional differentials in the speed of adjustments during the business cycle.

Of course, as Bolton (1980), Adams and Glickman (1980b), and Courbis (1980) point out, there are no "pure" TD or BU models that are realistic. Practitioners must try to construct models that successfully integrate TD and BU elements, as in Figure 6.1(c). In such models, activity that is determined in national markets (for example, interest rates) would be estimated from national data in a TD fashion, while local activity (such as retail sales) could be forecasted from local (BU) sources.

6.2.3. Uses of models

Models have been used in urban impact analysis and more general regional policy analysis in interesting, yet limited, ways. For example, Danziger et al. (1980), using an updated version of the Golladay–Haveman microsimulation model (see Holmer, 1980, for a discussion of the changes in the model), have analyzed welfare reform proposals and other income distribution policies. Ballard et al. (1980a) have calculated the effects of a redistribution of the federal budget: the NRIES multiregion econometric model was used for that study as well as several others, including energy policy. Polenske's input–output model has been used for a wide variety of experiments, including the regional impacts of disarmament. Some of these models will be reviewed in Section 6.3.

6.2.4. Characteristics of models

Whatever the model chosen (e.g., input–output, econometric, microsimulation, land use, etc.), several model characteristics should be present. First, the model should be multiregional so that interregional policy differentials and interregional feedbacks can be measured. Also, the model's regional components should consistently add to the national totals for each variable. Second, not only should the model depict interregional feedbacks, but it must recognize that local activity has effects on national events; that is, there is feedback from the regions to the nation. Third, a model must show the important differences and similarities among regions and recognize the hierarchical nature of interregional development. Fourth, the model should be built to be used by policy-makers as well as academic researchers. Therefore, a wide range of policy "levers" should be embedded in the model; these policy instruments should be at both the national and regional levels. Furthermore, the model should include a large number of variables which can be analyzed, such as output, employment, income, and others. I am a supporter of those who advocate the building of large model systems.

However, it must be recognized that formal empirical models have important limitations. First, they impose tremendous data requirements on the model-builder. Second, such models cannot meaningfully analyze structural change in the economy since the models are estimated using fixed coefficients. Third, many of the characteristics noted in my "ideal" model (e.g., it is multiregional and includes feedback mechanisms and policy levers) are usually not present in operational models. Therefore the usefulness of models must be understood by both model "producers" and their "consumers".

Having discussed some important modeling issues, I will now review some operational models and their policy applications.

6.3. Using models in regional policy studies

6.3.1. Econometric models

I will briefly summarize the nature of two multiregion models, one top down and the other bottom up. Some examples of policies that can be studied using these models will also be indicated.

A top-down model. The top-down model discussed here is that developed by Milne, Adams, and Glickman (Milne et al., 1978, 1980).* This model (MAG) is built to be linked to the Wharton long-term model and takes most of its exogenous inputs from the national model, distributing them to the nine regions represented in MAG. The linkages are mainly through the national markets that dominate manufacturing output, although there are also others. Each region contains equations for output, employment, and wages. In turn, the variable types are disaggregated by industry: durable and nondurable manufacturing, nonmanufacturing, farming, mining, and government. In addition, there are five types of nonwage income, unemployment rates, a large energy-demand submodel, and a demographic submodel. In all, there are 1600 endogenous variables in the model.

The MAG model has been used for several sets of regional policy simulations; see, in particular, Milne et al. (1978). First, the model was employed to gauge the long-term growth of the component regions given different assumptions about the future of the national economy. For example, the growth of the northern and southern regions is compared under assumptions of "slow" versus "fast" national growth. The simulations show that the North grows faster relative to the nation as a whole (although more slowly in absolute terms) under a fast-growth scenario. In addition, the energy submodel was employed to examine the effects of deregulation of natural gas and domestic oil prices. It was shown that there ought to be a trend toward equalization of energy prices among the regions** and, therefore, the North improves its position relative to the control solution. Another simulation shows the effects of a federal government expenditure redistribution in favor of the North.

A bottom-up model. One of the best examples of the bottom-up approach to regional econometric modeling is the national–regional impact evaluation system (NRIES)

---

* As noted in Section 6.2, most models have both TD and BU elements. Those discussed in this section contain a mixture of both, but are distinguished here by the major thrusts of their structures. MAG, for example, is primarily, but not completely, TD. The NRIES model, described later in this section, has a largely BU structure.

** The equalization occurs because as well-head prices increase, transportation costs decrease in proportion to total energy costs, and because the southern states start with lower energy prices.

developed at the U.S. Department of Commerce.* NRIES comprises 51 state-area**
econometric models that are integrated into a model of the United States economy.
The national model is the summation of the 51 independently constructed state
models. First, each state model generates individual growth patterns as if there were
51 separate single-region econometric models. National growth trends in this bottom-
up approach are therefore determined by regional growth patterns, as opposed to the
top-down approach of MAG.

Since regions do not grow independently, NRIES links them through an explicit
set of interregional "interaction variables" that are similar to gravity models. The
interaction variables are derived for each state and represent distance-deflated
economic activity in all other states, and are included in the state variable estimations.
The interaction variables (g) are calculated individually for each state (r), variable
(x^j), and period (t) based upon the following formula:

$$r_g x_t^j = \sum_{\substack{k=1 \\ k \neq r}}^{r} {}^k x_t^j / {}^{rk} d$$

The activity levels of the variables x^j in all other states are scaled by the distance
${}^{rk}d$ from the base state r and summed. The distance scalar ${}^{rk}d$ used is the distance
between the population centroids of each region. Since the interaction variables are
distance-deflated, the linkages are also spatially proportioned.***

The NRIES behavioral equations are econometrically estimated by ordinary least-
squares regression for each variable in each region using annual time-series data from
1955 to 1976. This results in a total of 11,730 equations (230 variables × 51 regions).
All equations in a given region are aggregated to form the state "model"; the state
models are then aggregated to form the complete NRIES model.

The NRIES model has been simulated to measure the impacts of several kinds
of policies. For example, Ballard et al. (1980b) have examined the spatial impacts
of housing and community-development programs. That study also considered the
redistribution of federal government grants-in-aid to local governments and the impact
of this spatial shift in spending on local economies. Ballard et al. (1980a) have extended the analysis to about three-quarters of the federal budget: they have calculated
the direct and indirect effects of a fiscally-neutral redistribution of spending (that
is, one in which expenditures in a given state are proportional to the taxes collected

---

* For a full description of NRIES, see Ballard et al. (1980c). For applications, see Glickman and
Jacobs (1979b), Gustely and Ballard (1979), Ballard and Wendling (1980), and Ballard et al.
(1980a, b).
** Washington D.C. is treated as a "state" in NRIES.
*** Thus, for example, while economic activity in California affects the economies of both Nevada
and New Hampshire, the influence on Nevada is greater because of its closer proximity to California.
The interaction variables are also weighted by the "mass" (levels of economic activity) of the
states. Here, for example, New York and Connecticut are nearly the same distance from Louisiana,
but the influence of New York on the Louisiana economy is much greater than that of Connecticut,
because of New York's greater size.

there). Comparisons are made between the artificially reallocated funding scheme and the one that actually took place between 1969 and 1977. Ballard et al. find that the northern tier states received less funds than they would have had a fiscally-neutral policy been in effect. NRIES was then used to calculate the total impact (direct and indirect) of this synthetic policy. The authors show that, over the period studied, the Mideast and Great Lakes regions "lost" (that is, they received less under the current system than they would have under a fiscally-neutral one) a total of $380 billion as measured by gross state product (GSP); New York and Illinois "lost" about half of this total, or about 9% of their GSP.

Of particular interest is the result that the present system results in a net loss in United States GNP over the eight-year period compared to the fiscally-neutral system. That is, even though there is no difference in overall federal spending or revenues (merely a spatial redistribution of them), GNP is $70 billion less under the present system. This occurs because the current system allocates greater expenditures to low-multiplier regions than does the fiscally-neutral system. This change in GNP with no change in total federal expenditures has been termed the "spatial balanced budget multiplier". This simulation illustrates an advantage of a BU system over the TD approach, as noted in Section 6.2. Here, the total spending remains unchanged, yet the spatial redistribution causes a change in national-level figures. This could not be traced were a TD model used.

6.3.2. Microsimulation models

At the opposite end of the modeling spectrum from macroeconometric models are microeconomic-based microsimulation models. This class of models can be characterized by a set of data on individual economic units (for example, households) and models that simulate the behavior of the units. Pioneered by Orcutt et al. (1961), these models have been used primarily to gauge the distributional effects of national programs.* In addition, a set of models developed by Golladay and Haveman (1977), Holmer (1980), and Betson et al. (1979) have added a spatial dimension to micro-simulation models. This set of models will be described next.

Essentially two models with several component submodels are used in this system. First, the "KGB" model developed by Kasten, Greenberg, and Betson (Betson et al., 1979) is used to trace the first-round effects of a public policy on households in different regions and of different races, family sizes, and income levels. Then, the RESIND model (Golladay and Haveman, 1977) determines the induced effects stemming from the first-round impacts described by the KGB model.

The KGB model first takes a representative sample of families drawn from the

---

* For example, federal distribution policy has been studied by Moeller (1973), Beebout et al. (1976), Hollenbeck (1976), Wertheimer and Zedlewski (1976), and Golladay and Haveman (1977); medical care has been evaluated by Holahan and Wilensky (1972) and Yett et al. (1978); and energy demand studied by Hollenbeck (1979).

Survey of Income and Education (SIE) to characterize the population by various socioeconomic variables before the policy change is simulated. Other information is obtained from tax schedules and predictive equations. Next, the values of net wage rates and disposable income are adjusted according to their post-change levels. Earnings are then adjusted to take account of changes in the supply of labor that result from wage and income changes.

The RESIND model then traces the effects of policy-induced changes in house-hold disposable income through the economic system and incorporates consumption responses, industry-based output responses (both direct and indirect), employment estimates by detailed occupational category, and changes in earnings by earnings class. And, in turn, all of these estimates are disaggregated into values for the component regions.

The structure of the RESIND model includes five component modules. Changes in disposable income from the income transfer programs (and the taxes required to finance them) lead to changes in the level and composition of consumption expenditures for those affected by the policy measures. These expenditures, in turn, affect the demands experienced by (and hence the output of) the various industries that supply consumers. Such changed output patterns will alter the demands placed on supplying industries. Because of the interdependence of the industries in the national economy, all sectors throughout the economy will experience such changes in gross outputs. In response to increased or decreased output levels, production sectors will alter the demand for labor of various occupational groups. These changes in occupational demands imply changes in the distribution of earnings and income, to the extent that the relative change in demand for highly skilled workers differs from that for unskilled workers.

Thus, in the first module, the microdata of the Survey of Income and Education are adjusted for underreporting and the net cost or benefit impact on each of the households is calculated. This first-round impact can be shown for various regions of the country and for various income classes. The second module simulates the changes in the level and pattern of consumption spending induced by the policy, for each of these families. This consumption demand simulation is obtained by applying the relevant expenditure sector coefficients to families distinguished by a variety of economic and demographic traits. The coefficients were estimated by fitting a 56-sector log-linear consumption expenditure system to the microdata of the 1972–1973 Consumer Expenditure Survey. From this simulation, the change in consumer demand for production from 56 sectors in all the regions is obtained.

The third module transforms this final demand sector into an estimate of changes in gross output required of all production sectors in various regions of the economy by incorporating the indirect demands placed by industries on each other. This is accomplished by means of the multiregion input–output table (MRIO) developed by Polenske (1975) and discussed in Section 6.3.3. From this module, the changes in gross outputs required of each of 79 industries in each of the regions by the policy change are estimated.

In the last two modules, the simulated changes in sectoral outputs are transformed into estimates of the changes in labor demand by occupation (module 4) and into dollars of induced earnings by income class (module 5). In the final module, the simulated estimates of occupational man-hour demands are also combined with occupational earnings data to estimate changes in earned income for each occupation in each region.

As a final step, the regional–occupational earnings estimates are mapped into incremental size distributions (consisting of 15 earnings and income classes) for each region and for the nation as a whole. The coefficients for this last mapping are from special tabulations employing the 1-in-1000 data tapes of the 1970 census. Hence, the final output of the model displays the distribution of policy-induced earnings by 15 income classes in each region and in the nation as a whole.

This combination of models is capable of a wide variety of policy simulations involving both spatial and interpersonal distributions of income. Among the more interesting policy simulations was the analysis of the welfare reform measure known as the Program for Better Jobs and Income (PBJI), undertaken by Danziger et al. (1980). Using the models described above, they found that the urban impacts of the PBJI (which was never enacted) would have been relatively small, but would have reduced the overall incidence of poverty. Most of the funds from PBJI would have gone to regions with low per capita income and high poverty levels, largely those in the South. Therefore, PBJI would have reinforced current growth trends by giving funds to the fast-growing but relatively low-income South, rather than the higher-income but slow-growing North. In addition, PBJI would have reduced income inequality among urban dwellers within metropolitan areas. Also, the PBJI aimed to distribute public service jobs to the South, but did not target these jobs on high-unemployment metropolitan areas. Finally, Danziger et al. concluded that fiscal relief would have been concentrated in states presently making relatively high welfare payments to large numbers of recipients.

Chernick and Holmer (1979), using the same models, analyzed a later welfare reform proposal, and the same microsimulation model was used to analyze the effects of a significant reduction in federal income tax rates by Holmer (1980).*

6.3.3. Input–output models

A third type of major empirical model used in regional policy studies uses the input–output approach.** The work on regional and multiregional input–output draws on the early work of Leontief (1941). As is the case in many econometric models,

---

* Work currently underway at the University of Wisconsin by R. Haveman, S. Danziger, and others seeks to analyze both the distributional effects and urban and regional impacts of programs involving housing subsidies for the poor, medical assistance, wage subsidies to counter youth unemployment, public service employment, and others.
** For reviews and summaries of this well-known technique, see Miernyk (1965, 1972), Richardson (1972), Riefler (1973), and Glickman (1977a).

input–output is essentially a demand-driven construct but is derived from the behavior of individual firms (similar to microeconomic simulation models). As with econometric models, input–output has been applied to both single-region and multiregion problems and for a wide variety of policy studies, particularly those. involving economic impact analysis.

The advantages of a multiregional input–output system over a single-region model are significant. Obviously, the single-region model must ignore interregional trade and feedback effects among regions; only interindustry feedbacks within regions will be accounted for in a single-region model. Furthermore, multiregion models provide a consistent accounting framework for checking data consistency (that is, interregional exports and imports must be equal). Finally, multiregion models have far more policy applications than the single-region variety. On the other hand, multiregion models are more difficult and costly to construct.

A variety of actual policy applications will now be considered in order to indicate the types of use of input–output. Single-region models have been used for many economic impact studies by changing the final demand components and calculating the resulting multipliers. For example, Isard and Langford (1969) used the single-region Philadelphia input–output table to study the effects of the Vietnam War on the local economy. First, they calculated that $284 million was the direct amount of extra spending in Philadelphia due to the war. Using this figure, the final demand elements were adjusted to take account of this change and the model calculated the indirect and induced effects of the war-related spending. In all, $996.6 million in extra output was generated in the economy, thus dwarfing the direct effects. Isard and Langford then applied the $284 million expenditure to education and low-income housing uses and calculated the direct and indirect effects; the total impact was some $40 million less in this "peace dividend" simulation. The Philadelphia table was later extended and used to study pollution problems.

Input–output models have been constructed for several regions, and have been used for a variety of policy studies. The studies have involved economic forecasting, the impacts of defense spending (see, for instance, Leontief et al., 1965), environmental problems (Cumberland, 1966; Isard, 1969; Leontief, 1970), and other subjects.

However, multiregion models enable analysts to study a far greater number of applications, as Polenske (1969) has argued. These include studies of regional differences in production techniques, regional accounting systems, transportation planning, and industrial location analysis, among others. Two multiregional models are reviewed below.

IDIOM. The model IDIOM (income determination input–output model) is an interregional model of the "balanced" or "intranational" type; see Leontief (1953) and Isard (1960, pp. 345, 346). This model has been developed by Dresch and others; see, for instance, Dresch (1980) and Dresch and Updegrove (1979). Such an intranational model disaggregates a national model into its regional components. It is therefore a top-down, demand-driven model, consisting of a primary national model and a secondary regional model. The national component is itself demand-driven and determines the national control totals for the regional model. The regional model takes no account of regional differences in the distribution of output, employment, or prices, and assumes invariant technologies across regions. The regional model divides the economy into "national" and "local" sectors; the former consists of those industries with national markets, the latter of those with local markets. The national industries are assumed to have no barriers to interregional trade and the regional distribution of output is made exogenously. The local industries have no interregional trade; all local demands are assumed to be satisfied within the region. In all, IDIOM is specified for 86 industries and 50 states.

IDIOM has been used for several regional policy simulations. One involved an assessment of U.S. military export policy: the effects of a possible reduction of $4.77 billion in military exports (Dresch and Updegrove, 1979). The policy was examined in the light of two possible compensatory policies. The first was a public works program designed to minimize regional employment effects; the second policy was a reduction in payroll taxes.

IDIOM showed that the regional distribution of export-related arms employment was fairly uniform, ranging between 0.2% and 0.5% of employment among the regions. However, within the larger regions, some states (for example, Connecticut) had more significant military-related employment concentrations. The multiplier effects (direct, indirect, and induced) of the export reduction were also calculated. Here, southern New England was seen to be most negatively affected, with total employment falling by 1.0% (compared to the national average of 0.7% and a decline in the southeast of only about 0.5%). Connecticut, a state with extensive military-related production, lost 1.6% of its employment.

One of IDIOM's attractive features is its ability to analyze the effects of two policies simultaneously and therefore to investigate policy tradeoffs. For example, Dresch and Updegrove (1979) reported the results of a regionally compensatory public works policy. A $4.9 billion national expenditure would offset the reduction in military exports. The regional distribution of net employment change would be fairly small (±0.1%). The policy of a regionally-based reduction of a labor tax rate yields a somewhat larger interregional variance in employment change.

IDIOM is a reasonable tool, not only for policy analysis, but also for policy selection. Although it has been subject to some criticism,* IDIOM is an interesting model.

MRIO. Polenske (1975) has developed the multiregional input–output model MRIO, which is fundamentally a bottom-up model. This model MRIO consists of 44 regional models, each with 79 industries. Unlike IDIOM, this model uses interregional trade data to link the regions: fixed, interregional trade coefficients were derived so that

---

* For instance, it does not consider factor supply elements and, once the regional distribution of the national industries is specified, the regional distribution of demand does not influence the regional distribution of output. The lack of factor cost considerations means that IDIOM cannot analyze changes in interregional competitive positions.

it is assumed that industry i in region j always imports a fixed proportion of its requirements of input k from each other region.

MRIO has been used for a variety of policy experiments, including studies of transportation, energy, and income distribution. (The income distribution application was adapted and employed as part of the Golladay–Haveman model discussed in Section 6.3.2.) The major advantage of MRIO over IDIOM – the use of interregional trade data – is also its main weakness. The model is very expensive to update and the 1963 trade data used are surely out of date.

6.4. Concluding remarks

This brief review of the use of empirical models in regional policy research provides an overview of some of their characteristics, strengths, and weaknesses. In the remaining paragraphs, I will summarize previous work done and suggest some possible future directions for work in this field.

The models reviewed here show a reasonable ability to handle a wide range of regional policy issues. These include forecasting experiments, income distribution and social service policies, and economic impact assessments, as well as studies of energy, environment, and federal government expenditure impacts. These models are, however, limited in several ways.

6.4.1. Data problems

Underlying many of the problems noted below is that of data. MRIO, for example, loses much of its usefulness because it uses data on interregional trade that are almost 20 years old.* Other data are also missing or of low quality in many countries: for example, quarterly time series for many key economic indicators, land-use and land-price data, migration statistics, energy and environmental data, and, in particular, capital stock and investment figures. Of course, quality and availability will vary among countries. The Japanese, for instance, have good migration and land-price data; the United States has rather poor data for both categories. Garnick (1980) has discussed data issues for the United States. One important item on a modeler's agenda should be to encourage data collection.

6.4.2. Spatial disaggregation

Few of the models go below the state level in defining their spatial units. This is a serious drawback for many types of policy study where knowledge of metropolitan

---

* The interregional trade data problem has been overcome in Japan, where a series of interregional tables has been built. See, for example, Japan Ministry of International Trade and Industry (1970) and Ihara (1979).

areas or individual jurisdictions is important. Models of smaller areas, such as MULTIREGION (Olsen et al., 1977) should be built.

6.4.3. Modeling priorities

Most of the models have been built for nonpolicy purposes. For example, many econometric models have conditional forecasting as their major purpose. Input–output, while useful for understanding the industrial structure of a region (or regions), has few, if any, policy levers.

Lyall (1980) takes a skeptical view of the use of models in federal policy-making. She argues that the structure, spatial scale, and cost of models have limited their use. Clearly models with more policy levers, which can be manipulated for a variety of policy experiments and which decision-makers can understand, are of great importance.*

Furthermore, regional models need to be built for important types of variables that have largely been ignored, at least in part because of lack of data. Migration, energy, and environment are of particular importance. Although there have been some models built in these areas (see Rogers, 1975, 1976; Rogers et al., 1978; US Department of Energy, 1978; Knox and Sandoval, 1980; Menchik, 1980; Willekens, 1980), more activity should be encouraged.

6.4.4. Model orientation

Almost all regional modeling to date has been demand-oriented. Clearly, factor supply considerations are crucial in trying to understand long-term relations among regions. Although some scholars have dealt with labor supply issues, I consider that it is equally important to understand interregional (as well as international) capital movements. As Bluestone and Harrison (1980) have pointed out, capital stock data are sparse in the United States and the effects of capital movements on local economies can be serious. Little work on this important subject has been done to date; Harris (1973) has modeled the investment process but his model does not really make use of that feature. Supply-side models should be built because demand models may be inconsistent. That is, the latter determine regional demand, but cannot determine whether supply-side conditions (e.g., an adequate labor supply) are met.

These are some of the issues to which I hope model-builders will turn. Although the work reported here is largely related to the United States, it should be noted that excellent work is going on elsewhere — see Courbis (1979, 1980) and Baranov et al. (1980) for some examples. This book includes some discussion of the models developed in other parts of the world.

---

* It is hard for me to underestimate the role of modelers in interacting with and educating public-policy-makers in the uses and limitations of models. Too often, people who make policy decisions do not grasp the nature of models or how they may be used.

References

Adams, F. G. and Glickman, N. J. (Editors) (1980a). Modeling the Multiregion Economic System: Theory, Data, and Policy. Heath/Lexington Books, Lexington, Massachusetts.

Adams, F. G. and Glickman, N. J. (1980b). Perspectives in multiregion modeling. In F. G. Adams and N. J. Glickman (Editors), Modeling the Multiregion Economic System: Theory, Data, and Policy. Heath/Lexington Books, Lexington, Massachusetts.

Adams, F. G., Brooking, C., and Glickman, N. J. (1975). On the specification and simulation of a regional econometric model: a model of Mississippi. Review of Economics and Statistics, 57: 286–298.

Ballard, K. and Glickman, N. J. (1977). A multiregional econometric forecasting system: a model for the Delaware Valley. Journal of Regional Science, 17: 161–177.

Ballard, K. and Wendling, R. M. (1980). The national–regional impact evaluation system: a spatial model of U.S. economic and demographic activity. Journal of Regional Science, 20: 143–158.

Ballard, K., Glickman, N. J., and Gustely, R. (1980a). A bottom-up approach to multiregional modeling: NRIES. In F. G. Adams and N. J. Glickman (Editors), Modeling the Multiregion Economic System: Theory, Data, and Policy. Heath/Lexington Books, Lexington, Massachusetts.

Ballard, K., Glickman, N. J., and Wendling, R. M. (1980b). Using a multiregion econometric model to measure the spatial impacts of federal policies. In N. J. Glickman (Editor), The Urban Impacts of Federal Policies. Johns Hopkins University Press, Baltimore, Maryland, pp. 192–216.

Ballard, K., Gustely, R., and Wendling, R. M. (1980c). The National–Regional Impact Evaluation System: Structure, Performance, and Applications of a Bottom-Up Interregional Econometric Model. Bureau of Economic Analysis, Washington, D.C.

Baranov, E. F., Martin, I. S., and Koltsov, A. V. (1980). Multiregional and regional modeling in the USSR. In F. G. Adams and N. J. Glickman (Editors), Modeling the Multiregion Economic System: Theory, Data, and Policy. Heath/Lexington Books, Lexington, Massachusetts.

Beebout, H., Doyle, P., and Kendell, A. (1976). Estimation of Food Stamp Participation and Cost for 1977: A Microsimulation Approach. Mathematical Policy Research, Washington, D.C.

Bell, F. W. (1967). An econometric forecasting model for a region. Journal of Regional Science, 7: 109–127.

Betson, D., Greenberg, D., and Kasten, R. (1979). A microsimulation model for analyzing alternative welfare reform proposals: an application to the program for better jobs and income. In R. Haveman and K. Hollenbeck (Editors), Microeconomic Simulation Models for Policy Analysis. Academic Press, New York.

Bluestone, B. and Harrison, B. (1980). Capital and Communities: The Causes and Consequences of Private Disinvestment. The Progressive Alliance, Washington, D.C.

Bolton, R. (1980). Multiregion models in policy analysis. In F. G. Adams and N. J. Glickman (Editors), Modeling the Multiregion Economic System: Theory, Data, and Policy. Heath/Lexington Books, Lexington, Massachusetts.

Bretzfelder, R. D. (1973). Sensitivity of state and regional income to national business cycles. Survey of Current Business, 53: 22–27.

Chernick, H. and Holmer, M. (1979). The Urban and Regional Impact of the Carter Administration's 1979 Welfare Reform Proposal. Mimeograph. U.S. Department of Health, Education, and Welfare, Washington, D.C.

Congressional Budget Office (1977). Troubled Local Economies and the Distribution of Federal Dollars. Congressional Budget Office, Washington, D.C.

Congressional Budget Office (1979a). The Effects of the Tokyo Round of Multilateral Trade Negotiations on the U.S. Economy: An Updated View. Congressional Budget Office, Washington, D.C.

Congressional Budget Office (1979b). U.S. Trade Policy and the Tokyo Round of Multilateral Trade Negotiations. Congressional Budget Office, Washington, D.C.

Courbis, R. (1979). The REGINA model: a regional–national model for French planning. Regional Science and Urban Economics, 9: 117–139.

Courbis, R. (1980). Multiregional modeling and interaction between regional and national development: a general theoretic framework. In F. G. Adams and N. J. Glickman (Editors), Modeling the Multiregion Economic System: Theory, Data, and Policy. Heath/Lexington Books, Lexington, Massachusetts.

Crow, R. T. (1973). A nationally linked regional econometric model. Journal of Regional Science, 13: 187–204.

Cumberland, J. H. (1966). A regional interindustry model for analysis of development objectives. Papers of the Regional Science Association, 17: 65–94.

Danziger, S., Haveman, R., Smolensky, E., and Taeuber, K. (1980). The urban impacts of the Program for Better Jobs and Income. In N. J. Glickman (Editor), The Urban Impacts of Federal Policies. Johns Hopkins University Press, Baltimore, Maryland, pp. 219–242.

Dresch, S. P. (1980). IDIOM: a sectoral model of the national and regional economies. In F. G. Adams and N. J. Glickman (Editors), Modeling the Multiregion Economic System: Theory, Data, and Policy. Heath/Lexington Books, Lexington, Massachusetts, pp. 161–165.

Dresch, S. P. and Updegrove, D. A. (1979). IDIOM: a disaggregated policy-impact model of the U.S. economy. In R. Haveman and K. Hollenbeck (Editors), Microeconomic Simulation Models for the Analysis of Public Policy. Academic Press, New York.

Engle, R. (1974). Issues in the specification of regional econometric models. Journal of Regional Science, 1: 250–267.

l'Esperance, W. L., Nestel, G., and Fromm, D. (1968). Gross state product and an econometric model of the state. American Statistical Association Journal, 44: 787–807.

Garnick, D. H. (1980). The regional statistics system. In F. G. Adams and N. J. Glickman (Editors), Modeling the Multiregion Economic System: Theory, Data, and Policy. Heath/Lexington Books, Lexington, Massachusetts.

Glickman, N. J. (1971). An econometric model of the Philadelphia region. Journal of Regional Science, 11: 15–32.

Glickman, N. J. (1977a). Econometric Analysis of Regional Systems: Explorations in Model Building and Policy Analysis. Academic Press, New York.

Glickman, N. J. (Editor) (1977b). Econometric Models and Methods in Regional Science I. Special Double Issue of Regional Science and Urban Economics, Vol. 7, Nos. 1 and 2.

Glickman, N. J. (Editor) (1979a). Econometric Models and Methods in Regional Science II. Special Double Issue of Regional Science and Urban Economics, Vol. 9, Nos. 2 and 3.

Glickman, N. J. (1979b). Urban impact analysis: its role in federal decisionmaking. Paper presented at a Conference on Public Policy and Management and Regional Science Association Mimeograph. Forthcoming in J. Crecine (Editor), Public Policy and Management Conference Proceedings.

Glickman, N. J. (1980a). Methodological issues and prospects for urban impact analysis. In N. J. Glickman (Editor), The Urban Impacts of Federal Policies. Johns Hopkins University Press, Baltimore, Maryland, pp. 3–32.

Glickman, N. J. (Editor) (1980b). The Urban Impacts of Federal Policies. Johns Hopkins University Press, Baltimore, Maryland.

Glickman, N. J. and Jacobs, S. S. (1979a). The urban impacts of HUD's Section 312 program. Urban Impacts Analysis 4. Department of Housing and Urban Development, Washington, D.C.

Glickman, N. J. and Jacobs, S. S. (1979b). The Urban Impacts of the Budget of the Department of Housing and Urban Development. Department of Housing and Urban Development, Washington, D.C.

Golladay, F. and Haveman, R. H. (1977). The Economic Impacts of Tax-Transfer Policy. Academic Press, New York.

Gustely, R. D. and Ballard, K. P. (1979). The Regional Macroeconomic Impact of Federal Grants: An Empirical Analysis. Mimeograph. Paper presented at a Conference on Municipal Fiscal Stress – Problems and Potentials, held at Miami, Florida.

Hall, O. P. and Licari, J. A. (1974). Building small region econometric models: extensions of Glickman's structure to Los Angeles. Journal of Regional Science, 14: 337–353.

Harris, C. C. (1973). The Urban Economies, 1985. Heath/Lexington Books, Lexington, Massachusetts.

Havemann, J. et al. (1976). Federal spending: the North's loss is the sunbelt's gain. National Journal, 8: 878–891.

Holahan, J. and Wilensky, G. (1972). National health insurance: costs and distributional effects. Urban Institute Paper 957-1. The Urban Institute, Washington, D.C.

Hollenbeck, K. (1976). An analysis of the impact of unemployment and inflation on AFDC costs and caseloads. MPR Project Report 76-12. Mathematical Policy Research, Washington, D.C.

Hollenbeck, K. (1979). A preliminary design for a microanalytic model of the indirect impacts of energy policy. Draft report submitted to the Energy Information Administration. Mathematical Policy Research, Washington, D.C.

Holmer, M. (1980). Urban, regional, and labor supply effects of a reduction in federal individual income tax rates. In N. J. Glickman (Editor), The Urban Impacts of Federal Policies. Johns Hopkins University Press, Baltimore, Maryland, pp. 494–511.

Ihara, T. (1979). An economic analysis of interregional commodity flows. Environment and Planning A, 11: 1115–1128.

Isard, W. (1960). Methods of Regional Analysis. MIT Press, Cambridge, Massachusetts.

Isard, W. (1969). Some notes on the linkage of the ecologic and economic systems. Papers of the Regional Science Association, 22: 85–96.

Isard, W. and Langford, T. W. (1969). Impact of Vietnam War expenditures on the Philadelphia economy: some initial experiments with the inverse of the Philadelphia input–output table. Papers of the Regional Science Association, 23: 217–265.

Japan Ministry of International Trade and Industry (1970). Input–Output Table of Japan 1965. Ministry of International Trade and Industry, Tokyo.

Knox, H. and Sandoval, A. D. (1980). Regional economic modeling for energy forecasting. In F. G. Adams and N. J. Glickman (Editors), Modeling the Multiregion Economic System: Theory, Data, and Policy. Heath/Lexington Books, Lexington, Massachusetts.

Leontief, W. (1941). The Structure of the United States Economy, 1919–39. Harvard University Press, Cambridge, Massachusetts.

Leontief, W. (1953). Studies in the Structure of the American Economy. Oxford University Press, New York.

Leontief, W. (1970). Environmental repercussions and economics: an input–output approach. Review of Economics and Statistics, 52: 262–271.

Leontief, W. et al. (1965). The economic impact – industrial and regional – of an arms cut. Review of Economics and Statistics, 47: 217–241.

Lyall, K. C. (1980). The role of models in federal policymaking. In F. G. Adams and N. J. Glickman (Editors), Modeling the Multiregion Economic System: Theory, Data, and Policy. Heath/Lexington Books, Lexington, Massachusetts.

Markusen, A. R. and Fastrup, J. (1978). The regional war for federal aid. Public Interest, 53: 87–99.

Menchik, M. D. (1980). Movement of jobs and people: ideas for regional modeling and related research. In F. G. Adams and N. J. Glickman (Editors), Modeling the Multiregion Economic System: Theory, Data, and Policy. Heath/Lexington Books, Lexington, Massachusetts.

Miernyk, W. (1965). The Elements of Input–Output Analysis. Random House, New York.

Miernyk, W. (1972). Regional and interregional input–output models: a reappraisal. In M. Perlman, C. L. Leven, and B. Chinitz (Editors), Spatial Regional and Population Economics. Gordon and Breach, New York, pp. 263–292.

Milne, W. J., Glickman, N. J., and Adams, F. G. (1978). A Framework for Analyzing Regional Decline: A Multiregion Econometric Model of the United States. Paper presented at a Conference on Spatial Development Under Stagnating Growth, held at Siegen, FRG, 6–19 August 1978. Revised version in Journal of Regional Science, 20: 173–189 (1980).

Milne, W. J., Adams, F. G., and Glickman, N. J. (1980). A top-down multiregional model of the United States economy. In F. G. Adams and N. J. Glickman (Editors), Modeling the Multiregion Economic System: Theory, Data, and Policy. Heath/Lexington Books, Lexington, Massachusetts.

Moeller, J. F. (1973). Development of a micro-simulation model for evaluating economic implications of income transfer and tax policies. Annals of Economic and Social Measurement, 2: 183–187.

Olsen, R. J. et al. (1977). MULTIREGION: A Simulation–Forecasting Model of BEA Economic Area Population and Employment. Oak Ridge National Laboratory, Oak Ridge, Tennessee.

Orcutt, G., Greenberger, M., Korbel, J., and Rivlin, A. (1961). Microanalysis of Socioeconomic Systems: A Simulation Study. Harper and Row, New York.

Perry, D. C. and Watkins, A. J. (Editors) (1977). The Rise of the Sunbelt Cities. Sage, Beverley Hills, California.

Polenske, K. R. (1969). A Multiregional Input–Output Model – Concept and Results. Harvard Economic Research Project, Cambridge, Massachusetts.

Polenske, K. R. (1975). The United States Multiregional Input–Output Model. Heath/Lexington Books, Lexington, Massachusetts.

Richardson, H. W. (1972). Input–Output and Regional Economics. Halsted Press, New York.

Riefler, R. F. (1973). Interregional input–output: a state of the art survey. In G. G. Judge and T. Takayama (Editors), Studies in Economic Planning Over Space and Time. North-Holland Publishing Co., Amsterdam, pp. 133–162.

Rogers, A. (1975). Introduction to Multiregional Mathematical Demography. Wiley Publishing Co., New York.

Rogers, A. (1976). Shrinking large-scale population projection models by aggregation and decomposition. Environment and Planning A, 8: 515–541.

Rogers, A., Raquillet, R., and Castro, L. (1978). Model migration schedules and their applications. Environment and Planning A, 10: 475–502.

Sakashita, N. (1974). Systems analysis in the evaluation of a nation-wide transport project in Japan. Papers of the Regional Science Association, 33: 77–98.

Salamon, L. and Helmer, J. (1980). Urban and community impact analysis: from promise to implementation. In N. J. Glickman (Editor), The Urban Impacts of Federal Policies. Johns Hopkins University Press, Baltimore, Maryland, pp. 33–66.

Saltzman, S. and Chi, H.-S. (1977). An explanatory monthly integrated regional/national econometric model. Regional Science and Urban Economics, 7: 49–81.

Stevens, B. H. (Editor) (1980). Symposium on multiregional forecasting and policy simulation models. Journal of Regional Science, 20: 129–206.

U.S. Department of Energy (1978). Annual Report to Congress: Volume Two. U.S. Government Printing Office, Washington, D.C.

U.S. Department of Housing and Urban Development (1978). The President's National Urban Policy Report. Department of Housing and Urban Development, Washington, D.C.

Vernez, G., Vaughan, R., Burright, B., and Coleman, S. (1977). Regional Cycles and Employment Effects of Public Works Investment. The Rand Corporation, Santa Monica, California.

Wertheimer, R. F. and Zedlewski, S. R. (1976). The Impact of Demographic Change on the Distribution of Earned Income and the AFDC Program: 1975–1985. Urban Institute Paper 985-1. The Urban Institute, Washington, D.C.

Willekens, F. (1980). Regional demographic modeling. In F. G. Adams and N. J. Glickman (Editors), Modeling the Multiregion Economic System: Theory, Data, and Policy. Heath/Lexington Books, Lexington, Massachusetts.

Yett, D. E., Drabek, L., Intrilligator, M. D., and Kimbell, L. J. (1978). A Forecasting and Policy Simulation Model of the Health Care Sector: The HRRC Prototype Microeconometric Model. Heath/Lexington Books, Lexington, Massachusetts.

Regional Development Modeling: Theory and Practice
M. Albegov, A.E. Andersson and F. Snickars (editors)
North-Holland Publishing Company
© IIASA, 1982

Chapter 7

ECONOMIC MODELS OF MIGRATION

Åke E. Andersson* and Dimiter Philipov
International Institute for Applied Systems Analysis, Laxenburg (Austria)

7.1. Introduction: economics and demography

Economic development is to a large extent determined by demographic changes. It is, for instance, obvious that the level and structure of consumption is determined by the size and structure of the population. Consumption functions often contain population factors as explanatory variables (Andersson and Lundqvist, 1976).

It is also obvious that the size and composition of the population has an important influence on the volume of production through the influence of the population on the labor supply. Most econometric studies include analyses of population and labor supply as well as the econometric human capital equations (Glickman, 1977). In other words, it is clear that the connection between economic development and demographic change is of interest to economic-model builders.

However, it should be noted that, until recently, most demoeconomic models have been too aggregated to be of any real interest to either demographers or regional policy-makers.

Many studies of demographic–economic interactions make no distinction between the sizes of different age groups and do not divide the area being studied into regions. This makes it difficult for demographers to interpret the results because fertility and migration patterns are compatible with demoeconomic predictions only at the national level. This chapter represents a moderate step in the direction of a more disaggregated analysis of the interactions between demographic and economic development processes. The main emphasis is placed on the one-way dependence of demographic variables on the state of the economic system, although we also suggest methods of including the mutual dependence of economic and demographic processes. The analysis also concentrates on migration rather than fertility or mortality, since

---

* Currently at the Department of Economics, University of Umeå, Umeå, Sweden.

migration is the variable of greatest interest to most demographers and economists.

We begin by studying the problem of local labor markets. This can be approached as an optimization problem in which migration is used as an instrument to maximize some national goal function; this approach is discussed in Section 7.2. A behavioral approach to the same problem is outlined in Section 7.3, and the institutional problems associated with labor market analysis are examined.

Both the optimization and labor market analyses are carried out at rather a high level of aggregation. In contrast, migration decisions are best analyzed at the microlevel. Section 7.4, which deals with migration analysis, therefore considers individual migrants rather than policy-makers. The aim is not to derive a general theory of migration, but rather to find a workable scheme in which the noneconomic and economic factors which affect migration can be empirically related to the observed migration frequencies. This section suggests that the multinomial logit model is a suitable tool for applied mobility analysis. Section 7.5 then discusses the problem of integrating the results obtained at different levels of aggregation, and the conclusions of this chapter are summarized in Section 7.6.

The best approach to use in an economic analysis of migration is by no means obvious. The mobility of the labor force and of the population as a whole has been a subject of interest to economists for many years, and has been studied using many different techniques. The most common of these approaches are classified according to type of model and level of aggregation in Table 7.1, and are discussed in more detail in later sections.

Table 7.1. The various approaches to economic modeling, classified by the type of model and the level of aggregation.

Level of aggregation	Model type		Solution criteria
	Deterministic	Stochastic	
Microlevel	Neoclassical optimal choice (Section 7.4)	Random utility (Section 7.4.2)	Optimal decisions
	Bounded rationality	"Elimination by aspects", information theory	Other solution criteria
Macrolevel	Macroeconomic planning (Section 7.2)	Stochastic macroplanning	Optimal decisions
	Equilibrium and disequilibrium (Section 7.3)	Information theory (Section 7.4.1)	Other solution criteria

7.2. Macroeconomic model of optimal migration of labor between regions

It is possible to view the migration process as a means of obtaining the macropolitically optimal population distribution. The criteria of optimality are necessarily many and can be represented in a goal function or as constraints. It is customary to represent the most important goals as constraints, which cannot be violated, while the less important goals may be traded off against each other in the goal function.

In this example we shall use the gross national product Q as an indicator of optimality. The value of Q will vary since it depends on labor mobility. We therefore shall not consider the labor used in households or in other nonmarket activities since they are not included in the measurement of Q.

Further, we shall assume that full employment is a political requirement, at both the regional and occupational levels. This means that the number of jobs in each region must be greater than or equal to the number of persons in the labor supply. To make political sense, this requirement must be supplemented by some wage regulation. For simplicity, we shall assume that the wage rates paid for each job are the same in all regions.

The socioeconomic background to the problem will be considered to be fixed over the period under study. Thus, it will be assumed that material capital, $\bar{K}_j^s$, and the public goods environment, $\bar{a}_j$, enter the production function of sector s in region j as parameters. We also assume that the number of workers employed in every occupation is fixed by the policy-makers. Finally, it is assumed that the institutional framework, consumption infrastructure, and so on, associate a certain mobility cost with migration between regions.

Note that mobility between regions may take the form of commuting or migration (or both). In this discussion it is not necessary to specify the type of mobility as long as the only factor that can be changed is the amount of labor in occupation o moving from region i to work in sector s in region j, M_{ij}^{os}. Changes in M_{ij}^{os} are induced by the social need to maximize the national product and are independent of the enterprises' perceived needs for labor. The latter will be considered later.

The model will therefore be a simple static social optimization model. The additional variables and parameters used are listed below (exogenous variables are denoted by overbars):

$\bar{b}_{ij}^{os}$ is the social mobility cost of the moves M_{ij}^{os} (here assumed to be independent of M_{ij}^{os})

$\bar{p}_j^s$ is the net price of a commodity produced by sector s in region j (assumed to be fixed)

Q_j^s is the level of production of a commodity by sector s in region j

$\{M_j^s\}$ is a vector with a typical element M_{ij}^{os}, for fixed j and s

K is capital

The optimal distribution of labor is given by

$$\underset{\{M_j^s\}}{\text{maximize}}\ Q = \sum_{s,j} \bar{p}_j^s Q_j^s(\{M_j^s\}, \bar{K}_j^s, \bar{a}_j) - \sum_{o,s,i,j} \bar{b}_{ij}^{os} \cdot M_{ij}^{os} \tag{7.1}$$

subject to

$$\sum_{i,o,s} M_{ij}^{os} = \bar{M}_j \qquad \text{(level of employment in region } j,\ j = 1, 2, \dots, n)$$

and

$$\sum_{j,i,s} M_{ij}^{os} = \bar{M}^o \qquad \text{(level of employment in occupation } o,\ o = 1, 2, \dots, m)$$

Note also the following natural constraints:

$$\sum_{i,j,s,o} M_{ij}^{os} = \sum_{j} \bar{M}_j = \sum_{o} \bar{M}^o \qquad \text{(total national labor force)}$$

$$M_{ij}^{os} \geqslant 0$$

Additional constraints of the type $\bar{M}_j = \bar{M}^o = 1$, for all j and o, will turn the model into a nonlinear optimal allocation model.

The conditions for optimality can be derived using Lagrangian techniques. Let ω^o and ω_j be Lagrangian multipliers corresponding to the o and j constraints, respectively. Providing Q_j^s has the appropriate properties, one condition for social optimality can be written

$$\frac{\partial L}{\partial M_{ij}^{os}} = \bar{p}_j^s \left(\frac{\partial Q_j^s}{\partial M_{ij}^{os}}\right) - \bar{b}_{ij}^{os} - \omega^o - \omega_j = 0 \tag{7.2}$$

where L is the Lagrangian for (7.1).

The Lagrangian multipliers give the value of the marginal product of labor, net of mobility costs, in occupation o (ω^o) or in region j (ω_j). They can then be used to trace changes in the maximum "net national product" induced by altering the amount of labor by one unit (e.g., man-hour). Equation (7.2) states that the value of the marginal product should compensate for the shadow price of full employment in each occupation, plus the shadow price of full employment in each region, plus the social cost of mobility. Note that because the Lagrangian multipliers are independent of the sectors, the value of the marginal product at the sectoral level depends only on the costs of moving labor.

In the above model the allocation of labor is discussed simply in terms of social benefit, and the wishes of the enterprises have been neglected. It is necessary, therefore, to consider the problem of labor allocation from the point of view of individual enterprises. We use the following notation:

π_j^s is the profit made by sector s in region j
$\bar{c}_{ij}^{os}$ is the private cost of the moves M_{ij}^{os}
$\bar{w}^o$ is the private price of labor in occupation o

Note that $\bar{c}_{ij}^{os}$ are costs paid by the enterprise for the moves and $\bar{w}^o$ is the salary paid by the enterprise to the employees.

The optimal mobility of labor as far as the employers are concerned may be obtained by solving the following problem with fixed s and j:

$$\underset{\{M_{ij}^{os}\}}{\text{maximize }} \pi_j^s = \bar{p}_j^s Q_j^s(\{M_j^s\}, \bar{K}_j^s, \bar{a}_j) - \sum_{i,o} (\bar{w}^o + \bar{c}_{ij}^{os}) M_{ij}^{os} \tag{7.3}$$

It is assumed that no other constraints are needed.

The condition for optimal labor mobility from the industrial point of view is represented by the first derivative of the profit function:

$$\frac{\partial \pi_j^s}{\partial M_{ij}^{os}} = \bar{p}_j^s \left(\frac{\partial Q_j^s}{\partial M_{ij}^{os}} \right) - \bar{w}^o - \bar{c}_{ij}^{os} = 0 \tag{7.4}$$

Optimization of labor mobility for the enterprises is thus assumed to be a question of maximizing short-term profits in a situation where the wage rate within a given occupation is uniform for all regions. To this wage rate should be added the cost of moving people to work in the given region.

Thus, there are at least two conflicting approaches to optimizing migration: one optimizes migration to maximize social welfare, the other optimizes migration to maximize industrial profits. Suppose that each approach has a unique solution – the two optima will then be equal only by chance. In order to arrive at consistent ideas for the optimal allocation of labor, it is necessary to reconcile these two approaches. (In a perfect market economy, the social welfare approach could be neglected; in a perfect planned economy, the industrial profit condition could be neglected. However, no perfect market or planned economy actually exists.)

One way of reconciling the approaches is to equate eqns. (7.2) and (7.4). This gives

$$\omega_j + \omega^o + \bar{b}_{ij}^{os} = \bar{w}^o + \bar{c}_{ij}^{os} \tag{7.5}$$

Since ω_j and ω^o are determined by the model for a given $\bar{b}_{ij}^{os}$, and we would like the social optimum to remain unchanged, we should alter the predetermined parameters on the right-hand side so that eqn. (7.5) is true. It is therefore necessary to modify the earnings of the worker by introducing regionally differentiated subsidies (or taxes); this changes the values of $\bar{c}_{ij}^{os}$ and $\bar{w}^o$. Measures of this type can be introduced by the state.

Parameters $\bar{c}_{ij}^{os}$ and $\bar{w}^o$ can therefore be treated as exogenous variables and used to generate different patterns of migration. This may be useful to policy-makers in both planned and mixed economies.

7.3. Equilibrium in labor markets

There is some disagreement among economists as to whether the existence of unemployment causes a state of disequilibrium in the labor market. The following quotation presents one point of view.

Before we proceed, I must make it clear why general equilibrium analysis is a proper approach to the study of unemployment. Indeed, most economists nowadays strongly object to such an idea. The objection comes from a misunderstanding of what general equilibrium analysis really is. Economists have been brought up to think that the very notion of equilibrium implies that, for each commodity, supply must equal demand, which of course cannot be the case for labour if some involuntary unemployment remains. But a general equilibrium is an abstract construct that has no logical obligation to assume equality between supply and demand. (Malinvaud, 1978, pp. 4, 5.)

We agree with Malinvaud that the concept of equilibrium is useful even in situations of unemployment. We also believe that equilibrium techniques may be used to analyze situations in which there are jobs vacant and no unemployment, such as can be observed in most planned economies. It can therefore be assumed that the concept of an equilibrium state may be usefully applied in market, planned, and mixed economies.

In order to explain these ideas in more detail, we will divide economies into three classes (Kornai, 1979), assuming a fixed price for labor:

1. Demand-constrained (Keynesian) economies in which there is a certain amount of unemployment, but no vacancies exist.
2. Supply-constrained (Kornaian) economies in which there are jobs vacant, but no unemployed workers.
3. Balanced (Walrasian) economies in which there is no unemployment and no vacancies exist.

By considering segments (sectors or geographical regions) of the economy we can generate various mixed economies containing different proportions of these three types as shown in Figure 7.1.

Consider point D. It represents an economy in which labor use in some segments is demand-constrained, in other segments is supply-constrained, and is balanced in the remaining segments. In the discussion that follows we shall consider these segments to be geographical regions. Point D then represents an economy with unemployment in some regions, job vacancies in others, and a balanced labor market in the remainder.

Point D is generally in a state of temporary disequilibrium. We assume that unemployed people will move to the regions with vacant jobs, and that companies will search for labor in different regions if they are unable to fill all their positions locally — this should lead to a state of equilibrium.

Once the system has moved from point D to one of the (possibly temporary) equilibrium positions, the economy can be classified as Keynesian, Kornaian, or Walrasian.* Thus, if any equilibrium criteria characteristic of the three classes of

---

* The definition of these three classes does not presuppose stability. If unstable situations occur, the dynamic trajectories of the processes must be studied. The dynamic properties of the three types of equilibrium are formally analyzed in Andersson and Batten (1979).

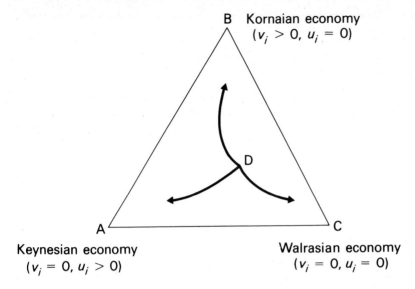

B Kornaian economy
$(v_i > 0, u_i = 0)$

D

A

Keynesian economy
$(v_i = 0, u_i > 0)$

C

Walrasian economy
$(v_i = 0, u_i = 0)$

Figure 7.1. Possible combinations of three types of economy in terms of the unemployment rate (u_i) in segment i and the proportion of jobs vacant (v_i) in segment i.

economy are known, they could be used to identify the equilibrium of any mixture of classes.

A Walrasian economy is characterized by a complete match between the supply of and demand for labor and commodities. The following two conditions must be satisfied if an equilibrium point is to exist:

1. Labor supply and demand must depend continuously on the price of labor (wages, salaries).
2. Let $\{p\} = \{p_1, p_2, \ldots, p_i, \ldots, p_n\}$ be a vector of prices. Then there must exist at least one vector $\{p\}$ such that supply matches demand in all segments i of the economy. It is also required that $\{p\} \geqslant \{\bar{p}_{min}\} \geqslant \{0\}$, where $\{\bar{p}_{min}\}$ is a vector of the lowest admissible prices.

The last inequality can be reformulated as $\{0\} \leqslant \{p\} \leqslant \{\bar{p}_{max}\}$, where $\{\bar{p}_{max}\}$ denotes a vector of the highest allowable prices.

One essential characteristic of a Kornaian economy is that the first condition and possibly also the last part of the second condition will not be satisfied. The need for unconditional full employment of certain resources means that supply is independent of prices at the national level, though this is not necessarily true at all regional levels. In a Kornaian economy, the demand for labor may also be independent of the price if prices are generally fixed by the state.

One important characteristic of a Keynesian economy is that the equilibrium price-vector, $\{p^*\}$, may include a price lower than the lowest admissible price; it may even be negative. In such a case unemployment must increase, unless $\bar{p}_{\min}$ is lowered below p^*. Similar arguments may be used if the equilibrium price exceeds the price ceiling $\bar{p}_{\max}$.

In the next three sections we describe Walrasian, Keynesian, and Kornaian equilibria and their relation to migration theory. The economies are represented at the macroeconomic level. The basic idea is that migration is a part of the aggregated plan for the supply of labor, and hence only labor migration will be considered. Demand is assumed to be derived from independent minimization of costs or maximization of profits in individual companies or over whole sectors or industries.

7.3.1. A Walrasian economy

Let $S_i(0)$ denote the supply of labor in region i at the beginning of a period during which migration takes place. Let S_i be the supply of labor at the end of this period, and M_{ij} be the migration from region i to region j. The supply of labor in region i at the end of the period is then given by

$$S_i = S_i(0) + \sum_j (M_{ji} - M_{ij}) \tag{7.6}$$

ignoring workers entering and leaving the work force through processes other than migration, e.g., death.

The migration flow M_{ij} depends on the relative socioeconomic standards of living in regions i and j. For simplicity, we will use the wage rate as a composite indicator to represent standard of living. This will be denoted by w_i for all regions i. Thus

$$M_{ij} = f_{ij}(w_i, w_j) \tag{7.7}$$

Substituting (7.7) into (7.6) yields

$$S_i = S_i(0) + \sum_j [f_{ji}(w_i, w_j) - f_{ij}(w_i, w_j)]$$

or

$$S_i = S_i(\{w\}) \tag{7.8}$$

where $\{w\} = (w_1, w_2, \ldots, w_n)$. Thus, supply is a function of the wages in all regions. We assume that all commodity prices are given, for instance, by the world market.

The demand for labor in region i is also a function of the vector $\{w\}$:

$$D_i = D_i(\{w\}) \tag{7.9}$$

However eqn. (7.9) does not adequately represent the demand for labor by the various enterprises. Their major aim is to maximize profits through the redistribution of the labor force (see our earlier analysis). Hence, the following optimization procedure can be used to derive the industrial demand in region j:

$$\text{maximize} \atop \{M_j^s\} \quad \pi_j^s = \bar{p}_j^s Q_j^s(\{M_j^s\}, \bar{K}_j^s, \bar{a}_j) - \sum_j \bar{w}_j M_j^s \tag{7.10}$$

This is essentially the same as eqn. (7.3), though the occupation indices have been suppressed. The notation used here is explained in more detail in Section 7.2.

Equations (7.9) and (7.10) are linked through the requirement that

$$D_j = \sum_s M_j^s \tag{7.11}$$

Thus, solving the optimization problem (7.10) gives the vector $\{M\}$, which can be used in eqn. (7.11) to generate demand.

Equilibrium in the classical Walrasian economy requires a simultaneous solution to the following equations:

$$S_i(\{w\}) - D_i(\{w\}) = 0 \qquad (i = 1, 2, \ldots, n) \tag{7.12a}$$

$$\sum_i [S_i(\{w\}) - D_i(\{w\})] w_i = 0 \qquad (i = 1, 2, \ldots, n) \tag{7.12b}$$

where $\{w\}$ is a vector of wage rates, the variables in this equilibrium problem.

If a solution exists, eqns. (7.12) give a set of equilibrium wage rates (w_i) for the regions. It is then possible to calculate the equilibrium migration patterns. Note that they will be given as a net-flow matrix due to the form of eqn. (7.6).

A solution to the equilibrium problem (7.12) can be found under fairly weak conditions. If supply is larger (smaller) than demand, it is sufficient to show that raising (lowering) the wage rate vector, $\{\bar{w}\}$, will equalize supply and demand. A continuous mapping can generally be assumed. If boundedness is also assumed, then a nonempty compact set is mapped onto itself. Under these conditions, the Brouwer theorem states that a fixed point, the equilibrium solution to the wage vector, will exist.

7.3.2. A Keynesian economy

The supply of labor in region i is given by eqn. (7.6) for both Walrasian and Keynesian economies. However, in a Keynesian economy the migration flow depends not only on wages but also on unemployment, measured by the regional unemployment rate u_i:

$$M_{ij} = f_{ij}[\bar{w}_i(1 - u_i), \bar{w}_j(1 - u_j)]$$
$$= f_{ij}(\{\bar{w}\}, \{u\}) \tag{7.13}$$

where the overbars indicate that wage rates are determined outside the model.

The supply equation corresponding to eqn. (7.8) will therefore be

$$S_i = S_i(\{\bar{w}\}, \{u\}) = S_i(0) + \sum_j [f_{ji}(\{\bar{w}\}, \{u\}) - f_{ij}(\{\bar{w}\}, \{u\})]$$

The demand equation remains unchanged:

$$D_i = D_i(\{\bar{w}\})$$

Thus, it is assumed that the maximization of profits is not constrained by the existence of unemployment.

One requirement for solution is given by

$$S_i(\{\bar{w}\}, \{u\}) - D_i(\{\bar{w}\}) = U_i > 0 \qquad (i = 1, 2, \ldots, n) \qquad (7.14)$$

where U_i is the number of unemployed workers. The wage rates are held fixed in eqn. (7.14), and hence it is only the unemployment rates defined as $u_i = U_i/D_i$ that can change. Therefore, eqn. (7.15) can be used to define the risk of unemployment:

$$\frac{S_i(\{\bar{w}\}, \{u\})}{D_i(\{\bar{w}\})} - 1 = u_i \qquad (i = 1, 2, \ldots, n) \qquad (7.15)$$

If we define the excess E of supply over demand by the ratio of supply to demand, $E_i = S_i/D_i$, eqn. (7.15) may be rewritten

$$E_i - 1 = u_i$$

In order to discuss the equilibrium at the interregional level, it is necessary to specify a certain relationship between the regional wage rates. This stems from the fact that at equilibrium the argument of the function f_{ij} from (7.13) ought to satisfy the equilibrium condition

$$\bar{w}_i(1 - u_i) - \bar{w}_j(1 - u_j) = 0 \qquad \text{for all } i \text{ and } j \qquad (7.16)$$

This condition reflects the equalization of the regional standards of living, $\bar{w}_i(1 - u_i)$, and underlies the planned supply and demand, which are quantitatively equalized in eqn. (7.15).

Case 1: $\bar{w}_i = \bar{w}_j$. It immediately follows from eqn. (7.16) that $u^* = u_i = u_j$ is an equilibrium condition. Then eqn. (7.15) gives

$$\frac{S_i(\{\bar{w}\}, \{u\})}{S_j(\{\bar{w}\}, \{u\})} = \frac{D_i(\{\bar{w}\})}{D_j(\{\bar{w}\})} = \gamma_{ij}$$

The quantities γ_{ij} represent the predetermined ratio of demand for labor in region i to demand for labor in region j. The above equation has a very simple interpretation: given that the wage rates are the same in each region, the unemployment rates are assumed to adjust until the relative supply of labor in each region is proportional to the predetermined demand ratios, γ_{ij}.

Case 2: $\bar{w}_i \neq \bar{w}_j$. Given that the wage rates are not the same in all regions, it follows that $u_i \neq u_j$ in order that the equality in (7.16) should hold, i.e., such that

$$\frac{1 - u_j}{1 - u_i} = \frac{\bar{w}_i}{\bar{w}_j}$$

Thus, if the regional wage rates are not equal, they can be used to define weighted regional differences in the excess of supply over demand. This simply means that the higher the wages in a certain region at equilibrium, the higher the unemployment in this region will generally be.

Keeping the wage rates fixed is a strong but not necessary condition. By relaxing it one can study situations of two main types: $|w_i - w_j| \leqslant \delta$ (the solidarity principle), in which the government restricts the regional wage differentials to a certain level; and $w_i \geqslant \min \bar{w}_i$ (the minimum-wage principle), in which the government imposes a certain minimum wage.

7.3.3. A Kornaian economy

The price system $\{w\}$ in a Kornaian economy is not generally flexible, as it is usually determined by the state (Kornai, 1979). The economy is such that the total demand for labor is usually greater than the total supply, for any supply strategy.

Let v_i denote the ratio of vacant jobs to total jobs in region i. The supply equation is again given by (7.6). The migration flow now depends on wage rates and vacancy rates

$$M_{ij} = f_{ij}(v_i, w_i, v_j, w_j)$$

and hence

$$S_i = S_i(\{\bar{w}\}, \{v\}) = S_i(0) + \sum_j (f_{ji} - f_{ij})$$

The demand for labor can also be assumed to depend on the vacancy rate

$$D_i = D_i(\{\bar{w}\}, \{v\})$$

One solution requirement is given by

$$D_i(\{\bar{w}\}, \{v\}) - S_i(\{\bar{w}\}, \{v\}) > V_i \qquad (i = 1, 2, \ldots, n)$$

where $v_i = V_i/D_i$ and V_i is the total number of vacancies.

As in case 1 of the Keynesian economy, equilibrium is given by the point at which the vacancy rate is the same in all regions, $v_i = v_j = v^*$. Hence,[†]

$$\frac{S_i(\{\bar{w}\}, \{v\})}{S_j(\{\bar{w}\}, \{v\})} = \frac{D_i(\{\bar{w}\}, \{v\})}{D_j(\{\bar{w}\}, \{v\})}$$

It is evident that the relative demands for labor in this case are not predetermined.

[†] It should be noted that the occurrence of a very large number of vacancies may encourage enterprises to reduce their search for labor, and hence no equilibrium will exist. In the Keynesian economy, an analogous situation occurs when the number of unemployed is too large – migration to search for a job may become meaningless. Both cases are extreme but real, and are not considered any further in this discussion.

If equilibrium is to be achieved, it is necessary that workers and jobs should be transferred between regions until the same relative vacancy rates are found in all regions. Case 2 of the Keynesian economy can also be given an analogous Kornaian interpretation.

The description of the three economies given above demonstrates that the lower-level policy-makers are faced with more difficult problems in a Kornaian, supply-constrained economy than in a Keynesian, demand-constrained economy. This stems from the fact that in the latter only labor is to be redistributed among regions while in the former both labor and jobs must be relocated. On the other hand, the same policy-makers in a Kornaian economy may have a larger number of acceptable solutions to their problems. Policy-makers are irrelevant to optimal migration patterns in a Walrasian economy.

We believe that it is possible to formulate a joint condition for microeconomic equilibrium in the labor market for both Keynesian and Kornaian economies (for simplicity, prices are assumed to be uniform over all dimensions). Let X denote a decision-maker (an enterprise in a Kornaian economy, and a labor supplier in a Keynesian economy). Let Π denote the probability of finding a contractor to supply (Kornaian economy) or demand (Keynesian economy) the labor. Let i denote segments of the economy. The microeconomic equilibrium is then given by

$$\Pi_n^k = \Pi_m^h \qquad (\forall\, k, h \in X; \quad n, m \in i)$$

In a Keynesian economy this condition can be interpreted as the requirement that the probability of finding a job should be the same in all segments of the economy; this implies that the risk of unemployment should also be the same throughout the economy.

7.4. Gross migration and decision-making

In this section, we will discuss a model that can be used to estimate gross migration rates. The migration rates are assumed to depend on a number of socioeconomic characteristics. The general formulation is

$$P_{ij}(x) = f_{ij}(x, E_k, E_{kl}) \qquad (k, l = 1, 2, \ldots, n) \tag{7.17}$$

where $P_{ij}(x)$ denotes the gross age-specific rate of migration from region i to region j for age group x, and E_k is a vector of socioeconomic and other characteristics whose values are exogenously defined and refer to region k. E_{kl} is a vector which represents the interaction between regions k and l.

A simple version of eqn. (7.17) that is very often used in practice is the linear function

$$P_{ij} = \sum_{m=1}^{M} \alpha_{ij}^m \cdot E_{ij}^m \tag{7.18}$$

To simplify the notation, the age groups x are not included; M is the number of exogenously given characteristics, and the α_{ij}^m are coefficients that can be determined by regression analysis. Note that the constraint that the estimated P_{ij} values must be in the interval $(0, 1)$ is not necessarily obeyed here.

Equation (7.18) is also often unsuitable for other reasons. It is likely that migration from region i to region j will depend to a certain extent on the characteristics of regions other than i and j, and this aspect is not considered in the equation. The assumption that increasing E will produce a linear increase in migration rates also seems doubtful. All in all, this approach is generally used more for its practical simplicity than for its theoretical content.

A model that avoids some of the disadvantages mentioned above is the multinomial logit model

$$P_{ij} = e^{V_{ij}} \Big/ \sum_{j=1}^{n} e^{V_{ij}} \tag{7.19}$$

where V_{ij} is a function that can be defined analogously to P_{ij} in (7.18). However, it also has some disadvantages of its own: since the model is nonlinear, it is difficult to estimate the unknown coefficients; the denominator does not depend on region j and hence the relative chance of choosing a specific region of destination does not depend on the number of regions available, i.e.,

$$P_{ij}/P_{ik} = e^{V_{ij}}/e^{V_{ik}}$$

This property is known as "independence of irrelevant alternatives".

Different ways of deducing the logit model are discussed in Sections 7.4.1 and 7.4.2. We consider a number of approaches to show that the derivation of the logit model does not depend on one specific set of assumptions. Among these approaches are models based on constant utility theory and random utility theory, as well as different models founded in information theory. The factor common to all these approaches is the explicit consideration of stochastic phenomena.

7.4.1. Derivation of the logit function from information theory

Consider the fundamental equation

$$w_i(X) = \sum_{j=1}^{n} m_i(x_j) \cdot \ln m_i(x_j) \qquad X = (x_1, x_2, \ldots, x_n)$$

where, in microinformation theory, $m_i(x_j)$ is the probability of an individual migrating to region j, given that the individual lives in region i; in macroinformation theory, $m_i(x_j)$ is the migration flow, or absolute number of migrants going from region i to region j; and, in both theories, X is the set of possible destinations (n in all) and $w_i(X)$ is the expected value of the system of destinations with respect to a fixed origin. By summing over the initial region i, the value of $w(X)$, called the entropy of the set X, may be obtained:

$$w(X) = \sum_{j=1}^{n} \sum_{i=1}^{n} m_i(x_j) \cdot \ln m_i(x_j) \qquad (7.20)$$

It is then necessary to find the most probable configuration of $m_i(x_j)$, i.e., to maximize entropy:

$$\text{maximize } w(X) = \sum_{i=1}^{n} \sum_{j=1}^{n} m_i(x_j) \cdot \ln m_i(x_j)$$
$$\scriptstyle m_i(x_j)$$

subject to system-wide constraints. In the macrocase these could be

$$\sum_{j=1}^{n} m_i(x_j) = P_i$$

which is a natural constraint, and

$$\sum_{i=1}^{n} c_{ij} m_i(x_j) \leqslant C_j \qquad (j = 1, 2, \ldots, n)$$

which is a constraint on the maximum possible flow of migrants into region j. Here c_{ij} could be a measure of the housing required by migrants to region j from region i.

The following necessary condition for a maximum can be derived using the Lagrangian for this program:

$$\frac{\partial L[\mu_i, \nu_j, m_i(x_j)]}{\partial m_i(x_j)} = -\ln m_i(x_j) - 1 - \mu_i - \nu_j c_{ij} = 0$$

where μ_i and ν_j are Lagrangian multipliers. Hence

$$m_i(x_j) = \exp[-(1 + \mu_i + \nu_j c_{ij})]$$

$$\sum_{j=1}^{n} m_i(x_j) = \exp[-(1 + \mu_i)] \sum_{j=1}^{n} \exp[-\nu_j c_{ij}] = P_i$$

$$\frac{m_i(x_j)}{P_i} = \frac{\exp[-(1 + \mu_i + \nu_j c_{ij})]}{\exp[-(1 + \mu_i)] \sum\limits_{j=1}^{n} \exp[-\nu_j c_{ij}]} = \frac{\exp(-\nu_j c_{ij})}{\sum\limits_{j=1}^{n} \exp(-\nu_j c_{ij})} \qquad (7.21)$$

Function (7.21) is of the same type as (7.19). Note that the value of ν_j can be derived using

$$\frac{\sum\limits_{j=1}^{n} c_j \exp(-\nu_j c_{ij})}{\sum\limits_{j=1}^{n} \exp(-\nu_j c_{ij})} = C_j \qquad (j = 1, 2, \ldots, n)$$

The following extension of (7.20) is very often useful:

$$I(M/M_0) = \sum_{j=1}^{n} \sum_{i=1}^{n} m_i(x_j) \cdot \ln \left[m_i(x_j)/m_i^0(x_j) \right]$$ (7.22)

where $m_i^0(x_j)$ gives some a priori information on the migration flow from region i to region j. The problem is then to estimate $m_i(x_j)$ such that the values are in some sense "correct" and as close as possible to the a priori information (see, for instance, Willekens et al., 1981). The quantity $I(M/M_0)$ is known as the information divergence.

7.4.2. Derivation of the logit model from probabilistic decision theory

Consider the one-dimensional problem of selecting region x_j as a destination from the set of regions $X = \{x_1, x_2, \ldots, x_n\}$. The region x_j is then described as the preferred alternative from the set of alternatives X.

Each alternative is described by a finite set of attributes, which can be arranged as the elements of a vector. Let z_i be the vector of attributes of region x_i. These attributes may be economic characteristics (such as the number of people employed, the average wage, the average dwelling area, or the type of services provided), geographic characteristics (such as the population density), or demographic characteristics (such as the mean age of the population or the average number of marriages in region x). Some of the attributes may be defined but not observed.

In probabilistic decision theory, an individual is assumed to select region x_i with a certain probability that depends on the observable attributes. There are two main arguments for this type of approach: (i) the individual may observe the attributes properly, but either his decision-making process is to some degree stochastic, or he does not consistently maximize his utility function; (ii) the individual acts rationally, but some of the attributes may be unobserved, or are observed with certain errors. A probability approach is thus required to deal with the difficulties in observation.

These arguments give rise to at least two theoretical approaches, often referred to as constant utility theory and random utility theory, respectively. The derivation of the logit model is different in each case and therefore the two theories will be considered separately.

Constant utility theory. This derivation is based on the axiom of choice introduced into probabilistic decision theory by Luce (1959); see also Luce et al. (1965), Volume I. The axiom is as follows:

Let $x_i \in U$, where U is a subset of X. Let $P_U(x_j)$ be the probability that a particular element (region) x_j is chosen from U, and let $P_X(U)$ be the probability that any element (region) in U is chosen from X. Then

$$P_X(U) = \sum_{x_j \in U}^{n} P_X(x_j)$$

Also, let

$$P_X(V|U) = P_X(V \cap U)/P_X(U)$$

where V is a subset of X, given that $P_X(U) \neq 0$. Then

$$P_X(x_j|U) = P_U(x_j)$$

or

$$P_X(x_j \cap U) = P_U(x_j) \cdot P_X(U)$$

The axiom states, in essence, that if some alternatives are removed from consideration, then the relative probabilities of the remaining alternatives will be preserved. In other words, the presence or absence of an alternative is irrelevant to the relative probabilities of two other alternatives, although, of course, the absolute values of these probabilities will generally be affected (Luce et al., 1965, p. 218). The axiom of choice is obviously a probabilistic version of the principle of independence of irrelevant alternatives.

Assuming that this axiom holds, Luce has shown that

$$P_X(x_j) = v(x_j) \bigg/ \sum_{i=1}^{n} v(x_i) \qquad (7.23)$$

where $v(x_j)$ is a ratio scale on X (a measure of cardinal utility).

Since the attributes of the alternatives are properly observed, there exists a measure $n(z_j)$ such that

$$n(z_j) = u(z_j) - c(z_j)$$

where $u(z_j)$ represents the benefit (generalized gain) of selecting region x_j, and $c(z_j)$ is the generalized cost (for instance distance, friction, or disutility) of this choice. This means that $n(z_j)$ can be defined as the net utility of selecting region x_j. The functions n, u, and c are all deterministic. It is the process of choosing the element, i.e., the region of destination, that is stochastic.

The probability $P_X(x_j)$ is the proportion of the population choosing to migrate to region x_j, and is determined by the vector z_j of observed attributes. The link between the model that describes this stochastic process and the net utility $n(z_j)$ appears in the analytical expression for $v(x_j)$:

$$v(x_j) = \exp[n(z_j)] = \exp[u(z_j) - c(z_j)] \qquad (7.24)$$

Substituting (7.24) into (7.23) gives

$$P_X(x_j) = \frac{\exp[u(z_j) - c(z_j)]}{\sum_{i=1}^{n} \exp[u(z_i) - c(z_i)]}$$

Note that the discussion in this section is at the level of an individual mover, and for this reason the utility $u(z_j)$ and the costs $c(z_j)$ are sometimes referred to as perceived utility and perceived costs.

The function $n(z_j)$ can also be defined by

$$n(z_j) = \sum_{k=1}^{m} \alpha_k z_{jk}$$

where z_{jk} is the kth coordinate of the m-dimensional vector z_j of attributes of region x_j.

Random utility theory. It was noted earlier in this section that the use of random utility theory is based on the assumption that some of the attributes are unobservable, and hence ought to be treated as stochastic variables. It is further assumed that the choice is made under the condition of bounded rationality. This means that the individual has his own well-defined structure of preferences, which can be represented numerically by a utility function. Given one and the same set of alternatives, described by one and the same set of attributes, a number of choices is possible; i.e., the choice has to be treated as a random variable.

These ideas are linked theoretically by assuming the utility function to be a random function of the attributes; i.e., $u(z_i)$ is a random variable for a fixed vector of attributes. In addition, the individual acts such that he maximizes his utility function; i.e., region x_i will be chosen if

$$u(z_i) > u(z_j) \qquad \text{for any } j \neq i$$

Since $u(z_i)$ is a stochastic function, there is a certain probability that this will occur:

$$P(x_i) = \text{prob}\left[u(z_i) > u(z_j)\right] \qquad \text{for } j \neq i, \; j = 1, 2, \ldots, n$$

It is always possible to represent $u(z_i)$ as

$$u(z_i) = v(z_i) + \epsilon(z_i)$$

where $v(z_i)$ is a deterministic function and $\epsilon(z_i)$ is a stochastic function. Here $v(z_i)$ is the mean, or strict, utility function and can be defined as $v(z_i) = E[u(z_i)]$, or the expectation of $u(z_i)$. The stochastic function $\epsilon(z_i)$ represents either the errors in observing the attributes or the effects of unobserved attributes. The expression for the probability then becomes

$$P(x_i) = \text{prob}\left[\epsilon(z_j) - \epsilon(z_i) < v(z_i) - v(z_j)\right] \tag{7.25}$$

In order to obtain the probabilities $P(x_i)$ explicitly, it is necessary to define a probability distribution for the differences $\epsilon(z_j) - \epsilon(z_i)$ and to express $v(z_i)$ analytically. This probability distribution must behave consistently under maximization, i.e., if $\epsilon(z_j)$ and $\epsilon(z_i)$ have one distribution (not necessarily with the same parameters), then $\max[\epsilon(z_j), \epsilon(z_i)]$ must have the same distribution.

McFadden (1978) shows that the extreme value (or Weibull, or Gnedenko) probability distribution

$$P[\Gamma \leqslant \gamma] = \exp[\exp(\gamma + \alpha)]$$

where α is a parameter, is one possible distribution satisfying the requirements stated above and yielding the logit model. An additional assumption is that the unobserved elements of the vector z_i should independently follow the extreme value distribution.

A theoretical proof (due to Holman and Marley) that the assumption of the axiom of choice is equivalent to the assumption of an extreme value distribution can be found in Luce et al. (1965, Vol. III, p. 338).

The strict utility function $v(z_j)$ can be defined as a linear combination of the observed (deterministic) attributes:

$$v(z_j) = \sum_{k=1}^{m} \alpha_k z_{jk}$$

Thus, the logit model is given by

$$P(x_j) = \frac{\exp\left(\sum_{k=1}^{m} \alpha_k z_{jk}\right)}{\sum_{j=1}^{n} \exp\left(\sum_{k=1}^{m} \alpha_k z_{jk}\right)}$$

7.5. The problem of aggregation

It is obvious from the preceding sections that there are many models that can be used in the analysis of migration problems. Each one of these models has been formulated to give answers to specific questions. Macroeconomic labor migration models are, for instance, designed to give some structure to net migration flows, while the micromodels introduce regularities into the behavior of certain finely disaggregated demographic groups. It is very difficult to obtain a consistent aggregation when coordinating these separate models, especially when the primary requirement is to generate gross migration flows between regions. These gross flows are needed for the multiregional population projection models.

In this section we suggest that the results obtained from economic models at the aggregate level and at various levels of disaggregation should be used as constraints on the estimation of gross migration probabilities. The approach is based on information theory and thus belongs to the class of stochastic macromodels. The general idea is to use the economic results, for instance, in the form of net migration constraints, as a priori information on migration and then to search for the most probable pattern of migration flows.

At this stage in our research, we can only outline one possible heuristic procedure which combines the microeconomic, stochastic utility theory with the labor-market-equilibrium or optimization theories presented above.

There are three types of a priori information on migration flows: microbehavioral decision functions of the logit type; macroequilibrium (Kornaian, Keynesian, Walrasian) conditions or social optimality conditions; and previously observed behavior. It is possible to introduce all of this a priori information into a theoretical framework, either in two stages or in one simultaneous calculation. We intend to try both procedures but the simpler and thus preliminary approach is to work in stages.

In the first stage, we assume that the probability of migration can be estimated using microdata in a logit model; for instance

$$P_{ij} = \frac{\exp(V_{ij})}{\sum\limits_{i=1}^{n} \exp(V_{ij})}$$

where P_{ij} is the probability of an individual moving from region i to region j, and V_{ij} is an indicator of the differences between the two regions. This indicator is assumed to be measured statistically using some econometric technique.

The macroequilibrium conditions affecting net flows between the regions must also be taken into account. The exact nature of these conditions depends on the economic system.

In a Keynesian economy, for example, these equations would be

$$N_{ij} = M_{ij} - M_{ji} = f[\bar{w}_j(1-u_j), \bar{w}_i(1-u_i), \dots]$$

The function f is assumed to have been measured statistically (eqn. 7.13). We propose to estimate a gross migration structure $\{M_{ij}\}$ that would be consistent with macroconstraints of this form while giving a microsolution as close as possible to the pattern generated by the logit function.

If we use the information divergence criterion (eqn. 7.22), the procedure may be outlined as follows:

$$\text{minimize} \sum_{i=1}^{n} \sum_{j=1}^{n} M_{ij} \ln \left(\frac{M_{ij}}{\left[\exp(V_{ij}) \Big/ \sum\limits_{j=1}^{n} \exp(V_{ij}) \right] \bar{P}_i} \right)$$

subject to one of the following four sets of constraints:

Keynesian	$M_{ij} - M_{ji} = f(u_i - u_j, w_i - w_j, \dots)$
Kornaian	$M_{ij} - M_{ji} = g(v_i - v_j, w_i - w_j, \dots)$
Walrasian	$M_{ij} - M_{ji} = h(w_i - w_j)$
Social optimality conditions	$M_{ij} - M_{ji} = $ optimal net migration as computed in a planning model of type (7.1)

The following constraints also have to be included for consistency:

$$M_{ij} \geq 0$$

$$\sum_{i=1}^{n} \sum_{j=1}^{n} (M_{ij} - M_{ji}) = 0$$

$$\sum_{j=1}^{n} M_{ij} = \bar{P}_i$$

where $\bar{P}_i$ is the total population of region i.

Thus, in the Keynesian case, the two-step procedure generates the Lagrangian optimization problem:

$$\text{minimize } G = \sum_{i=1}^{n} \sum_{j=1}^{n} M_{ij} \ln \left(\frac{M_{ij}}{\left[\exp(V_{ij}) \Big/ \sum_{i=1}^{n} \exp(V_{ij}) \right] \bar{P}_i} \right) - \sum_{i=1}^{n} \lambda_i (M_{ij} - M_{ji} - f)$$

λ can be interpreted here as the effect of macroeconomic conditions on decisions to migrate to and from region i.

A more ambitious procedure would be to estimate the V_{ij}, and f, g, or h functions simultaneously. In this case the parameters of these functions become the unknowns, and the values of M_{ij} are treated as observations. This gives rise to a nonlinear optimization criterion and a nonlinear set of constraints. The solution of this type of program requires powerful numerical methods.

7.6. Conclusion

The discussion in the previous sections describes methods which could be used to carry out an economically consistent forecast of a multiregional population. We have concentrated on migration as the demographic variable of primary importance in the spatial distribution of the population, at least in the short term and possibly over the medium term. The functions used in these models are specified in accordance with socioeconomic theories; unspecified functions are left only in the equilibrium and optimization models, and may easily be deduced from standard theories described elsewhere in the economics literature.

References

Andersson, Å. E. and Batten, D. F. (1979) An Interdependent Framework for Integrated Sectoral and Regional Development. WP-79-97. International Institute for Applied Systems Analysis, Laxenburg, Austria.

Andersson, Å. E. and Lundqvist, L. (1976). Regional analysis of consumption patterns. Papers of the Regional Science Association, 36.

Glickman, N. (1977). Econometric models and methods in regional science. Regional Science and Urban Economics, 7.

Kornai, J. (1979). Resource-constrained versus demand-constrained systems. Econometrica, 47.

Luce, R. D. (1959). Individual Choice Behavior. Wiley Publishing Co., New York.

Luce, R. D., Bush, R., and Galanter, E. (1965). Handbook of Mathematical Psychology. Volumes I, II, and III. Wiley Publishing Co., New York.

Malinvaud, E. (1978). The Theory of Unemployment Reconsidered. Blackwell Publishing Co., Oxford, pp. 4, 5.

McFadden, D. (1978). Modeling the choice of residential location. In A. Karlqvist, L. Lundqvist, F. Snickars, and J. Weibull (Editors), Spatial Interaction Theory and Planning Models. North-Holland Publishing Co., Amsterdam.

Willekens, F., Pór, A., and Raquillet, R. (1981). Entropy, multiproportional, and quadratic techniques for inferring detailed migration patterns from aggregate data. In Andrei Rogers (Editor), Advances in Multiregional Demography. RR-81-6. International Institute for Applied Systems Analysis, Laxenburg, Austria.

Regional Development Modeling: Theory and Practice
M. Albegov, A.E. Andersson and F. Snickars (editors)
North-Holland Publishing Company
© IIASA, 1982

Chapter 8

TECHNOLOGICAL TRANSFER OF REGIONAL ENVIRONMENTAL MODELS*

Oscar Fisch
Department of City and Regional Planning, College of Engineering, Ohio State University, Columbus, Ohio (USA)

8.1. Introduction

The Research Applied to National Needs–Regional Environmental Systems (RANN–RES) program supported a highly diverse set of projects over the five-year period from 1971 to 1976. Approximately 8.5 million dollars was allocated for research on a total of 18 projects. One of the main links between these projects was the development of large computer-based environmental models to be used as planning and decision-making aids at the state and regional levels. In the fiscal year 1976, the National Science Foundation (NSF) started to assess the relative success of these projects and began to establish guidelines for future work in this field.

In an early assessment, it was found that few of the project teams had had the opportunities and/or the funds to test the transferability, general applicability, and utility of the models. The project described in this chapter (Fisch and Gordon, 1979) was designed to overcome this problem by independently developing sites for screening and testing models that are likely to be improvements on those currently in use in planning agencies in Ohio. The following questions were considered:

1. Which needs of regional planning agencies can better be met by the use of RANN–RES models or submodels than by current methods?
2. Which modeling projects have produced tools that can be easily transferred and used?
3. Which models applicable to Ohio present no calibration difficulties?
4. What would be the cost of further extensions to successful models within Ohio and elsewhere?
5. What are the main computer requirements of Ohio regional planners?

* This work was supported by National Science Foundation Grant NSF/ENV77-15020. Any opinions, findings, conclusions, or recommendations expressed here are those of the author and do not necessarily reflect the views of the National Science Foundation.

The main aims of the project were: to assess the utility and transferability of previous RANN–RES models by providing the necessary test facilities; to determine the needs of a typical cross-section of users; and to transfer selected modeling tools to regional planning agencies and provide on-line capacity for testing and use.

It was initially thought that the main task would be to act as a clearing-house. The technical assessment of the models to be transferred was considered to be a minor function.

The clearing-house carried out two main operations: the technical and physical transfer of the models, and the implementation of the models for use by local public agencies.

The first operation involved gathering information about the models (i.e., documentation, software, and original testing data) from the NSF–RANN investigators and working closely with them, where necessary, to complete the documentation, define input requirements, redefine proxy variables, and so on.

The second operation involved close contact with three regional planning agencies in the state (direct users) and with a special board of representatives from different levels of government in the state (advisory committee). These three regional planning agencies assigned staff to help in setting up some of the models that supposedly fitted their short-term and/or long-term planning needs, provided data input, and cooperated on the implementation, testing, operation, and evaluation of the models. Workshops with regional staff were to be held at an early stage to clarify and correct first estimates of needs, and to determine the capacity of the agency to use new tools. It will be shown later that this initial concept was altered by the force of circumstances, changing the project originally planned into something slightly different. Before going any further, however, it is necessary to present the RANN–RES program that generated these projects against the historical and philosophical background of the USA in the last decade.

The beginning of the 1970s saw the demise of the huge space research program in the USA. Congress was under political pressure to reallocate federal funds to alleviate the urban crisis and to respond to increasing public concern about environmental issues. In addition, Congress decided against development of the supersonic transport-plane, thereby rendering thousands of engineers and scientists redundant. Summer crash programs were initiated at schools such as MIT and Berkeley, with the aim of converting engineers and scientists into regional and urban planners. The implicit assumption was that these people would apply the sophisticated systems approach learned in their previous professions to the problems of urban and regional planning.

The philosophy behind the program has been summarized by Mar et al. (1977, p. 69) as follows:

A fundamental goal of all RANN/RES projects was to employ the systems approach to regional environmental-systems analysis. The RES projects sought to demonstrate that (1) the computer and model technology associated with system analysis could perform existing analysis more effectively and comprehensively than existing manual, discipline-oriented processes and (2) the formulation of any analysis as a component of a larger system analysis incorporating complex feedback and inter-action mechanisms would reveal unanticipated outcomes, thereby reducing the surprises often encountered in management decision making. Implicit in the effort to accomplish these two goals were the necessity to (1) merge the knowledge and methods of many disciplines associated with the study of various components of RES and (2) unite the efforts of the analyst/modeler with those persons in policy-making positions.

Bertalanffy (1968) has long advanced the necessity and feasibility of the system approach, recognizing the value inherent in the whole being more than the sum of its parts theory and alleging that the reductionalist approach provides information from only the parts of the problems it addresses.

This statement seems to consider the systems approach in two different lights: first, from the epistemological viewpoint, the systems approach is presented as a new cognitive theory; and second, it is shown as an all-embracing mix of two branches of metaphysics — cosmology, in which the systems approach considers the totality of parts and their interlinking laws, and ontology, in which the systems approach is presented as a theory of reality. This statement was made in 1977 and ignores an early warning (Weber, 1972) of the risks associated with the application of systems analysis to sociotechnical problems.

Scientists and engineers generally concentrate upon problems with clearly defined objectives and draw upon their experience in the physical sciences to determine whether a problem has been solved or whether the solution is technically feasible. In contrast, government research into social planning or policy planning must deal with problems that are ill-defined, and that can only be resolved through careful political judgment.

The 18 RANN–RES projects were funded with the aim of encouraging the develop-ment of large-scale computer models, with particular emphasis on the use of existing knowledge to solve regional environmental problems.

Most of the projects try to include factors such as the economic and ecological impacts of different types of land use, the urban development of rural land, the management of residuals, and the interaction between urban and agricultural environ-ments in their computer models. Their basic aim is to integrate the most recent develop-ments in the social sciences (economics, demography, sociology of human migrations, development of human resources, and growth and development of human settlements), engineering and planning (transportation and land-use planning and development, resource consumption analysis, disposal of residuals, soil fertility analysis, and com-munity development and planning), and geography and ecology (analysis of resource availability and ecosystem response).

Thus, the projects being evaluated were basically trying to obtain quite specific results and it seems reasonable to expect that problem-solving former scientists would be well suited to this type of work.

The 18 models cover quite a large number of topics, as shown by the titles listed in Appendix 8A, and treat these subjects at a wide variety of spatial and temporal

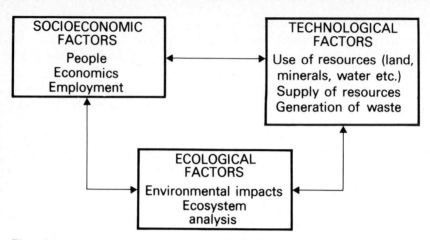

Figure 8.1. The main factors considered in regional environmental projects.

levels with varying degrees of complexity. The original scope of the models was outlined by Mar et al. (1977).

Figure 8.1 outlines the procedure used in the RANN–RES projects. The principal investigator was supposed to be familiar with the most recent theoretical and methodological advances in each of the three fields denoted by boxes in the figure. His main task was to develop links between the squares to produce a computer model able to solve problems not only in the original geographical setting but in any area with similar problems.

8.2. Model transfer: ex ante

We initially defined model transfer as the process whereby a technical model originating in one institutional setting (e.g., a university) is adopted for use in another institutional setting (e.g., a regional planning agency).

Three important assumptions were made at the start of our project: the end product of the 18 projects, a soft technology, was at the working prototype stage; our task would be mainly that of screening the models to match them with the requirements of the regional agencies; and the agencies were in real need of at least some of the models evaluated. The fundamental belief underlying the last two assumptions was that the design of the soft technology was prompted by a real regional need of high national and regional priority and that implementation of this technology was the driving force of the overall research process. However, this took no account of the fact that the organizations developing the models were not necessarily going to be the final users.

The federal programs of technology transfer in use at the time relied on the simple strategy of disseminating technical information and did not result in the adoption of the models by large numbers of regional agencies. This failure to adopt models that had already been built was explained as follows:

1. The destination agencies (final users) could not design the models because they do not have the necessary "brain" power.
2. The originating agencies (designers) have the necessary "brain" power and know about the users' needs and were therefore responsible for the design of the models.
3. The intended destination agencies did not adopt the models because they were not reached by the dissemination program, or because the information provided was either insufficient or too sophisticated for them. Too much detail discourages potential users as much as too little.

A RANN–RES workshop was held, with the 18 principal investigators participating. A special group reported on the transfer and utilization of models, and recommended to the plenary session (Mar et al., 1977, p. 51):

> The users must be encouraged to identify with a technology because acceptance is not possible without identification. *User commitment* is the *key* to successful transfer. . .

In order to establish that commitment it was recommended, among other things, that the user be convinced of the value of the technology in meeting his particular needs.

In brief, we initially understood that the task of model transfer would involve: advocation of the systems approach to solve the problems faced by the regional agency; validation of the structure, theory, and methodology used in the model as the most advanced available at the time; introduction of the model for use in the regional agency, after minor adjustment to the new operational situation.

We considered that we had a good start in our work because one of the models under investigation was already being implemented in one of the three regional planning agencies in the state of Ohio (a subcontractor of our project).

The quality of the projects to be transferred was not in doubt, given the summaries of the projects presented at the workshop. The national reputations of some of the scholars in the different teams and the number of journal articles generated by the projects reinforced our feeling that the 18 projects did indeed represent the best of the "state of the art" at that time.

8.3. Model transfer: ex post

In the private sector, research activities are governed by a combination of scientists, technicians, marketing production engineers, and regional sales managers. A clear corporate goal unifies and guides the research, and the technology transfer phase is reflected in the design of the technology.

In the public sector, finding a market for a newly developed technology is an after-thought. A potential user of the technology must be identified at the preliminary stages of the research so that the project may be endorsed; this is also a way of enhancing the relevance of the project to national and regional objectives. However, when funding is assured, the interaction between the research workers and the original endorser is totally informal, with the latter, in the majority of cases, playing the role of a guest whose next invitation is conditional on his being polite to his hosts.

In screening the 18 projects to be transferred by matching them to the needs of the local planning agencies, our ex ante assumptions about model transfer started to break down very rapidly.

Our first assumption was that the models would be at the working prototype stage when submitted to our clearing-house operation. This implies that the following items had already been prepared:

1. An "executive" report outlining the characteristics of the model.
2. An exhaustive description of the theory and methodology behind the model, and a discussion of their implicit claims to represent the most advanced state of the art.
3. An exhaustive report of the tests carried out at the development agency.
4. Complete documentation of the computer code, with a list of input requirements.
5. A set of computer tapes to be transferred, including the model and the original testing data.

The executive report is needed to persuade the regional agencies to adopt the model. The second and third items are required to validate the model, while the fourth and fifth are used to implement the model at the planning agency.

We found that the executive report was missing in almost all of the cases studied, and it was necessary to use a summary of the second and/or third items or the RANN–RES workshop papers to persuade the local agencies to adopt the model.

The other items appeared in every possible combination; in one case all five items were missing. We define an item as "missing" if it was not possible to obtain it within the first 18 months of our project; at that time, the projects had been completed for at least two years.

8.4. Advocacy stage

One-third of the models were immediately discarded because they focused on problems so unique to the area originally studied that nothing in the approach or results could possibly be applied to Ohio state. Since the models involved were very complex, it was necessary to break them down into a set of components to match them to the needs of the local planning agencies.

In a survey performed at the beginning of our project, the advisory committee and

Table 8.1. Components of the RANN–RES models ranked in first ten by destination agencies.[a]

Components ranked 1–10 by destination agencies	Percentage of agencies ranking component in first ten		
	⩽ 25%	25–75%	⩾ 75%
a		X	
b			X
c		X	
d		X	
e	X		
g	X		
h	X		
k		X	
l		X	
m		X	
n		X	
p	X		
q		X	
r	X		
t	X		
w			X
z		X	
aa		X	

[a]From a total of 27 components, nine were not ranked in the top ten by any destination agency. For a description of each component, see Appendix 8B.

Table 8.2. Components of the RANN–RES models ranked in first five by destination agencies.[a]

Components ranked 1–5 by destination agencies	Percentage of agencies ranking component in first five		
	⩽ 25%	25–75%	⩾ 75%
a		X	
b			X
c		X	
d		X	
g	X		
h	X		
k	X		
m	X		
n	X		
q	X		
w		X	
z	X		

[a]From a total of 27 components, 15 were not ranked in the top five by any destination agency. For a description of each component, see Appendix 8B.

the three regional planning agencies were asked the following question: would simulation models with any of the following outputs be useful to your agency? Later, they were asked to rank the components in order of usefulness to their agency (see Tables 8.1 and 8.2). The components ranked among the top ten by the various agencies can be given the following overall ranking: (1) b; (2) w; (3) a, c, d; (4) k, m, n, q, z; (5) g, h; (6) l, aa; and (7) e, p, r, t; for a description of each component see Appendix 8B.

Using these results, the degree of readiness of the models to be transferred, and our close contact with the three regional planning agencies (the destination agencies), we selected six models for transfer. Since one model was already being transferred, we now had a total of seven models to validate and implement in the local planning agencies.

8.5. Validation stage

The advocacy stage mainly involved putting forward the arguments for the systems approach and persuading the local agencies to use research funded by the NSF for regional planning purposes. The validation stage proved unexpectedly difficult in several ways; these are outlined below.

In the first part of the validation stage, it was necessary to establish a clear match between the capabilities of the model and the needs of the destination agency. If D is defined as the set of microservices demanded from the model by the local planning agency and S is the set of microservices that the originating agency claims the model could supply, the following combinations are possible: (1) $D \cap S = E$; (2) $D \cap S = D$; (3) $D \cap S = S$; (4) $D \cap S = M$; and (5) $D \cap S = D = S$. Here E is the empty set and M is the subset produced by a partial match between supply and demand. Case (1) represents a clear incompatibility between the services provided by a model and the needs of the potential destination agencies and case (3) represents a situation in which the destination agency requires all the services provided by the model and more. The worst case from our point of view is case (4), in which the partial match must be large enough to warrant the implementation of the whole model. This case also shares some of the problems of case (3), in that all the services provided by the model are required but not all the related needs of the destination agency are fulfilled. In such a situation the exogenous variables of the model may have to be forecast by the destination agency, possibly with the aid of a new model more costly than the one being transferred. Case (2) and particularly case (5) are the ideal situations, but neither was observed in practice.

The second part of the validation stage involved checking the consistency between claims made by the originating agency about the services provided by the model and the real ability of the software to deliver those services. In more than one case there were discrepancies between the documentation of the model and the results of the tests, and more dramatically, between the description of the methodology and the computer code of the model. We spent a considerable amount of time rebuilding the original logical and mathematical structure of such models from the FORTRAN code of the software.

The most difficult problem to deal with at this stage was associated with the quality of the research behind the models to be transferred. In the majority of the projects, there was no mention of a literature search or the integration of the various branches of science listed in Figure 8.1 that would justify the description of the model as the most advanced "state of the art". We expected to be able to compare the theoretical and methodological state of the models between 1970 and 1972 with the advances reported in the literature during the period of the original research (1971–1976). This would show us where the models were lacking in terms of technological development, given the natural delay between conception of the model and transfer of model technology. However, this last expectation was never fulfilled, given the observed differences between the theoretical and methodological state of the art in 1972 and the state of the art reported in the projects. These differences will become clearer in the discussion of the activation stage which follows.

8.6. Activation stage

It is commonly thought that the main reason for the underutilization of federal research is the lack of scientific and technical skills at the regional planning agencies. This lack of skills may frequently block the adoption of new models, either because the agency fails to realize the potential of the model or because it cannot meet the technical demands of the implementation–maintenance stage. However, we believe that this inability to evaluate models at the destination agency also frequently leads to the adoption of unsuitable models, a process that is reinforced by the prestige and credibility of the research agency. One example of this was the adoption of the community analysis model (CAM).*

CAM was an ambitious attempt to model the microbehavior of different classes of population in urban neighborhoods, in terms of residential choice by location, tenure, and price of housing. The community analysis model is based on nine sociodemographic and housing submodels, which are described in more detail later in this section.

The core of the methodological approach is the definition of a comprehensive list of neighborhood actors, together with their decisions, their social stratifications and interactions, and the major determinants of their social and spatial location at any particular moment in time.

There are 12 "neighborhood actors" who make "decisions", such as whether or not to have a child, whether or not to move into or out of the region, which neighborhood and type of housing unit to live in, mode of tenure, and so on. These actors are classified by age, ethnic–racial composition, and education, and there is a full list of factors or "determinants" which may affect the decisions they take. For example, migratory decisions are affected by determinants such as age, race, education, and racial changes in neighborhoods; moves may even be forced on a household by demolition.

The spatial unit considered in this approach is a "neighborhood" within a metropolitan region. A neighborhood is defined as a census area or combination of adjacent

---

* The remainder of Section 8.6 follows the description of the model presented in Fisch and Gordon (1979), Volume 3: An Evaluation of Birch's Community Analysis Model.

census areas that are relatively homogeneous in terms of types of housing available, social status, etc.

All of the basic transformations of data in the community analysis model are linear. They are therefore ideally suited for matrix specification, with the state being given by a vector and the operator transforming the data by an appropriate matrix. However, the author chose instead to represent the state by a cubic matrix (population state: $4 \times 3 \times 3$), and to describe, sometimes without specific equations, the transformations of these matrices. This meant than we had to spend considerable time in translating the nonmathematical explanations and computer code used in the model into more standard vector-state and matrix-operator representations. It was then possible to analyze the dimensions of the problem, the sequential steps in the transformation, and the assumptions and data requirements associated with the model. These analyses are summarized below. Each step is represented in standard matrix notation; see Rogers (1971) for a good introduction to this. The size of each matrix or vector is indicated below the equation in which it appears. The meaning of the equation is then explained nonmathematically, and this is followed by any observations or comments that should be noted at this stage.

Step 1

$$p(l, t-1) = \underset{36 \times 1}{S(l, t-1)} \underset{1 \times 1}{P(l, t-1)}$$
$$\scriptstyle 36 \times 1$$

Output. Population at location l at time $t-1$ (lagged).

Transformation. Total population to population classified by age–ethnicity–education.

Operator. State's sample tape $S(t-1)$ is adjusted to local conditions to give $S(l, t-1)$.

Observation. Transformation is exogenous; not included in the software provided. The main problem is the derivation of $S(l, t)$, $S(l, t+1)$, etc., for forecasting purposes.

Step 2

$$\underset{36 \times 1}{p^0(l, t)} = \underset{36 \times 36}{C(t-1)} \underset{36 \times 1}{p(l, t-1)}$$

Output. Population at location l at time t (current).

Transformation. Population at time $t-1$ to population at time t (cohort survival).

Operator. $C(t-1)$ is derived from fertility and mortality rates adjusted by ethnicity. No clear educational mobility adjustment.

Observation. Migration is not considered.

Step 3

$$\underset{27 \times 1}{h^0(l, t)} = \underset{27 \times 36}{HR(t-1)} \underset{36 \times 1}{p^0(l, t)}$$

Output. Number of households at location *l* at time *t* (current).

Transformation. Number of people (population) to number of households (headship rate **HR**).

Operator. The headship rates are derived nationally from data for previous periods.

Observation. No specific behavioral explanation of changes in household size, the most important factor in analyzing housing demand. Household size dropped approximately 25% during the 1970s. Data from 1979 show that 53% of households contained 1 or 2 people, whereas 10 years before this figure was only 45%.

Step 4

$$h^0(l, T, t) = \underset{54 \times 27}{\mathbf{TE}(t-1)} \underset{27 \times 1}{h^0(l, t)}$$
$$\underset{54 \times 1}{}$$

Output. Number of households at location *l* and time *t* with type of tenure *T*.

Transformation. Total households to households classified by type of tenure *T*.

Operator. **TE** is derived from the State's Public Use Sample Tape of the Census.

Observation. No consideration of local conditions, behavioral changes, etc.

Step 5

$$\underset{27 \times 1}{m^0(l, t)} = \underset{27 \times 54}{\mathbf{MT}(t-1)} \underset{54 \times 1}{h^0(l, T, t)} + \underset{27 \times 1}{fm(l, t)}$$

Output. Households willing to move out of location *l* at time *t*.

Transformation. Households classified by type of tenure to households willing to move at location *l* at time *t*.

Operator. Mobility rate of households with different types of tenure **(MT)** derived at the metropolitan level plus number of households forced to move by demolition (*fm*). Mobility rate derived from data for previous periods.

Observation. No local adjustment of this mobility rate. No clear indication of how to derive **MT**(*t*), **MT**(*t* + 1), and *fm*(*l, t* + 1), etc.

Step 6

$$\underset{27 \times 1}{M^0(t)} = \sum_{l=1}^{L} \underset{27 \times 1}{m^0(l, t)}$$

Output. Total regional pool of migrating households at time *t*. *L* is the number of neighborhoods in the region.

Step 7

$$\underset{27 \times 1}{X^0(l, t)} = \underset{27 \times 27}{\mathbf{PR}(l)} [\underset{27 \times 1}{M^0(t)} + \underset{27 \times 1}{I(t)}]$$

Output. Number of households moving to each neighborhood of a region at time *t*.

Transformation. Total number of household movers in region plus total households in-migrating to region, to number of households moving away to each neighborhood of region.

Operator. Diagonal matrix with elements derived from probability function generator (PFG):

$$\sum_i PR_{jj}(i) = 1.0; \quad jj = 1, 2, \ldots, 27.$$

Observation. The PFG requires that the user specify the structure of the 27 equations to be fitted. Some of the variables (filters) used in the model are: SLOTS, CONTIG, PMIN, PFOR, JOBACC, and CLASS. SLOTS is the number of vacancies created by one class of households for households of the same type. According to the author, this seems to be a "good" filter. But SLOTS is obtained from the number of households moving from location l generated in Step 5. In general, there is no documentation of any statistical fitting of observed data for the Miami Valley region. Thus, it is impossible to evaluate the reliability of the method or of the results obtained from it. In addition, the model does not calculate the net or gross migration into the region.

Step 8

$$\hat{x}^0\underset{162 \times 1}{(l, T, P, t)} = \mathbf{TEP}\underset{162 \times 27}{(t)} \hat{X}^0\underset{27 \times 1}{(l, t)}$$

Output. Number of households choosing housing with different types of tenure at different price levels.

Transformation. Total number of households moving to region, to number of households choosing each type of housing (according to tenure and price).

Operator. Matrix of regional split rates classified according to price and tenure, with local adjustments. Derived from data for previous periods.

Observation. No rationale behind changes in either tenure or prices.

For the whole region

$$\hat{X}^0(T, P, t - 1) = \sum \hat{x}^0(l, T, P, t - 1)$$

$$TENDM(T, t - 1) = \frac{\sum_P x(T, P, t - 1)}{\sum\sum \hat{x}^0(T, P, t - 1)}$$

$$TENDM\underset{54 \times 1}{(T, t)} = \mathbf{ST}\underset{54 \times 54}{(t)} TENDM\underset{54 \times 1}{(T, t - 1)}$$

$\mathbf{ST}(t)$ is a shift tenure operator (no rationale given).

$$TPDM(P, t-1) = \frac{\hat{x}^0(T, P, t-1)}{\sum_P \hat{x}^0(T, P, t-1)}$$

$$\underset{81\times1}{TPDM(P, t)} = \underset{81\times81}{SP(t)}\underset{81\times1}{TPDM(P, t-1)}$$

$SP(t)$ is a shift price operator (no rationale given).

$TEP(t)$ is the deconsolidation operator obtained by merging ST and SP.

$TEP(t) - TENDM(T, t)*TPDM(P, t)$ is a constant characteristic of the occupying households.

Step 9

$$\underset{162\times1}{\hat{x}^0(T, P, t)} = \sum_l \underset{162\times1}{\hat{x}^0(l, T, P, t)}$$

8.7. Integrating applied research and technology transfer

Research and development in the private sector is ruled by profitability; the product must be marketable, and its quality is the result of a tradeoff between marketability and profit margin. However, federally sponsored applied research designed to fill high priority needs in regional and local government cannot be assessed simply by its profitability. Federal programs generally operate in a top-down mode, without any effective measure of failure or success that takes transferability into account. Would programs designed to act in a bottom-up mode be more likely to succeed?

A successful historical example of applied research was produced by the 1862 Morril Act, which created the Land-Grant Universities. The original Agricultural and Mechanical Colleges, with heavy support from the U.S. Department of Agriculture, successfully merged the top-down and bottom-up approaches in the design and implementation of new agricultural techniques. The Agricultural Experimentation Stations, with their close contact with the destination agencies (County Agricultural Extension Services or even the farmers themselves) created an efficient means of transferring existing technology from research laboratories to local agencies and supported the testing and minor adjustments necessary to tailor it to specific local requirements (top-down approach). At the same time, all the local needs not met by existing technologies passed directly into the program of basic and applied research at the college laboratories (bottom-up approach). This dialectical interface between theoretical and practical demands in research and development has resulted in an extraordinarily high productivity growth rate in the agricultural sector over the last 100 years.

To conclude, our experience of model transfer to regional and local destination agencies suggests that the process could be improved in four important ways:

1. A bottom-up approach is needed to shape the program of basic and applied research in this field. Some needs of destination agencies cannot be fully met either because basic research is lacking, or, if a model has ostensibly been designed

for use in a particular region, because the needs of the region were not fully investigated at the inception of the program.

2. Permanent technical clearing-houses are needed to extract the models from the originating agencies and transfer them to destination agencies. They must perform three important functions: advocation, validation, and implementation.

3. A mechanism for peer group review of the models is needed, especially with regard to their potential transferability; peer group evaluation of the transferability of applied research would also be useful.

4. Funds should be diverted to destination agencies to create a buyers' market for technologies. This mechanism should mean that scarce resources are spent on the main priorities of the destination agencies, and applied research grants will then be allocated to the agencies, when a research proposal is accepted by a destination agency. It is hoped that this would reduce the mismatch between the needs of the local planning agencies and the services provided by the existing models.

References

Bertalanffy, L. V. (1968). General Systems Theory. George Brazilier, New York.

Fisch, O. and Gordon, S. (1979). Evaluation and Testing of NSF–RANN Sponsored Land-Use Modeling Projects with Ohio as a Test Case. Seven volumes. Ohio State University Research Foundation, Columbus, Ohio.

Mar, B. et al. (1977). Regional Environmental Systems. Department of Civil Engineering, University of Washington, Seattle, Washington.

Rogers, A. (1971). Matrix Methods in Urban and Regional Analysis. Holden-Day, San Francisco, California.

Weber, M. (1972). Dilemmas in a General Theory of Urban Planning. Institute of Urban and Regional Planning, University of California, Berkeley, California.

Appendix 8A. List of RANN–RES projects

Birch, D. L., Environmental Models for Planning and Policy Making, Harvard School of Business Administration.

Cox, D. C., Hawaii Environmental Simulation Model, University of Hawaii.

Fruh, E. G., Methodology to Evaluate Alternative Coastal Zone Management Policies: Application in the Texas Coastal Zone, University of Texas.

Heady, E. O., National Environmental Models of Agricultural Policy, Land Use and Water Quality, Iowa State University.

Jameson, D. A., Regional Analysis and Management of Environmental Systems, Colorado State University.

Kadlec, R. H., Feasibility of Utilization of Wetland Ecosystem for Nutrient Removal from Secondary Municipal Waste Water Treatment Plant Effluent, University of Michigan.

Koenig, H. E., Design and Management of Rural Ecosystems, Michigan State University.

Meadows, D. L., Natural Resource Availability and Policy Implications in the United States, Dartmouth College.

Northorp, G. M., Impact of Economic Development and Land Utilization Policies on the Quality of the Environment, The Center for Environment and Man.

Odum, H. T., Feasibility of Utilizing Cypress Wetlands for Conversion of Water and Nutrients in Effluents from Municipal Waste Water Treatment Plants, University of Florida.

Reid, G. W., An Urban Model for the Evaluation of Alternative Growth Policies, University of Oklahoma.

Rowe, P. G., Environmental Analysis for Development Planning; Chambers County, Texas, Rice Center for Community Design and Research.

Schrueder, G. F., Forest Land-Use Allocation and Environmental Systems Evaluation, University of Washington.

Steinitz, C., The Interaction between Urbanization and Land: Quality and Quantity in Environmental Planning and Design, Harvard Graduate School of Design.

Thompson, R. G., National Economic Models of Industrial Water Use and Water Treatment, University of Houston.

Tolley, G. S., Environmental Control and Land Use Interactions in the Chicago Region, University of Chicago.

Voelker, A., Regional Environmental Systems Analysis, Oak Ridge National Laboratory.

Watt, K. E. F., Land Use, Energy Flow and Decision Making in Human Society, University of California at Davis.

Appendix 8B. List of components of RANN–RES models

(a) A detailed regional input–output table at the SMSA or county level derived from the input–output table for the state.

(b) Spatial allocation of future population to census areas.

(c) Determination of future wholesale and retail activity at the neighborhood level, based on the predicted population distribution and present neighborhood characteristics.

(d) An overlay of the spatial allocation of future industries, utilities, retail and commercial facilities and residences with maps delineating land capability.

(e) A biological data retrieval system which relates water-quality changes to effects on organisms.

(f) Energy-demand forecasts for census areas, districts, and SMSAs for both rented and owner-occupied housing units. Energy-use categories include central heating, cooking, water heating, dryer heat, air conditioning, space heating, and kitchen energy.

(g) Forecasts of school-age population in each grade, land use, vacant land status, construction type, and assessed value of each subzone in townships.

(h) Estimates of township budget and tax rates.

(i) List of economic consequences to multicounty regions of state restrictions on the use of agricultural fertilizers.

(j) List of economic consequences to multicounty regions of state restrictions on soil loss.

(k) Analysis and ranking of citizens' political preferences concerning planning-related issues in a given county or region.

(l) List of the combination of planning decision variables that best satisfies established goals under specified regional priorities.

(m) Projection of economic activity for a multicounty region, based on historical growth.

(n) Projection of economic activity for a multi-county region, based on interactions of economic sectors and growth rates derived from external systems.

(o) Simulation of the pastureland ecosystem through consideration of herbivore and plant-growth subsystems. The simulation can be carried out for individual farms.

(p) Classification of land in a county or multicounty region into regional response units, i.e., units of land that react in specific ways to development or environmental change.

(q) Socioeconomic and land-use projections in which population, employment, industrial development, and labor-force participation rates for a region are allocated to 40-acre cells.

(r) Ground-level concentrations of air pollutants (gases and particulates).

(s) Estimates of the deposition rates of air pollutants through fallout and washout.

(t) Hydrologic outputs for a watershed, including total annual runoff, annual surface runoff, maximum discharge, and sediment production.

(u) Prediction of the number of fish in a reservoir or group of reservoirs, based on the concentration of total dissoved solids and mean depth of the reservoir.

(v) Estimates of the effects of sulfur dioxide on a forest ecosystem.

(w) Forecasts of public expenditures and revenues for major political jurisdictions within a region.

(x) Simulation of the formation of coalitions by groups concerned with local planning issues.

(y) Effects of air pollutants on urban areas, in terms of economic costs.

(z) Analysis of the effects of future urban development on air quality.

(aa) Analysis of the effects of future urban development on water quality in a particular river basin.

PART IV

MULTIOBJECTIVE ANALYSIS IN REGIONAL PLANNING

Regional Development Modeling: Theory and Practice
M. Albegov, A.E. Andersson and F. Snickars (editors)
North-Holland Publishing Company
© IIASA, 1982

Chapter 9

MULTIPLE OBJECTIVES IN MULTILEVEL MULTIREGIONAL PLANNING MODELS

Peter Nijkamp and Piet Rietveld
Faculty of Economic Sciences, Free University, Amsterdam (The Netherlands)

9.1. Introduction

The aim of multiregional policy analysis is to provide tools for spatial conflict resolution. To develop harmonized planning strategies for a system divided into a set of subsystems (for example, districts within a metropolis, regions within a country, branches within an industry, or sectors within a national economy) requires methods for the resolution of conflicts of goals or interests arising from the interdependence of the various components of the system. Hence, collective decision-making should guarantee an allocation of resources (money, commodities, investments, etc.) such that the final state of the system reflects a meaningful compromise between the various policy options.

The complex interactions between both the components of the system and the policy or decision levels of the system can only be analyzed properly if information is provided on the structure of all the components. This information comprises, amongst other things: (a) interdependencies between the components of the system (for example, overspill effects and externalities); (b) conflicts between various priorities, objectives, or targets within one component of the system (for example, friction between equity and efficiency at the intraregional level); and (c) conflicts between the priorities, objectives, or targets set by the various components of the system (for example, competition between various cities for federal funds).

To provide the information in category (a) it is necessary to construct a structure model which describes all interactions within and between the components of the system (for example, an interregional model describing the functional economic relationships within and between regions in a national economy).

The policy conflicts inherent in (b) require the use of multiobjective programming theory in which a vector optimization problem reflects the conflicts between a set of different objectives (for example, the friction between the aim of maximum production and maximum environmental protection within an area).

Finally, the conflicts between the various components of the system require a coordinating mechanism at a higher level to provide a meaningful compromise between the various aims at the lower levels (for example, the allocation of investment funds by central government in order to stimulate the regional potential for industrial growth). This category of coordination problems is best studied using multilevel programming theory.

Thus, the interdependencies, interactions, and causal–functional relationships within and between the components of a system are described by means of a structure model. This model and its constraints define the feasible area for both a multiobjective programming analysis and a multilevel programming analysis. Multiobjective programming deals with the multidimensional nature of choices and conflicting options in real-world policy problems. Multilevel programming concentrates on coordinating different decision levels in an attempt to encourage the best social choices. The aim of the present study is to provide a synthesis between multiobjective and multilevel programming. This implies some sort of nested hierarchy between a coordination center (or central policy unit) and the components of the system. Thus, it will be assumed that there are several decision units (in particular, regions) that are competing with each other in external conflicts caused by spatial overspills and environmental externalities. Furthermore, the policies of the various components are assumed to be coordinated by a central planning unit [using either a top-down (centralized) or bottom-up (decentralized) approach]; this is important because each component has its own interests, leading to multilevel conflicts. Finally, each component is assumed to have a set of multiple conflicting objectives leading to internal conflicts.

This chapter contains a brief introduction to multiobjective programming and multilevel programming. Special attention will be paid to the problems of coordination in a combined multiobjective–multilevel programming model. The computational aspects will be discussed, and the conflicts inherent in the possibility of the parties forming coalitions will be examined. After a formal comprehensive analysis of the problems of conflict resolution, an empirical application based on a multiregional policy model will be presented incorporating, amongst other things, pollution, employment, and a two-level policy structure.

9.2. A multiobjective programming framework

Traditional policy models are generally based on the assumption that individual decision units are trying to achieve one-dimensional objectives (maximum revenues, utility, social welfare, etc.). These policy analyses are often developed for a wonderland containing no other decision units, while external overspill effects are frequently excluded by assumption.

Recently, however, there has been a growing awareness of the existence and relevance of overspill effects in the larger decision-making units (such as regions or countries). A simultaneous analysis of all relevant policy objectives (implying a multidimensional objective profile) and of all relevant decision units (implying a multicomponent profile)

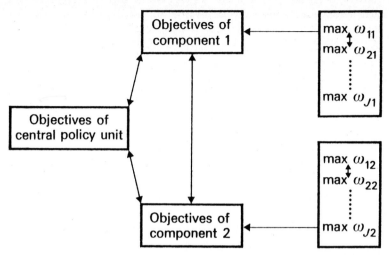

Figure 9.1. Multiobjective framework for planning in a two-component system. The J objective functions of each component (e.g., region) r are denoted by ω_{jr}, where $j = 1, 2, \ldots, J$ and $r = 1, 2$.

would complicate the traditional policy and programming models considerably. Figure 9.1 illustrates a double multidimensional policy framework which reflects the interdependencies and interactions among the various components of the policy structure under analysis. The various objective functions are denoted as $\omega_1, \omega_2, \ldots,$ ω_J. The framework in Figure 9.1 can easily be extended to a structure with three or more levels. The figure clearly reflects a double choice conflict, i.e., between objectives and between components. The conflict between components is a result of both the effects of overspill between these components (for example, spatial externalities and input–output linkages in a multiregional system) and competition for scarce resources allocated by a higher decision or policy level. The conflicts between objectives emerge from the divergence between objectives within a certain component, and may be studied using multiobjective programming (see, for example, Johnsen, 1968; Fandel, 1972; Cochrane and Zeleny, 1973; Hill, 1973; Guigou, 1974; Zeleny, 1974, 1976; Haimes et al., 1975; Wallenius, 1975; Wilhelm, 1975; Keeney and Raiffa, 1976; Thiriez and Zionts, 1976; Bell et al., 1977; van Delft and Nijkamp, 1977; Nijkamp, 1977, 1979; Starr and Zeleny, 1977; Cohon, 1978).

The presence of several competing policy units introduces an additional complication because it involves a double choice conflict. This problem can be described formally using adjusted multiobjective programming models. Consider, for instance, a system with two components (for example, two regions in a national system). Then the intracomponent and intercomponent structure of component 1 can be formally represented by means of the following model:

$$x_1 = f(x_1, x_2, e_1) \tag{9.1}$$

Table 9.1. A component interdependence matrix.

		Component 1 x_1	Component 2 x_2
Component 1	x_1	x_{11}	x_{12}
Component 2	x_2	x_{21}	x_{22}

where x_1 is a vector of relevant variables for component 1 (for example, employment level, sectoral production levels, levels of pollution, energy consumption, etc.), x_2 a vector of the same variables for component 2, and e_1 a vector of exogenous variables for component 1. An analogous model may obviously be assumed for region 2. Examples of intercomponent relationships in a multiregional system include input–output linkages, environmental externalities, transportation flows, and migration and commuting flows. Table 9.1 illustrates these interdependencies. The diagonal blocks of Table 9.1 represent intracomponent relationships while the off-diagonal blocks represent intercomponent relationships (including any conflicts between the components).

Beside the structural relationship reflected by eqn. (9.1), it is also necessary to consider the set of technical, economic, environmental, and institutional constraints that limit the sphere of action of decision-makers. Given eqn. (9.1), the feasible area of x_1 may be represented by K_1, i.e.,

$$x_1 \in K_1 \tag{9.2}$$

This leads to the following multicomponent multiobjective programming structure for the whole system:

$$
\left.
\begin{aligned}
&\max \omega_{11}(x_1) \\
&\max \omega_{21}(x_1) \\
&\quad\quad \text{for component 1} \\
&\max \omega_{J1}(x_1)
\end{aligned}
\right\}
$$

$$
\left.
\begin{aligned}
&\max \omega_{12}(x_2) \\
&\max \omega_{22}(x_2) \\
&\quad\quad \text{for component 2} \\
&\max \omega_{J2}(x_2)
\end{aligned}
\right\} \tag{9.3}
$$

$$\left. \begin{array}{l} \max \omega_{1c}(x_1, x_2) \\[6pt] \max \omega_{2c}(x_1, x_2) \\[4pt] \quad \cdot \\ \quad \cdot \\ \quad \cdot \\[4pt] \max \omega_{Jc}(x_1, x_2) \end{array} \right\} \quad \text{for the whole system}$$

subject to $x_1 \in K_1, x_2 \in K_2$

and $\quad (x_1, x_2) \in K$

Table 9.2 shows the multiobjective matrix obtained by combining both the components of the system and the system as a whole. Each column of the table represents the structure of a certain objective function over the various components, while each row represents the profile of all relevant objectives for a given component. Thus, a vertical maximization would maximize the value of one objective function for all components, while a horizontal maximization would maximize the value of all objective functions within one component.

Table 9.2. A multiobjective matrix profile.

Components	Objective functions			
	ω_1	ω_2	$\cdot\ \cdot\ \cdot$	ω_J
1	ω_{11}	ω_{21}	$\cdot\ \cdot\ \cdot$	ω_{J1}
2	ω_{12}	ω_{22}	$\cdot\ \cdot\ \cdot$	ω_{J2}
System	ω_{1c}	ω_{2c}	$\cdot\ \cdot\ \cdot$	ω_{Jc}

Clearly, both directions involve serious policy conflicts between both components and objectives. Therefore, an appropriate compromise framework that leads to a satisfactory result for both components and objectives has to be devised. In our view, an interactive learning procedure is extremely useful in that it guarantees coordination at the multicomponent level and compromise at the multiobjective level (for a formal exposition of interactive programming, see Nijkamp and Rietveld, 1976, 1978, 1979). The two parties in the interaction are the policy-maker, who is responsible for the decision but who lacks a clear insight into the problem, and the analyst, who possesses this insight and who is available to aid the policy-maker in his decision. The interactive programming procedure used in conflict resolution is basically as follows. First, the analyst produces a provisional compromise solution (see below) which is then assessed by the policy-maker(s). The policy-maker indicates the proposed values of the objective functions that are not satisfactory to him, and these preferences are incorporated by the analyst in the next stage of the analysis. The procedure is then repeated until a converging satisfactory compromise solution is finally obtained.

A general outline of the interactive procedure is given in Figure 9.2.

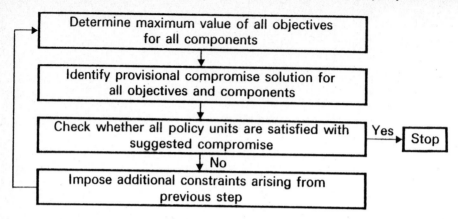

Figure 9.2. Flow diagram illustrating how a compromise solution acceptable to all policy units may be obtained through an interactive procedure.

There are various methods of calculating provisional compromise solutions (see Nijkamp and Rietveld, 1979). In this chapter we will make use of a method proposed by Theil (1964), which consists of the following steps:

1. Find the solution vector x_j that yields the maximum attainable value for each objective ω_j, using mathematical programming.
2. Given the solution vectors x_j $(j = 1, 2, \ldots, J)$, construct the payoff matrix $\mathbf{P}$ with a typical element $p_{jj'}$ defined as the value of objective ω_j when $x_{j'}$ is the decision vector. The elements p_{jj} of the main diagonal are clearly the maximum attainable values of the objectives ω_j $(j = 1, 2, \ldots, J)$.
3. Given $\mathbf{P}$, construct the loss matrix $\mathbf{Q}$ with a typical element $q_{jj'}$ defined as $p_{jj} - p_{jj'}$. Thus $q_{jj'}$ represents the difference between the maximum attainable level of objective j, and its value when objective j' is maximized. Obviously, $q_{jj'} \geqslant 0$ and $q_{jj} = 0$.
4. Find compromise weights $(\lambda_1, \lambda_2, \ldots, \lambda_J)$ such that for all pairs of objectives j, j' the weighted loss for objectives j caused by the maximization of objective j' $(\lambda_j q_{jj'})$ is equal to the weighted loss for objective j' when objective j is maximized $(\lambda_{j'} q_{j'j})$. Hence, the number of conditions imposed on the λ_j is equal to $\frac{1}{2} J(J-1)$, which is larger than the number of unknowns $(J-1)$. Theil concludes that exact equality between pairs of weighted losses is in general unattainable. He therefore proposes to approximate these equalities by determining weights $(\lambda_1, \lambda_2, \ldots, \lambda_J)$ such that

$$\sum_j \lambda_j q_{jj'} = \sum_j \lambda_{j'} q_{j'j} \quad \text{for } j' = 1, 2, \ldots, J \tag{9.4}$$

Theil shows that weights determined in this way have the attractive properties of being positive and unaffected by linear transformations.

5. Determine the compromise solution by maximizing $\Sigma_j \lambda_j \omega_j$ under certain constraints, where the λ_j are the compromise weights determined in step 4.

Before examining the interactive framework for a multilevel multiobjective programming model in detail, it is first necessary to study some of the important features of multilevel programming models in general.

9.3. Multilevel programming: an introduction

Multilevel programming models provide a framework for the coordination of decisions made in the various components of a system. It is assumed that the coordination has to be accomplished by a central unit that has the authority to give certain directives to the components. However, the central policy unit has only fragmentary knowledge about the structure of the problem. Hence, in addition to the learning process described in Section 9.2, another learning process has to be introduced to gather sufficient information about the structure of the problem.

Multilevel programming models can be described as follows. Assume that there are R components $(r = 1, 2, \ldots, R)$. Every component r has a series of J objectives $\omega_r = (\omega_{1r}, \omega_{2r}, \ldots, \omega_{Jr})'$ that have to be maximized and that depend on I instrument and state variables $x_r = (x_{11}, x_{21}, \ldots, x_{I1})'$. If we assume that the system has a linear structure, then $\omega_r = C_r x_r$, where C_r is a $(J \times I)$ matrix of impact coefficients. If we also assume that the components or subdivisions have solved their internal goal conflicts, we may take for granted the existence of a welfare function $\omega_r = \Sigma_j \lambda_{jr} \omega_{jr}$ for each component, where λ_{jr} is the weight attached to objective j by component r.

The restrictions faced by the components may be divided into two classes: internal restrictions

$$B_r x_r \leqslant b_r \tag{9.5}$$

and joint restrictions

$$\sum_r A_r x_r \leqslant a_r \tag{9.6}$$

It is clear that when there are only internal restrictions, the components are independent of each other and no coordination is therefore needed. The need for coordination consequently stems from the occurrence of joint restrictions.

In order to coordinate the decisions, the center has to formulate an objective. In many multilevel studies it is assumed that the central objective ω_c is simply the sum of the objectives of the components: $\omega_c = \Sigma_r \omega_r$. This assumption will be followed in the rest of this section and in Section 9.4.

The central planning problem can now be formulated as:

$$\max \lambda_1' \, C_1 x_1 + \ldots + \lambda_R' \, C_R x_R$$

$$\text{subject to} \quad A_1 x_1 + A_2 x_2 + \ldots + A_R x_R \leqslant a$$

$$B_1 x_1 \qquad\qquad\qquad \leqslant b_1$$

$$B_2 x_2 \qquad\qquad \leqslant b_2 \tag{9.7}$$

$$\cdot \qquad\qquad \cdot$$

$$B_R x_R \leqslant b_R$$

$$x_r \geqslant 0 \qquad r = 1, 2, \ldots, R$$

Problem (9.7) cannot be solved immediately if it has been assumed that the center has only limited information about the matrices B_r. The answer is for the components r to determine the optimal values of x_r themselves, guided by some coordination from the center. This coordination can be accomplished either directly or indirectly.

In the *direct method* the center begins by distributing the common resources a among the components according to a provisional distribution $(a_1, a_2, \ldots, a_R)$ satisfying $\Sigma_r a_r = a$. Then each component r solves

$$\max \omega_r = \lambda_r' \, C_r x_r$$

$$\text{subject to} \quad B_r x_r \leqslant b_r$$

$$A_r x_r \leqslant a_r \tag{9.8}$$

$$x_r \geqslant 0$$

and reports the shadow prices π_r of the common resources back to the center. Given this information about prices, the center revises the distribution of resources to increase efficiency. When the shadow prices are equal in all regions, a redistribution does not increase ω_c, so that the optimum has been attained. Examples of direct methods can be found in Schleicher (1971), ten Kate (1972a), Kornai (1975) and Johansen (1978).

A characteristic feature of *indirect methods* is that the center computes the distribution of resources only *after* the optimal prices of resources have been determined. The center begins by generating provisional prices π for the common resources. The components then solve

$$\max \lambda_r' \, C_r x_r - \pi' A_r x_r$$

$$\text{subject to} \quad B_r x_r \leqslant b_r \tag{9.9}$$

$$x_r \geqslant 0$$

and report to the center the optimal amounts a_r required. If $\Sigma_r a_r = a$, the overall optimum has been attained. If not, the center has to adjust the prices such that this condition will ultimately be fulfilled. See ten Kate (1972b) for a more precise statement of the conditions for optimality in the direct and indirect methods, and Dantzig

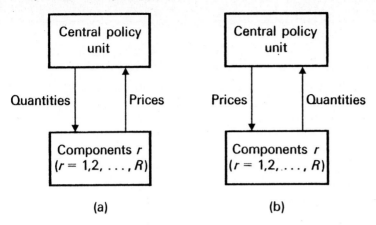

Figure 9.3. The two methods of coordinating resource allocation and resource prices between the central policy unit and the R components of the system: (a) direct method; (b) indirect method.

and Wolfe (1960), Dantzig (1963), Baumol and Fabian (1964), and Johansen (1978) for examples of indirect methods.

The main distinction between direct and indirect methods is that in the former the center provides information about quantities to the components and receives information about prices from the components, while in the latter the situation is reversed (see Figure 9.3).

In the next section we will discuss the way in which the center determines prices or quantities so that the process will converge.

9.4. Computational aspects of multilevel programming

One of the first algorithms of use in indirect methods of distribution was developed by Dantzig and Wolfe (1960), and can be outlined as follows. Let S_r denote the set of all x_r satisfying the constraints for each component in (9.7):

$$S_r = \{x_r | \ B_r x_r \leqslant b_r ; x_r \geqslant 0\} \tag{9.10}$$

At the beginning of the procedure, B_r (and consequently also S_r) is unknown to the center. The Dantzig–Wolfe algorithm carries out a stepwise determination of the relevant part of S_r ($r = 1, 2, \ldots, R$), which is a convex polyhedron with a finite number of extreme points. Let T_r be the matrix of extreme points of S_r. Then every $x_r \in S_r$ can be written as a convex combination of the columns of T_r, i.e., $x_r = T_r \mu_r (\iota' \mu_r = 1, \mu_r \geqslant 0)$. Consequently, (9.7) is equivalent to:

$$\max \lambda_1' \ C_1 T_1 \mu_1 + \ldots + \lambda_R' \ C_R T_R \mu_R$$

subject to $A_1 T_1 \mu_1 + \ldots + A_R T_R \mu_R \leqslant a$

$$\iota' \mu_r = 1 \qquad r = 1, 2, \ldots, R$$
$$\mu_r \geqslant 0 \qquad r = 1, 2, \ldots, R \tag{9.11}$$

The difference between (9.7) and (9.11) is that in the former $x_1, x_2, \ldots, x_R$ are the unknowns to be determined, whereas in the latter the unknowns are $\mu_1, \mu_2, \ldots, \mu_R$. The question is, obviously, how to determine $T_1, T_2, \ldots, T_R$. To solve this problem Dantzig and Wolfe (1960) proposed the following algorithm:

1. In the first two steps T_r is completely unknown. The center provides information about arbitrary prices π to the components.
2. The components solve (9.9) and each reports one column of T_r to the center.
3. The center solves (9.11) for the T_r matrix, as far as its columns are known. The dual variables related to the common resources a are reported to the components.

The iterations are then repeated until the recurrent solution of (9.11) converges. See Malinvaud (1972) and Johansen (1978) for related algorithms.

A very attractive property of this algorithm is that it guarantees convergence in a finite number of steps. The information reported by the components of the system is apparently handled in a very efficient way.

An equally attractive procedure has been developed by ten Kate (1972a) for methods of direct distribution. In fact, his proposal is based on a dualization of the Dantzig–Wolfe algorithm. In its original formulation, the Dantzig–Wolfe algorithm tries to determine in a stepwise fashion the relevant part of the feasible region implied by (9.9). The algorithm of ten Kate attempts to identify the relevant part of the set of feasible dual variables implied by (9.8) in a similar way.

The central programming problem implied by the ten Kate method is as follows. Let π_{kr}^i represent the dual variable related to the kth common resource of component r during iteration i. Let v_r^i be the vector of dual variables related to the resources b_r of component r during iteration i. Then $\eta_r^i = b_r' v_r^i$ represents the value of the resources b_r during iteration i. The center has to solve the following problem in step i':

$$\max z_c = \sum_r v_r$$

$$\text{subject to} \quad v_r \leqslant \eta_r^i + \sum_k \pi_{kr}^i a_{kr} \qquad i = 1, 2, \ldots, i' - 1$$
$$r = 1, 2, \ldots, R \tag{9.12}$$
$$\sum_r a_{kr} = a_k \qquad k = 1, 2, \ldots, K$$

The result of (9.12) is a new distribution of resources $(a_1, a_2, \ldots, a_R)$ based on the information about productivities per component deduced from all preceding steps.

The algorithm guarantees convergence within a finite number of iterations. This attractive property is not shared by some alternative direct methods, such as those described in Schleicher (1971), Kornai (1975), and Johansen (1978). These other methods, however, are more intuitively appealing. The central idea is that when a component r has reported a relatively high productivity π_{kr}^{i-1} in iteration $i-1$, the central objective function can be improved by reallocating, in iteration i, a certain amount of a_k to component r at the expense of less productive components. Obviously, this rule about the direction of a redistribution must be followed by a rule which determines the order of magnitude of the redistribution. However, it is not easy to find such a rule since a small step-size may imply a very slow rate of convergence, while a large step-size may imply no convergence at all. Schleicher (1971) solves this dilemma by introducing a variable step-size, which can be altered in each iteration as necessary.

Nijkamp and Rietveld (1981) contains a more extensive analysis of various aspects of multilevel planning methods, including amongst other things the amount of a priori information needed, the size of the information streams, the number of iterations, and the complexity of the computations to be performed. It seems to be impossible to find a single method which performs better than all other methods under all criteria, although within certain subsets of methods more definite conclusions can be drawn.

Nijkamp and Rietveld (1981) also contains some computational results obtained using the Schleicher and ten Kate methods, and it is clear that the latter is superior to the former as far as speed of convergence is concerned.

9.5. Multiobjective multilevel planning

Comparing Sections 9.3 and 9.4 with Section 9.2, it is clear that the multidimensional character of the objectives has not been recognized sufficiently in the multilevel planning methods discussed here. This is shown in at least two ways:

1. The conflicts between the components of the system are neglected — for example, the central objective is assumed to be simply the unweighted sum of the objectives of the components.
2. The components can easily determine the weights λ_r they attach to the various objectives they pursue. Internal conflicts consequently receive only minor attention.

9.5.1. Conflicts between components

Most multilevel planning studies consider that each component has a unidimensional objective, such as the maximization of income, which can be aggregated in a meaningful way. If other interpretations are placed on $\lambda_r' C_r x_r$ (for example, if it is assumed to represent welfare in component r), difficulties may arise. In this case an aggregation is only meaningful if a consensus can be reached about a common measure of welfare in the various components. Another problem arises in indirect methods, since the objective $\lambda_r' C_r x_r - \pi' A_r x_r$ in (9.9) essentially assumes the existence of welfare payments from component r to the center, which hardly seems very likely.

It will be shown in this section that it *is* possible to design multilevel programming models which take into account the various conflicts between objectives. Salih (1975) has given an example of an indirect method which does this; in the present section we will concentrate on direct methods, especially that of ten Kate.

As a first step, we assume that the central objective function is no longer $\omega_c = \Sigma_r \lambda'_r \omega_r$, where λ_r is the vector with weights determined by component r, but $\omega_c = \Sigma_r \gamma'_r \omega_r$, where the γ_r indicate the priorities of the center with respect to the objectives of component r.

Allowing this divergence of priorities gives rise to some very interesting phenomena. Assume, for example, that $a_1, a_2, \ldots, a_R$ is the optimal division of resources when there is no difference in political weights (i.e., $\gamma_r = \lambda_r, r = 1, 2, \ldots, R$); this would give rise to the solutions $x_1, x_2, \ldots, x_R$. What will happen if a difference in political weights does exist? Obviously, there is no guarantee that the resources will be used according to the intentions of the center. Hence, one may wonder whether it is possible for the center to determine a division $\mathring{a}_1, \mathring{a}_2, \ldots, \mathring{a}_R$ that yields the best feasible outcome, given the divergence of priorities.

We will show that a division $\mathring{a}_1, \mathring{a}_2, \ldots, \mathring{a}_R$ is possible in principle, although it implies considerably more computation at the component level. The information π_{kr} is useless in the new situation, since the center is no longer interested in the effect of resource k on the welfare of component r, but rather its effect on the various objectives $1, 2, \ldots, J$ of component r. The components must therefore provide information in the form of a $J \times K$ matrix Π_r with elements π_{jkr}, instead of a vector π_r with elements π_{kr}. Unlike π_r, Π_r cannot be derived directly from the simplex tableau; sensitivity analysis is necessary (see Wagner, 1975).

Given this new information, the ten Kate method for direct distribution can be adapted to compute $\mathring{a}_1, \mathring{a}_2, \ldots, \mathring{a}_R$ in an iterative fashion, while retaining its attractive convergence properties:

$$\max \sum_r v_r$$

$$\text{subject to} \quad v_r \leqslant \sum_{j,k} \gamma_{jr} \pi^i_{jkr} a_{kr} + \sum_j \gamma_{jr} \eta^i_{jr} \qquad \begin{array}{l} r = 1, 2, \ldots, R \\ \\ i = 1, 2, \ldots, i' - 1 \\ \\ k = 1, 2, \ldots, K \end{array} \qquad (9.13)$$

$$\sum_r a_{kr} = a_k$$

Note that the priorities of both levels play a role in (9.13), since both the central weights γ_{jr} and the weights λ_{jr} assigned by the components will affect the result. [Note that the dual variables π^i_{jkr} depend on the λ_{jr} values (9.8).] It is obvious that when $\gamma_{jr} = \lambda_{jr}$ for all j and r, (9.13) is equivalent to (9.12).

The multilevel planning method described above is characterized by a certain measure of decentralization. The center does not tell the components which decision x_r they should take; it only specifies the boundaries a_r within which the components must operate. This method generally yields a lower value for the central objectives when the wishes of central and component authorities do not coincide. Thus, if the objectives of central and component authorities conflict, the center will be aware of a certain cost of decentralization.

Section 9.6 is devoted to a numerical example using the multiobjective multilevel planning method implied by (9.7), and illustrates the cost of decentralization in a multiregional planning problem.

9.5.2. Conflicts between objectives

The second shortcoming of standard multilevel planning methods lies in their neglect of conflicting objectives: these methods take into account uncertainties in the decision structure, but not in the priority structure. If we combine the learning processes concerning both items, we obtain a double interactive structure containing $2R + 2$ participants of the following types: R policy-making bodies or components; R bodies for analytical aid to decision-makers at the component level; one central policy-making body; and one body for analytical aid to decision-makers at the central level. Figure 9.4 presents the ensuing communication network. It shows how the information exchange between the center and the components has been extended so that it now includes two phases of deliberation at both levels.

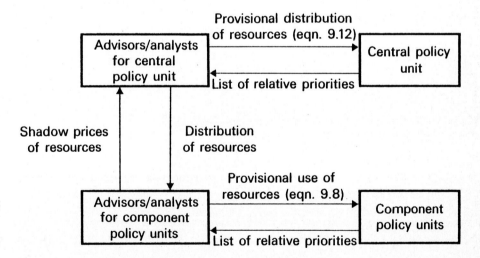

Figure 9.4. Communication between advisors and policy-makers in a direct multilevel multiobjective planning process.

The compromise procedure described at the end of Section 9.2 can easily be generalized to achieve a compromise solution in a multilevel framework. This approach will be adopted in Section 9.6.

9.6. A numerical application

Here we illustrate the ideas put forward in earlier sections in a numerical application of a multilevel multiobjective policy model with conflicting interests. We will consider a simple two-region input–output model for The Netherlands. The model is assumed to contain the following elements:

> 4 sectors: agriculture, industry, services, and transport ($s = 1, 2, 3, 4$)
> 2 regions: Rijnmond (the greater Rotterdam region) and the rest of The Netherlands ($r = 1, 2$)
> 2 objectives: maximization of regional income and minimization of regional pollution ($j = 1, 2$)

The model considers the following variables:

> $y_r' = (y_{1r}, y_{2r}, y_{3r}, y_{4r})$ are the production levels in sectors $1, 2, 3, 4$ in region r
> e_r is the emission of pollutants in region r
> m_r is the influx of pollutants into region r
> $i_r = (i_{1r}, i_{2r}, i_{3r}, i_{4r})$ is the productive investment in sectors $1, 2, 3, 4$ in region r
> v_r is the environmental investment in region r

The two objectives are

$$\text{max!} \ \omega_{1r} = c_r' y_r \tag{9.14}$$

and

$$\text{min!} \ \omega_{2r} = m_r \tag{9.15}$$

where c_r is the vector of value-added coefficients in region r.
The model contains 11 common constraints.

1. Eight input–output relationships for the intermediate deliveries between the four sectors in both regions:

$$\begin{bmatrix} -I + A_{11} & A_{12} \\ A_{2i} & -I + A_{22} \end{bmatrix} \begin{bmatrix} y_1 \\ y_2 \end{bmatrix} \leqslant a_1 \tag{9.16}$$

   where the matrices A are the input–output matrices and a_1 indicates the final demand in the various sectors.

2. One constraint for the limited amount a_2 available for total investment:

$$\iota' i_1 + v_1 + \iota' i_2 + v_2 \leqslant a_2 \tag{9.17}$$

3. Two constraints for the relationships between the influx of pollutants into each region and the emission in both regions.

$$m_1 - h_{11}e_1 - h_{21}e_2 + f_1v_1 = 0$$

$$m_2 - h_{12}e_1 - h_{22}e_2 + f_2v_2 = 0 \tag{9.18}$$

where the coefficients h denote the multiregional diffusion pattern of pollution and f_r represents the efficiency of pollution abatement investments in reducing the emission of pollutants.

In addition to these 11 constraints, there are six constraints for each region separately:

1. Four constraints for the restricted supply of capital available for each sector:

$$y_r - \hat{s}i_r < \hat{s}\bar{K}_r \tag{9.19}$$

where $\hat{s}$ is the diagonal matrix of capital productivities and $\bar{K}_r$ is the vector of the amounts of capital available at the beginning of the planning period.

2. One constraint for the limited amount of labor available:

$$\iota' \hat{L}_r y_r < b_{2r} \tag{9.20}$$

where $\hat{L}_r$ is the diagonal matrix of labor productivities and b_{2r} is the labor force in region r.

3. One relationship for the link between the level of production and the emission of pollutants in a region:

$$e_r - d'_r y_r = 0 \tag{9.21}$$

where d_r is a vector of emission coefficients.

Before we discuss a numerical application of the model[*] to multilevel planning, we should examine the way in which the decomposition has been accomplished (see Figure 9.5).

Figure 9.5(a) illustrates the decomposition proposed above. The zero matrices indicate that the two components (regions) are partly independent of each other. The multilevel planning procedure aims at a decision based on 11 common constraints. Figure 9.5(b) illustrates another method of decomposition, namely by classifying the variables according to whether they are "economic" or "environmental". If this method of decomposition is used, the multilevel algorithm has only three common constraints.

In general, it is not possible to predict the decomposition that will be preferred by the center, since although a small number of common constraints may be preferred from the point of view of computation, this type of decomposition will also reduce the influence of the center on the components in any priority conflicts.

One interesting implication of the existence of two methods of decomposition [into regions ($r = 1, 2$) or policy fields ($p = 1, 2$)] is the possibility of three-level planning models (see Figure 9.6). In Figure 9.6(a) the regions are responsible for

---

[*] This model is based on Mastenbroek and Nijkamp (1976). The precise data used in the numerical applications are available on request from the authors.

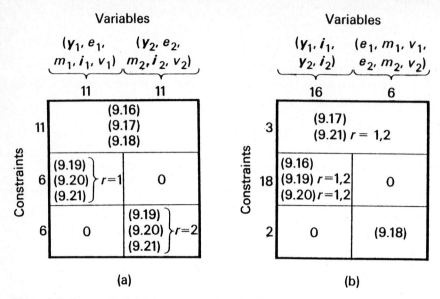

Figure 9.5. Two methods of decomposing the problem. In method (a) the variables are classified according to region; in method (b) economic variables are separated from environmental variables.

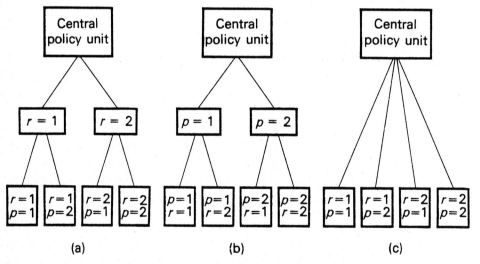

Figure 9.6. Three hierarchical approaches to multiregional multisectoral planning: (a) the regions r are responsible for sectoral planning; (b) the sectoral organizations p are responsible for regional planning; (c) regional and sectoral planning are integrated on one level.

economic and environmental (facet) policies, while in Figure 9.6(b) the economic and environmental organizations are responsible for regional policies. If the center is not content with either of these options, it may try to achieve a direct integration of

Table 9.3. Payoff matrix.

Objective maximized	Resulting values of objectives			
	ω_{11}	ω_{21}	ω_{12}	ω_{22}
ω_{11}	624	-402	686	-79
ω_{21}	63	0	654	-41
ω_{12}	105	-140	2863	-316
ω_{22}	63	-39	654	-32

regional and facet policy as in Figure 9.6(c). One disadvantage of decomposition (c) is that the decision problem is more complex than in (b) or (a), as shown by the number of common constraints; there are 13 common constraints in (c), as compared with 11 and 3 in (a) and (b), respectively. In practice, the policy-making structure is frequently a very complex mixture of (a), (b), and (c).

Let us now consider a numerical application of the model, using the decomposition into regions illustrated in Figure 9.5(a). The conflicting nature of the objectives can be represented by the payoff matrix (see Table 9.3), which indicates the values of the various objectives when one of these objectives is maximized. (Note that we have multiplied ω_{21} and ω_{22} by -1 so that min! ω_{2r} can be replaced by max! ω_{2r}.) It follows from the definition of the payoff matrix that the best attainable levels of each objective can be found on the main diagonal. With the exception of $(\omega_{21}, \omega_{22})$, all pairs of objectives show a considerable degree of conflict.

We then make the following assumptions:

1. The weights attached to the objectives by the center are $(\gamma_{11}, \gamma_{21}, \gamma_{12}, \gamma_{22}) = (1, 3, 1, 3)$. These weights are not known explicitly by the center, however, and hence an interactive multiobjective decision (MOD) procedure must be employed to find the desired outcomes. Furthermore, the center does not have any data on the regional decision structure $(\mathbf{A}_r, \mathbf{B}_r, \mathbf{b}_r, \lambda_r)$, so that an interactive multilevel planning procedure must be employed.

2. The weights attached to the objectives by the regions are not the same as the weights given by the center: $(\lambda_{11}, \lambda_{21}) = (1, 0)$ and $(\lambda_{12}, \lambda_{22}) = (1, 2)$. These weights are not known explicitly by the regions, so that an interactive MOD procedure must again be used. However, the regions have reliable information on the decision structure $(\mathbf{A}_r, \mathbf{B}_r, \mathbf{b}_r)$.

Tables 9.4 and 9.5 show how the two regions $(r = 1, 2)$ use the interactive MOD procedure to determine the optimal allocation of resources a_r^i during the first two iterations $(i = 1, 2)$ of the multilevel planning procedure.

The weights λ underlying these compromises are also given in the tables. The attractiveness of the compromise solutions to the regional policy-making bodies have been estimated such that they are consistent with the assumed values of the regional weights λ_r. The values that must be improved in a certain step have been indicated in the tables by underlining.

Table 9.4. Results of the interactive multiobjective decision procedure for region 1 during the first two iterations i of the multilevel planning procedure.[a]

i	Step	Type of solution	Values of objectives		Weights corresponding to compromise	
			ω_{11}	ω_{21}	λ_{11}	λ_{21}
1	1	Ideal	392.7	− 200.0		
		Minimum	351.6	− 224.0		
		Compromise	368.0	− 209.6	0.368	0.632
1	2	Ideal	392.7	− 209.6		
		Minimum	368.0	− 224.0		
		Compromise	377.9	− 215.4	0.368	0.632
1	3	Ideal	392.7	− 215.4		
		Minimum	377.9	− 224.0		
		Compromise	383.9	− 218.8	0.368	0.632
1	4	Ideal	392.7	− 218.8		
		Minimum	383.9	− 224.0		
		Compromise	387.4	− 220.9	0.368	0.632
1	5	Ideal	392.7	− 220.9		
		Minimum	387.4	− 224.0		
		Compromise	389.5	− 221.1	0.368	0.632
2	1	Ideal	323.9	− 106.4		
		Minimum	298.5	− 159.8		
		Compromise	323.6	− 121.1	0.677	0.323
2	2	Ideal	323.9	− 121.1		
		Minimum	323.6	− 159.8		
		Compromise	323.7	− 136.2	0.990	0.010

[a] The values underlined are those that must be improved in a given step.

Table 9.5. Results of the interactive multiobjective decision procedure for region 2 during the first two iterations i of the multilevel planning procedure.[a]

i	Step	Type of solution	Values of objectives		Weights corresponding to compromise	
			ω_{12}	ω_{22}	λ_{12}	λ_{22}
1	1	Ideal	1515.6	− 147.5		
		Minimum	1515.6	− 147.5		
		Compromise	1515.6	− 147.5		
2	1	Ideal	2793.1	− 237.9		
		Minimum	2787.8	− 247.1		
		Compromise	2791.0	− 243.4	0.633	0.367
2	2	Ideal	2791.0	− 237.9		
		Minimum	2787.8	− 243.4		
		Compromise	2789.8	− 241.2	0.633	0.367

[a] The values underlined are those that must be improved in a given step.

Given the absolute priority that the policy-making body of region 1 is assumed to attach to objective ω_{11}, only improvements in the value of this objective are considered in Table 9.4. For region 2 it is found that the objectives do not conflict during the first iteration. The main reason is that the resources a_2^1 available to region 2 are so restrictive that the region does not have the opportunity to develop a policy reflecting its own political views. This is not the case during the second iteration, when the value of a_2^2 is such that the region has (limited) scope to balance the relative importance of the objectives.

Having determined the desired allocation of resources, the regions r report the prices π_{jkr}^i to the center; these prices reflect the productivity of resource k with respect to objective j during iteration i. This information is used by the center during iteration $i + 1$ to determine a new distribution of resources. Table 9.6 shows how the center calculates the new distributions of resources for $i = 2$. The ω_{11}, ω_{12}, ω_{21}, and ω_{22} given in Table 9.6 are implicit in (9.13):

$$\omega_{jr} = \min_{i'=1,2,\ldots,i} \left(\sum_k \pi_{jkr}^{i'} a_{kr} + \eta_{jr}^{i'} \right) \tag{9.22}$$

It is important to note that these values are only approximations, owing to the limited information available to the center during iteration i. We may conclude that the range of the various objectives will decrease considerably when 4 or 5 steps of the interactive MOD procedure have been carried out.

Table 9.6. Results of the interactive multiobjective procedure for the center during the iteration $i = 2$ of the multilevel planning procedure.[a]

i	Step	Type of solution	Values of objectives				Weights corresponding to compromise			
			ω_{11}	ω_{21}	ω_{12}	ω_{22}	λ_{11}	λ_{21}	λ_{12}	λ_{22}
2	1	Ideal	544	0	28,138	23				
		Minimum	238	− 366	26,438	− 396				
		Compromise	392	− 32	27,118	− 152	0.36	0.36	0.05	0.23
2	2	Ideal	544	− 32	28,138	− 146				
		Minimum	392	− 336	27,118	− 396				
		Compromise	453	− 154	27,526	− 250	0.35	0.21	0.07	0.37
2	3	Ideal	453	− 51	28,138	− 241				
		Minimum	392	− 154	27,526	− 380				
		Compromise	416	− 104	27,771	− 301	0.30	0.21	0.04	0.45
2	4	Ideal	416	− 63	28,138	− 297				
		Minimum	392	− 104	27,771	− 380				
		Compromise	402	− 86	27,918	− 332	0.35	0.23	0.04	0.37
2	5	Ideal	402	− 70	28,138	− 331				
		Minimum	392	− 86	27,918	− 380				
		Compromise	395	− 80	28,006	− 351	0.41	0.26	0.03	0.30

[a] The values underlined are those that must be improved in a given step.

Table 9.7. Iterations of the multiobjective multilevel algorithm. Only a subset of the iterations is included.

i r		Common resources											ω_{jr}^{*}
		1	2	3	4	5	6	7	8	9	10	11	
1 1	a	-5.000	-80.000	-100.	-5.000	20.000	100.000	75.	5.000	200.000	4000.000	-6000.000	
	$\Pi(j = 1)$	0.000	0.000	0.	0.000	0.000	0.000	0.	0.000	0.206	0.000	-0.019	390.
	$\Pi(j = 2)$	0.000	0.000	0.	0.000	0.000	0.000	0.	0.000	0.000	-0.008	0.032	-221.
1 2	a	1.000	50.000	35.	3.000	-50.000	-650.000	-550.	-20.000	550.000	-4000.000	6000.000	
	$\Pi(j = 1)$	0.000	7.924	0.	165.979	0.000	0.000	0.	2.930	0.000	-0.170	-0.000	1516.
	$\Pi(j = 2)$	0.000	0.000	0.	0.000	0.000	0.000	0.	0.000	0.030	0.038	-0.002	-148.
2 1	a	-5.000	-146.667	-100.	-70.333	20.000	100.000	75.	2.778	444.444	6322.105	-3412.391	
	$\Pi(j = 1)$	0.000	1.099	0.	0.000	0.000	0.000	0.	49.795	0.001	-0.000	-0.101	324.
	$\Pi(j = 2)$	0.000	0.000	0.	0.000	0.000	0.000	0.	0.000	0.000	-0.008	0.032	-136.
2 2	a	1.000	116.667	35.	68.333	-50.000	-650.000	-550.	-17.778	305.556	-6322.105	3412.391	
	$\Pi(j = 1)$	0.000	0.000	0.	0.000	0.000	0.000	0.	0.000	0.000	-0.034	0.000	2790.
	$\Pi(j = 2)$	0.000	0.000	0.	0.000	0.000	0.000	0.	0.000	0.030	0.038	-0.002	-241.
5 1	a	-4.235	-186.365	-100.	-21.667	14.074	70.370	75.	6.431	710.181	3000.500	-3053.268	
	$\Pi(j = 1)$	1.178	3.461	0.	0.378	0.000	0.000	0.	0.000	-0.000	0.000	-0.277	188.
	$\Pi(j = 2)$	0.000	0.000	0.	0.000	0.000	0.000	0.	0.000	0.120	-0.008	0.032	-36.
5 2	a	0.235	156.365	35.	19.667	-44.074	-620.370	-550.	-21.431	39.819	-3000.500	3053.268	
	$\Pi(j = 1)$	325.833	0.000	0.	0.000	0.000	0.000	0.	0.000	0.135	-0.233	0.000	1966.
	$\Pi(j = 2)$	0.000	0.000	0.	0.000	0.000	0.000	0.	0.000	0.000	0.038	-0.002	-120.

													ω_{jr}
6 1	a	−5.760	−189.754	−100.	−6.908	14.074	70.370	75.	6.431	750.000	4090.035	−3133.789	
	$\Pi(j = 1)$	1.176	3.461	0.	0.378	0.000	0.000	0.	0.000	0.000	0.000	−0.277	202.
	$\Pi(j = 2)$	0.000	0.000	0.	0.000	0.000	0.000	0.	0.000	0.120	−0.008	0.032	−43.
6 2	a	1.760	159.754	35.	4.908	−44.074	−620.370	−550.	−21.431	0.000	−4090.035	3133.789	
	$\Pi(j = 1)$	0.000	0.000	0.	207.411	0.000	0.000	0.	3.266	0.133	−0.214	0.000	2108.
	$\Pi(j = 2)$	0.000	0.000	0.	0.000	0.000	0.000	0.	0.000	0.000	0.038	−0.002	−162.
10 1	a	−5.674	−160.455	−100.	−20.695	14.074	70.370	75.	6.840	750.000	5571.662	−2760.549	
	$\Pi(j = 1)$	1.178	3.461	0.	0.378	0.000	0.000	0.	0.000	−0.000	−0.000	−0.277	195.
	$\Pi(j = 2)$	0.000	0.000	0.	0.000	0.000	0.000	0.	0.000	0.120	−0.008	0.032	−43.
10 2	a	1.674	130.455	35.	18.695	−44.074	−620.370	−550.	−21.840	0.000	−5571.662	2760.549	
	$\Pi(j = 1)$	0.000	7.672	0.	0.000	0.000	0.000	0.	0.000	0.116	−0.189	0.000	2759.
	$\Pi(j = 2)$	0.000	0.000	0.	0.000	0.000	0.000	0.	0.000	0.000	0.038	−0.002	−217.
11 1	a	−4.855	−160.927	−100.	−20.695	14.074	70.370	75.	6.840	750.000	5571.662	−2763.138	
	$\Pi(j = 1)$	1.178	3.461	0.	0.378	0.000	0.000	0.	0.000	−0.000	0.000	−0.277	195.
	$\Pi(j = 2)$	0.000	0.000	0.	0.000	0.000	0.000	0.	0.000	0.120	−0.008	0.032	−43.
11 2	a	0.855	130.927	35.	18.695	−44.074	−620.370	−550.	−21.840	0.000	−5571.662	2763.138	
	$\Pi(j = 1)$	0.000	7.672	0.	0.000	0.000	0.000	0.	0.000	0.116	−0.189	0.000	2762.
	$\Pi(j = 2)$	0.000	0.000	0.	0.000	0.000	0.000	0.	0.000	0.000	0.038	−0.002	−217.

* The final column ω_{jr} gives the values of ω_{1r}^i and ω_{2r}^i for each i and r.

We will next consider how the results of the MOD procedures carried out by the center and the regions may be used in a multilevel planning algorithm to find an optimal distribution of resources. Table 9.7 shows the distribution of resources a^i_r between the regions and the corresponding productivities reported by the regions for a number of iterations i. The final column (ω_{jr}) gives the values of ω^i_{1r} and ω^i_{2r} for every i and r. The algorithm converges after 11 iterations, and we have two comments to make on the convergence. First, the reported shadow prices for several common resources are equal to zero for both regions. There is obviously no need to revise the distribution of such resources. Second, the algorithm does not guarantee that the process of convergence will be monotonic: the values of the central and regional objectives oscillate to a certain extent.

It is interesting to compare these results with those for a situation in which the central and regional weights are the same $(\gamma_{jr} = \lambda_{jr} \,\forall\, j, r)$. When $(\lambda_{11}, \lambda_{21}, \lambda_{12}, \lambda_{22}) = (1, 2, 1, 3)$, the multilevel planning algorithm yields as a solution for $\omega = (\omega_{11}, \omega_{21}, \omega_{12}, \omega_{22})'$:

$$\omega = (295, -59, 2762, -218)'$$

$$\sum_{j,r} \gamma_{jr}\,\omega_{jr} = 2226 \tag{9.23}$$

When $(\lambda_{11}, \lambda_{21}, \lambda_{12}, \lambda_{22}) = (1, 0, 1, 2)$, however, Table 9.7 shows that the optimal solution is

$$\omega = (195, -43, 2762, -217)'$$

$$\sum_{j,r} \gamma_{jr}\,\omega_{jr} = 2177 \tag{9.24}$$

We conclude that the difference in the weights produces a redistribution of resources. This redistribution affects region 2 only slightly, but has a stronger effect on region 1. In the example considered here, the relative autonomy of the regions (they are free to use the resources according to their own views) gives rise to a cost of decentralization equal to 49 units $(2226 - 2177)$.

9.7. Conclusion

Multilevel planning is an important concept in regional planning, since it explicitly recognizes that lack of information will induce communication flows in the form of hierarchical networks. This chapter has shown that the range of multilevel planning methods can be extended using MOD methods to include uncertainties about priorities and conflicts between divergent objectives.

References

Baumol, W. J. and Fabian, T. (1964). Decomposition, pricing for decentralization and external economies. Management Science, 11: 1–32.

Bell, D. E., Keeney, R. L., and Raiffa, H. (1977). Conflicting Objectives in Decisions. Wiley Publishing Co., New York.

Cochrane, J. L. and Zeleny, M. (Editors) (1973). Multiple Criteria Decision Making. University of South Carolina Press, Columbia, South Carolina.

Cohon, J. L. (1978). Multiobjective Programming and Planning. Academic Press, New York.

Dantzig, G. (1963). Linear Programming and Extensions. Princeton University Press, Princeton, New Jersey.

Dantzig, G. and Wolfe, P. (1960). The decomposition principle for linear programming. Operations Research Quarterly, 8: 101–111.

van Delft, A. and Nijkamp, P. (1977). Multicriteria Analysis and Regional Decision-Making. Martinus Nijhoff, The Hague.

Fandel, G. (1972). Optimale Entscheidung bei Mehrfacher Zielsetzung (Optimal Multicriteria Decisions). Springer Verlag, Berlin, GDR.

Guigou, J. L. (1974). Analyse des Données et Choix à Critères Multiples (Data Analysis and Multicriteria Decisions). Dunod, Paris.

Haimes, Y. Y., Hall, W. A., and Freedman, H. T. (1975). Multi-Objective Optimization in Water Resource Systems. Elsevier, Amsterdam.

Hill, M. (1973). Planning for Multiple Objectives. Monograph Series 5. Regional Science Research Institute, Philadelphia, Pennsylvania.

Johansen, L. (1978). Lectures on Macroeconomic Planning. North-Holland Publishing Co., Amsterdam, Chapter 2.

Johnsen, E. (1968). Studies in Multi-Objective Decision Models. Monograph 1. Lund Economic Research Center, Lund, Sweden.

ten Kate, A. (1972a). Decomposition of linear programs by direct distribution. Econometrica, 40: 883–898.

ten Kate, A. (1972b). A comparison between two kinds of decentralized optimality conditions in nonconvex programming. Management Science, 18: 734–743.

Keeney, R. L. and Raiffa, H. (1976). Decision Analysis with Multiple Conflicting Objectives. Wiley Publishing Co., New York.

Kornai, J. (1975). Mathematical Planning of Structural Decisions. 2nd edition. North-Holland Publishing Co., Amsterdam.

Malinvaud, E. (1972). Lectures on Microeconomic Theory. North-Holland Publishing Co., Amsterdam.

Mastenbroek, A. P. and Nijkamp, P. (1976). A spatial environmental model for an optimal allocation of investments. In P. Nijkamp (Editor), Environmental Economics. Volume 2. Martinus Nijhoff, The Hague, pp. 19–38.

Nijkamp, P. (1977). Theory and Application of Environmental Economics. North-Holland Publishing Co., Amsterdam.

Nijkamp, P. (1979). Multidimensional Spatial Data and Decision Analysis. Wiley Publishing Co., New York.

Nijkamp, P. and Rietveld, P. (1976). Multiobjective programming models, new ways in regional decision making. Regional Science and Urban Economics, 6: 253–274.

Nijkamp, P. and Rietveld, P. (1978). Impact analyses, spatial externalities and policy choices. Northeast Regional Science Review, 7: 20–37.

Nijkamp, P. and Rietveld, P. (1979). Conflicting social priorities and compromise social decisions. In E. G. Cullen (Editor), Analysis and Decision in Regional Policy. Pion Press, London, pp. 153–177.

Nijkamp, P. and Rietveld, P. (1981). Hierarchical multiobjective models in a spatial system. In P. Nijkamp and J. Spronk (Editors), Multicriteria Analysis: Operational Methods. Gower Publications, London, pp. 163–186.

Salih, K. (1975). Goal conflicts in pluralistic multi-level planning for development. International Regional Science Review, 1: 49–72.

Schleicher, S. (1971). Decentralized optimization of linear economic systems with minimum information exchange of the subsystems. Zeitschrift für National Ökonomie, 31: 33–44.

Starr, M. K. and Zeleny, M. (Editors) (1977). Multiple Criteria Decision Making. North-Holland Publishing Co., Amsterdam.

Theil, H. (1964). Optimal Decision Rules for Government and Industry. North-Holland Publishing Co., Amsterdam.

Thiriez, H. and Zionts, S. (Editors) (1976). Multiple Criteria Decision Making. Springer Verlag, Berlin, GDR.

Wagner, H. M. (1975). Principles of Operations Research. Prentice-Hall, Englewood Cliffs, New Jersey, p. 130.

Wallenius, J. (1975). Interactive Multiple Criteria Decision Methods. Helsinki School of Economics, Helsinki.

Wilhelm, J. (1975). Objectives and Multi-Objective Decision Making under Uncertainty. Springer Verlag, Berlin.

Zeleny, M. (1974). Linear Multiobjective Programming. Springer Verlag, Berlin.

Zeleny, M. (Editor) (1976). Multiple Criteria Decision Making. Springer Verlag, Berlin.

Regional Development Modeling: Theory and Practice
M. Albegov, A.E. Andersson and F. Snickars (editors)
North-Holland Publishing Company
© IIASA, 1982

Chapter 10

A MATHEMATICAL PROGRAMMING APPROACH TO LAND ALLOCATION IN REGIONAL PLANNING

Åke E. Andersson* and Markku Kallio**
International Institute for Applied Systems Analysis, Laxenburg (Austria)

10.1. Introduction

Workers in many disciplines (e.g., theoretical geography, economics, and operations research) have attempted to develop efficient methods for allocating land between different uses. Numerous approaches have been used, differing mainly in their treatment of space and time dimensions. The various possibilities are outlined in Table 10.1.

Approaches involving continuous time and/or continuous space have (up until now) proved to be of limited practical value, except in some qualitative analysis. We have chosen to use a discrete space and time model able to cope with the practical complications involved in generating policy alternatives in our long-term studies of the Skåne area in southern Sweden and the Silistra region in Bulgaria. In this chapter we will formulate the regional development program as a (dynamic) mathematical programming problem and outline a procedure for finding an optimal solution under various criteria.

After an introductory discussion of regional planning, the problem is formulated as a dynamic (nonconvex) quadratic programming problem with integer variables. A solution procedure based on the theory of optimization over networks is then developed and illustrated with a numerical example.

10.2. Characteristics of regional development

The regional development planning problem is basically the problem of finding a suitable trajectory for the spatial pattern of various activities in a region. The following

---

* Currently at the Department of Economics, University of Umeå, Umeå, Sweden.
** Currently at the Department of Management Science, Helsinki School of Economics, Helsinki, Finland.

Table 10.1. The various approaches to the land allocation problem.

Space	Time	
	Discrete	Continuous
Discrete	Most mathematical programming approaches (Andersson and La Bella, 1979)	Most optimal control models (Isard et al., 1979)
Continuous	New urban economics (Mills, 1972) Weber models (Cooper, 1967; Nijkamp and Paelinck, 1975)	Isard's dynamic transportation–location models (Isard et al., 1979) Beckmann–Puu transportation–location models (Beckmann, 1953; Puu, 1981)

three considerations must be taken into account: the current or initial spatial pattern of resources within the region; the expected (or planned) total volume of different types of activities in the future; and the criteria used to evaluate the possible future spatial patterns of these activities. Each of these considerations will now be discussed in some detail.

The current situation may be described by a map indicating the distribution of resources within different subregions. As an example, the regional subdivision and the main road network for the Skåne and Silistra regions are given in Figures 10.1 and 10.2, respectively. The term "resources" as used here is understood to include both natural resources and various types of capacity. Examples include industrial capacity, agricultural capacity, the capacity of the transportation system, water resources, renewable resources (such as forests), and nonrenewable resources (such as mineral deposits).

Resources are used over time for various activities, such as industrial or agricultural production. We shall assume that the total volume of these activities is known over the period under study. Forecasts of this type may be readily available in a centrally planned economy (as in Bulgaria) or they may have to be estimated using econometric techniques (as in Sweden). A spatial pattern is said to be feasible at a given time if it provides sufficient resources to achieve the estimated levels of regional activity at that time. This may require an increase over time in some of the resources (such as housing or industrial units) or it may suggest a decrease over time in others (such as the use of mineral resources).

In general, there is much freedom in designing feasible patterns: there are a number of possible locations for most of the activities and, furthermore, the activity at certain sites may change over time. We must then consider the criteria that should be taken into account when comparing alternative feasible land-use patterns over time. A single criterion is clearly insufficient. Over the whole planning period, we must consider simultaneously the investment costs involved in changing the capacity for various activities in each subregion, the operating costs of the production activities, communication within the region, and the environmental problems created by certain

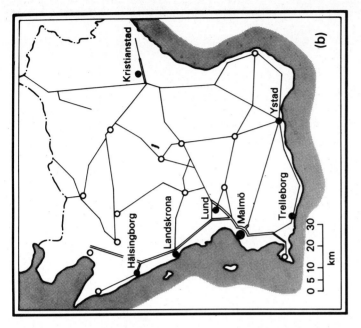

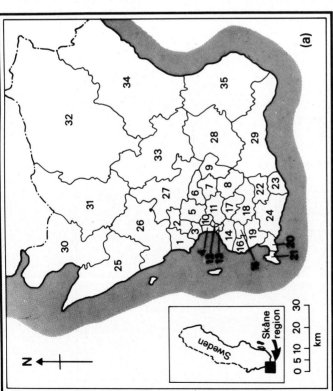

Figure 10.1. The Skåne region in southern Sweden: (a) subdivision into municipalities; (b) network of major roads in 1976 – the double lines represent motorways.

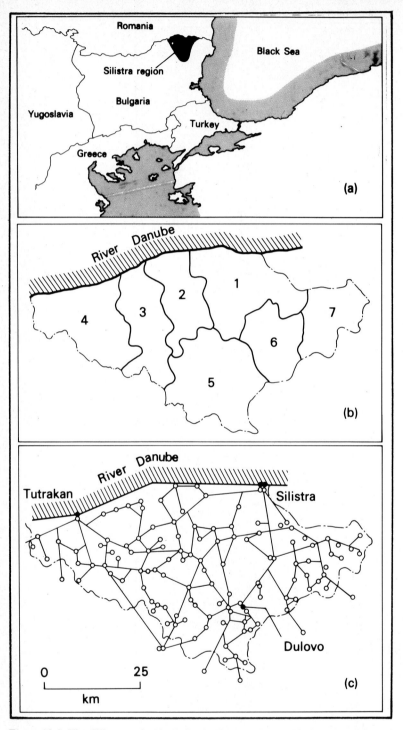

Figure 10.2. The Silistra region in Bulgaria: (a) position within southeastern Europe; (b) regional subdivisions; (c) network of major roads.

spatial patterns. Economies of scale are assumed to play an important role in determining production costs. Furthermore, the siting of a production unit relative to natural resources and other production units may, of course, have a significant effect on the operating costs.

10.3. A planning model

10.3.1. Feasible allocation of land

Our next task is to formulate the planning problem as a mathematical programming model. We shall first describe the set of alternative spatial patterns in terms of mathematical relations. We will then formulate a precise statement of the criteria proposed for evaluating the various plans.

As indicated above, we adopt a discrete time/discrete space approach, i.e., we consider the planning horizon to be divided into T periods ($t = 0, 1, 2, \ldots, T-1$) and the region to be divided into R subregions ($r = 1, 2, \ldots, R$). For instance, each period t may be five years long, in which case the planning horizon may consist of three, four, or five such periods. Each subregion r is associated with a land area L_r and an initial capacity x_{i0}^r for activity i, where $i = 1, 2, \ldots, I$. Thus, there is a total area L_r available for all these activities in subregion r during each period t.

We shall denote by x_0 the vector whose components are x_{i0}^r. The allocation of land for different activities i in the first period $t = 0$ is determined by the initial state x_0, though the land use in other periods may be altered through investment decisions. Let $y_i^r(t)$ be the increase in capacity i (for activity i) in subregion r during t, let $d_i^r(t)$ be the decrease in capacity i (demolition), and let $x_i^r(t)$ be the total capacity i in subregion r at the beginning of period t, for all i, r, and t. In this notation, we have, for all t,

$$x(t + 1) = x(t) + y(t) - d(t) \tag{10.1}$$

where $x(t) = \{x_i^r(t)\}$, $y(t) = \{y_i^r(t)\}$, and $d(t) = \{d_i^r(t)\}$ are nonnegative vectors with $I \times R$ components, and $x(0) = x_0$.

One way of handling economies of scale is to consider a set of indivisible production units. Assuming that these units correspond to real alternatives, the production-cost estimates can be obtained relatively easily. This approach leads to an integer programming formulation. In particular, for our present purposes it is sufficient to consider only one plant size – that which yields an average production cost close to the minimum possible and yet represents a relatively small unit compared with the total capacity increase required. Thus, $y(t)$ is a nonnegative integer vector.

Note that no physical depreciation is assumed in eqn. (10.1). Thus, the operating cost is assumed to cover the reinvestment cost needed to maintain the capacity at the same level over period t. For the amount $d(t)$ to be demolished we may set lower and upper bounds $V(t)$ and $U(t)$, respectively:

$$V(t) \leqslant d(t) \leqslant U(t) \tag{10.2}$$

We shall assume $d(t)$ and x_0 to be integer vectors, so that $x(t)$ is also an integer vector.

Let $z_i(t)$ be the total amount of capacity i at the beginning of period t; the total capacity is then $z(t) = \{z_i(t)\}$. Then

$$z_i(t) = \sum_r x_i^r(t) \tag{10.3}$$

The minimum amount of capacity required at time t is given by a vector $Z(t) = \{Z_i(t)\}$; this is basically an estimate of the total volume of activities expected within the region at time t. It is clear that

$$z(t) \geqslant Z(t) \tag{10.4}$$

The land availability (L_r) constraint can be taken into account through the following inequality:

$$\sum_i x_i^r(t) \leqslant L_r \qquad \text{for all } r \text{ and } t \tag{10.5}$$

Although this may seem quite restrictive, it is reasonable to assume that the same amount of land is needed for each unit of capacity of each industrial activity. The size of this unit is roughly determined by the chosen scale of the production units. For activities in which the economies of scale are less important, the unit of capacity is determined so that its land requirement is about the same as that for an industrial unit. The purpose of this slightly restrictive assumption is to obtain a network flow formulation which then greatly simplifies the analysis of our model.

An example of the network structure of our model for a two-period, two-subregion, and two-activity case is given in Figure 10.3. The nodes on the left refer to the land available and those on the right to the installed capacity. The vertically directed arcs on the left describe unoccupied land and those on the right the capacity carried over from one period to the next (for which we have a lower bound given by eqn. (10.4)). The other arcs, which are horizontal and not necessarily directed, refer to land allocated to the two activities or land made available through demolition. The conservation equations for each node, together with the lower bounds given by eqn. (10.4) (for the vertical flows on the right) and the need to consider integer quantities, constitute the constraints for possible land allocations.

10.3.2. Evaluation criteria

We evaluate the various allocations using the following criteria: investment, demolition, and operating costs (including the cost of transporting raw materials and industrial products); private communication costs (such as the costs of commuting and recreation); and environmental considerations, especially the problems caused by congestion and synergetic effects. The following paragraphs explain how these factors may be quantified.

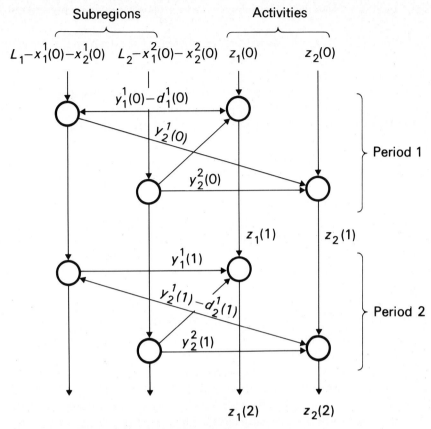

Figure 10.3. The network structure of a model with $T = I = R = 2$.

Investment, demolition, and operating costs. Let $B_i^r(t)$ and $D_i^r(t)$ be the unit cost of investment and demolition, respectively, for capacity i in subregion r during period t. Define $B(t) = \{B_i^r(t)\}$ and $D(t) = \{D_i^r(t)\}$. Then the total cost of investment and demolition in period t is $B(t)y(t) + D(t)d(t)$.

We divide the operating costs into the costs of interaction between the various activities for which we must find a location (such as the costs of transporting goods, or the costs of communication) and other operating costs (which may also be dependent on the locations of the production units). The costs of interactions between activities i and j located in subregions r and s, respectively, are denoted by $x_i^r(t)c_{ij}^{rs}(t)x_j^s(t)$, where $x_i^r(t)$ and $x_j^s(t)$ define the number of units and $c_{ij}^{rs}(t)$ is the cost of interaction per unit of activity i in subregion r and per unit of activity j in subregion s.

[It should be noted that, in order to take advantage of the network structure of the model, the subregional capacity levels $x_i^r(t)$ must be suppressed; i.e., it is necessary to solve for $x_i^r(t)$ from (10.1) and substitute elsewhere. While doing so, it is necessary to restrict the demolition activities in order to ensure that the $x_i^r(t)$ remain nonnegative.

For instance, one may allow demolition only in certain periods and for some initially existing capacity.]

This treatment of interaction costs was first proposed by Koopmans and Beckmann (1957). Similar formulations were later developed by Lundqvist and Karlqvist (1972), Andersson (1974), Snickars (1977), and Los (1978).

The interaction cost $c_{ij}^{rs}(t)$ may be interpreted as a potential transportation (communication) cost. The cost $c_{ij}^{rs}(t)$ is then a product of the frequency of interaction of one unit of activity i with activity j divided by $Z_j(t)$ (the estimated total number of units j), a nondecreasing function of the distance between subregions r and s, and the unit cost of interaction. The product of the first two factors yields an estimate of the number of interactions between one unit of activity i in subregion r and one unit of activity j in subregion s.

Let $F_i^r(t)$ represent all other operating costs for one unit of activity i in subregion r during t. This term may, for instance, include the costs of interaction between the industrial unit and site-specific utilities (such as mineral deposits, water sources, or ports). If we define a square matrix $C(t) = \{c_{ij}^{rs}(t)\}$ and a vector $F(t) = \{F_i^r(t)\}$, then the total investment, demolition, and interaction costs for period t, denoted by $I_1(t)$, can be written

$$I_1(t) = B(t)y(t) + D(t)d(t) + F(t)x(t) + x(t)C(t)x(t) \qquad (10.6)$$

Alternatively, the interaction costs may be interpreted as a measure of accessibility, which will now be defined. The accessibility A_{ij}^{rs} of a unit j in subregion s to a unit i in subregion r is defined as a product of the frequency of interaction of one unit of i with j, and a nonincreasing function of the distance between subregions r and s. Defining a square matrix $A = \{A_{ij}^{rs}\}$, the total system accessibility is given by $x(t)Ax(t)$. Because a high level of accessibility is desirable, unlike a high cost, we may replace the interaction cost $x(t)C(t)x(t)$ in eqn. (10.6) by the negative of the total system accessibility (possibly multiplied by a positive scalar since accessibility is not necessarily measured in monetary units).

Both potential transportation (communication) costs and accessibility are of fundamental importance in spatial planning problems. Accessibility in particular has been an important concept in recent developments in regional theory; it has been given an axiomatic foundation by Weibull (1976), and the definition above is consistent with his assumptions. Because accessibility adds to the dimensionality of our decision criteria, we shall consider the potential transportation costs to be a measure of communication costs.

Private communication costs. Private communication costs are treated in much the same way as other communication costs (see above). However, it is necessary to distinguish between private and other communication costs because these constitute two separate criteria for evaluating the various solutions to our planning problem. Private communication costs, denoted by $I_2(t)$, are then given as

$$I_2(t) = F_p(t)x(t) + x(t)C_p(t)x(t)$$

where $F_p(t)$ is a vector of unit communication costs between housing and site-specific recreation areas (i.e., lakes, rivers, forests, etc.) and $C_p(t)$ is a matrix of the potential communication costs involved in connecting the housing units to other (non-site-specific) facilities. Thus, any component of $F_p(t)$ which does not correspond to interaction with a housing unit is defined as zero. Similarly, components of $C_p(t)$ are equal to zero if they do not correspond to interaction with a housing unit.

Congestion. Excessive congestion of activities is the most obvious kind of environmental problem. We measure congestion by the density of capital allocated to each subregion (i.e., congestion in subregion r is defined as $\Sigma_i K_i^r(t)x_i^r(t)/L_r$, where $K_i(t)$ is the capital stock per unit of activity i). The average congestion in a regional system is defined as the weighted sum of the congestion in each subregion. If the ratio of capital stock in subregion r to the total capital stock within the region is used as the weighting coefficient, then the average congestion, denoted by $I_3(t)$, is given by

$$I_3(t) = \left\{ \sum_r \left[\sum_i K_i^r(t)x_i^r(t) \right]^2 \Big/ [K(t)L_r] \right\} = x(t)G(t)x(t)$$

where $G(t)$ is an appropriately defined square matrix and $K(t)$ is the total capital stock.

Environmental synergetic effects. Environmental problems are generally much more complicated than those described in the paragraph above. For example, a unit of heavy industry is of little environmental consequence if located with other units of heavy industry at a considerable distance from housing. On the other hand, if it has to be located close to housing or outdoor recreation centers, the disturbance can be enormous. Thus, one method of assessing this disturbance is to take into account the number of persons affected. An environmental interaction matrix $E = \{E_{ij}^{rs}\}$ would consequently measure the disturbance between different activities i and j located in subregions r and s, respectively. It is clear that it may be very difficult to assign numerical values to the parameters E_{ij}^{rs}; one possible solution is to use powers of ten (e.g., 0.1, 1, 10, 100, etc.). A measure $I_4(t)$ for environmental synergetic effects is then given by

$$I_4(t) = x(t)Ex(t)$$

For each of the four criteria c and for each period t, we define a weighting factor $\beta_c(t)$ that reflects the importance of the criterion over the period considered. Thus, our planning problem becomes a four-criteria optimization problem, where the criteria I_c are given by

$$I_c = \sum_t \beta_c(t)I_c(t) \qquad \text{for } c = 1, 2, 3, 4$$

Rather than using any particular multicriteria optimization technique we simply use nonnegative weights λ_c to form a linear scalarizing function that should be minimized. In this way, we obtain a linear approximation to the (negative) utility function g

$$g = \sum_c \lambda_c I_c \tag{10.7}$$

We may adopt different values for the parameters λ_c in order to generate a set of interesting development possibilities for the region.

10.3.3. Summary of the model

To summarize, the planning problem (P) is to find nonnegative integer vectors $x(t)$, $y(t)$, $d(t)$, and $z(t)$, for all t, that minimize g in (10.7) subject to (10.1)–(10.5) with the initial state $x(0) = x_0$.

The objective function of this problem is quadratic, though in general not convex. It is easy to see, for instance, that the potential transportation cost matrix $C(t)$ is normally not positive semidefinite. For a static one-activity, two-zone problem where transportation costs are equal to the distances d_{rs} between subregions r and s, the potential transportation cost is given by

$$xCx = (x^1, x^2) \begin{bmatrix} d_{11} & d_{12} \\ d_{21} & d_{22} \end{bmatrix} \begin{bmatrix} x^1 \\ x^2 \end{bmatrix}$$

where the diagonal elements d_{rr} are equal to zero. Clearly, if $d_{rs} > 0$ for $r \neq s$, matrix C is not positive semidefinite, since if $(x^1, x^2) = (1, -1)$ then $xCx < 0$. It can be shown that this is a general result for multiactivity multizone problems (see Snickars, 1977). Our planning problem will thus not necessarily have a unique optimum. Instead we expect a number of different spatial patterns to correspond to local optima, one or more of which will also be a regional optimum. This is illustrated by a numerical example in Section 10.5.

10.4. A solution technique

In this section we consider the network formulation of the problem (P), i.e., we assume that variables $x_i^r(t)$ have been obtained using (10.1), substituted elsewhere, and that their nonnegativity is guaranteed. Let x be a vector whose components are the decision variables $y_i^r(t)$, $d_i^r(t)$, and $z_i(t)$, for all i, r, and t. Denote the objective function in eqn. (10.7) by $g = g(x)$ and the set of all nonnegative vectors x satisfying the constraints (10.1)–(10.5) by S. In this notation, problem (P) may be restated as finding an integer vector x to minimize $g(x)$, where $x \in S$. The set S can be formally described as the set of feasible solutions to a transshipment network of the type illustrated in Figure 10.3.

We then make use of the fact that every linearized problem (L) (in which the objective function of (P) is replaced by a linear function) is a transshipment problem for which very efficient solution techniques exist (see, for example, Bradley et al.,

1977). This is because every extreme point of S is an integer solution provided that L_r, $D_j(t)$, and $Z_j(t)$ are integers for all r, j, and t (see, for example, Dantzig, 1963). Thus, the need for integral variables can be relaxed when solving the linearized problem.

We propose to solve (P) in the following way:

1. Choose an initial solution $x^0 \in S$, and set the iteration count k to 0.
2. Solve the linearized problem (L)

$$\underset{x \in S}{\text{minimize}} \ \nabla g(x^k)x$$

for an optimal solution $z^k \in S$. Here $\nabla g(x^k)$ denotes the gradient of $g(x)$ at $x = x^k$.
3. Solve the line search problem (Q)

$$\text{minimize} \ g(\alpha x^k + (1-\alpha)z^k)$$

for an optimal solution $\alpha^k \in [0, 1]$.
4. Stop if $\alpha = 0$, or if

$$\underset{i \leqslant k}{\min} \ g(z^i) - g[\alpha^k x^k + (1-\alpha^k)z^k] < \delta$$

where δ is a given tolerance, or if any other appropriate criterion is satisfied (e.g., computing time is exhausted). Otherwise replace x^k by $\alpha^k x^k + (1 - \alpha^k)z^k$, replace k by $k + 1$, and return to step 2.

As mentioned above, the linearized problem (L) is a transshipment problem which can be solved extremely efficiently. We shall follow the method reported in Kallio et al. (1979). The optimal (basic) solution for (L) satisfies the requirement that all variables $x_i^r(t)$ should be integral. Thus, z^k is a feasible solution of (P). We approximate the optimal solution of (P) by the best of the solutions z^k generated by the above procedure. This, of course, may not be an exact solution of (P).

Problem (Q) is a quadratic problem with one variable α and one constraint, $0 \leqslant \alpha \leqslant 1$. Thus, (Q) is extremely simple. Let (R) be the problem that is obtained by relaxing the integrality requirement on x in (P). Solution x^{k+1} is the best for problem (R) and can be found by moving from x^k in the direction z^k. Thus, the sequence $\{x^k\}$ generated by this procedure is exactly the same as that generated by the Frank–Wolfe method (Frank and Wolfe, 1956) when applied to problem (R). If x^k converges to an optimal solution x^* for (R) and $\bar{x}$ is optimal for (P), then

$$g(x^*) \leqslant g(\bar{x}) \leqslant \underset{i < k}{\min} \ g(z^i)$$

i.e., $g(x^*)$ is a lower bound on the optimal value of (P). Even though $g(x^*)$ may be unknown, the difference between $\underset{i<k}{\min} \ g(z^i)$ and $g(x^k)$, the best feasible values for (P) and (R) found so far, may still be used as a stopping criterion. This is illustrated in Figure 10.4. If g is a convex function, x^k converges to x^*. Otherwise the method may be applied several times, starting with different values of x^0.

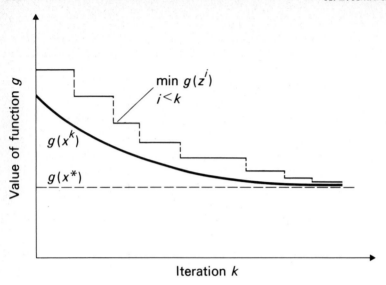

Figure 10.4. Values of $g(x^*)$, $g(x^k)$, and $\min_{i<k} g(z^i)$ as a function of the number of iterations k.

10.5. A numerical example[†]

As a simple example, we consider a static problem with four subregions and four activities. The activities j, their building requirements Z_j, the subregions r, and their land availability L_r are described in Table 10.2.

Let x_i^r be the number of units i to be located in subregion r, and let $x = (x_1^1, x_2^1, x_3^1, \ldots, x_1^4, x_2^4, x_3^4, x_4^4)$. If the investment costs are assumed to be independent of the subregion they can be considered as a constant term and omitted from further consideration. The linear term in the objective function will then consist only of the cost of communication between housing and recreation facilities. This linear term is then $cx = (0, 0, 0, 54{,}900, 0, 0, 0, 45{,}500, 0, 0, 0, 32{,}800, 0, 0, 0, 39{,}400)x$.

Table 10.2. Building requirements for four activities j and land availability in four subregions r.

Activity j	Building requirement Z_j	Subregion r	Land availability L_r
1 (agriculture)	5	A	1
2 (industry)	4	B	2
3 (service)	3	C	5
4 (housing)	6	D	10

[†] The authors wish to thank András Pór of the International Institute for Applied Systems Analysis for carrying out the computations reported in this section.

The quadratic term xQx consists of congestion and communication costs, where the matrix $Q = \{Q_{ij}^{rs}\}$ is given by

$$Q_{ij}^{rs} = \begin{cases} \alpha_{ij}d_{rs} + 1/L_r & \text{if } r = s \text{ and } i = j = 4 \\ \alpha_{ij}d_{rs} & \text{otherwise} \end{cases}$$

and

$$\{\alpha_{ij}\} = \begin{bmatrix} 2 & 3 & 1 & 0 \\ 1 & 5 & 3 & 1 \\ 1 & 4 & 3 & 10 \\ 1 & 4 & 6 & 8 \end{bmatrix} \qquad \{d_{rs}\} = \begin{bmatrix} 20 & 30 & 50 & 100 \\ 30 & 30 & 40 & 80 \\ 50 & 40 & 40 & 50 \\ 100 & 80 & 50 & 50 \end{bmatrix}$$

The objective function appears to be nonconvex. Thus, we ran our solution procedure from randomly generated solutions x^0. The procedure was repeated a number of times, each run taking a few seconds on the PDP 11/70 computer at IIASA. Two local optima were found. Both solutions appeared to be equally good, thus yielding two optimum regional solutions to our location problem.

These solutions are given in Table 10.3.

Table 10.3. The two optimum solutions to the land allocation problem.

j	r				Total	r				Total
	A	B	C	D		A	B	C	D	
1	1	0	0	4	5	1	2	0	2	5
2	0	0	0	4	4	0	0	0	4	4
3	0	2	0	1	3	0	0	0	3	3
4	0	0	5	1	6	0	0	5	1	6
Total	1	2	5	10		1	2	5	10	

10.6. Implementation of a plan

A plan generated with the aid of a model is of little value if it cannot be implemented. There are essentially four ways of implementing a plan:

1. To leave implementation to the market system without any constraints but with charges (rent) for land use.
2. To use central authorities to implement complete investment strategies.
3. To use the planning system to generate local constraints for activities and leave the detailed implementation to the market.
4. To organize negotiations between the allocators of land and the allocators of investment.

Each approach will now be discussed briefly.

10.6.1. Market implementation

This method involves determining rental values for land in different subregions. The decision-makers in each sector would then be allowed to choose the locations which would yield the location pattern best meeting their needs. The theory behind the use of such rental values is outlined below.

Consider first a simple case in which n activities i are to be sited in n available subregions r. Let b_i^r be the net benefit of activity i (excluding the rent for land) located in subregion r. Suppose that the locations are chosen so that the total net benefit is as large as possible. The optimal plan is then the solution of the following assignment problem (Dantzig, 1963):

$$\text{maximize} \sum_{ir} b_i^r x_i^r$$

$$\text{subject to} \sum_i x_i^r \leqslant 1 \tag{10.8}$$

$$\sum_r x_i^r = 1 \tag{10.9}$$

$$x_i \geqslant 0$$

For an optimal basic solution, x_i^r is equal to 1 if activity i is to be located in subregion r, and is zero otherwise.

Let p_r and π_i denote the optimal dual multipliers for constraints (10.8) and (10.9), respectively. If, according to the (optimal) plan, activity i is assigned to subregion r, then $b_i^r - p_r - \pi_i = 0$. Here p_r may be interpreted as the rent for subregion r. Thus, $\pi_i = b_i^r - p_r$ is the profit for activity i. Given the rental values p_r, another location k for activity i would yield a profit $\pi_i^k = b_i^k - p_k$. By the optimality condition, $b_i^k - p_k - \pi_i \leqslant 0$, or $\pi_i^k = b_i^k - p_k \leqslant \pi_i$; i.e., any other location k for i would yield a profit π_i^k less than or equal to π_i. Thus, maximizing the profit of each activity separately yields an optimal location pattern under this rental price system.

It is generally believed that a decentralized pricing system cannot be used to allocate a resource if there are economies of scale leading to difficulties in separating various functions. In fact, it has been shown by Koopmans and Beckmann (1957) for the above example that the optimal solution cannot be implemented in this way if the goal function is nonlinear — for example, if it is quadratic. The same is usually true when capacity for certain activities has to be built in units of a given size. In our problem, both of these factors are likely to prevent a market implementation of the plan. The pure market solution to the implementation problem can therefore be ruled out.

10.6.2. Centralized implementation

Another method for implementing a plan is to have it enforced by the regional authority. However, this procedure demands a great deal of information at the central planning level. A planning model such as that described in this chapter is, by necessity, of a highly aggregated nature. Aggregation of this type may rule out a centralized implementation scheme, which requires detailed information finely disaggregated into fairly homogeneous branches of industry. One might also argue that it is impossible, or at least uneconomical, to generate very disaggregated technological and administrative data at the central level.

10.6.3. Zoning

"Zoning" is a compromise between centralized and market implementation in which central authorities constrain the use of land in each subregion to fall within a certain class of activities, leaving all detailed decisions to the market. It is clear that a planning model could be used to generate such constraints on land use.

10.6.4. Negotiation

"Negotiation" may also be seen as a compromise between the pure planning and market approaches to implementation. This procedure, however, comes closer to the market approach. The allocation model is used to generate a reasonably representative set of Pareto-optimal location patterns. These solutions may then be used as reference points in the negotiation (on investment in new units of production and other activities) between the land-allocating authorities and the sectoral decision-makers.

It should be noted that the best implementation approach to use in a certain situation cannot be determined objectively; it depends to a great extent on the institutional structure of the country and region concerned.

References

Andersson, Å. E. (1974). Towards an integrated theory of intersectoral and interregional growth. In A. Karlqvist (Editor), Dynamic Allocation of Urban Space. Saxon House, Farnborough, England.

Andersson, Å. E. and La Bella, A. (1979). A System of Models for Integrated Regional Development. Proceedings of Task Force Meeting I on Regional Development Planning for the Silistra Region. CP-79-7. International Institute for Applied Systems Analysis, Laxenburg, Austria.

Beckmann, M. (1953). The partial equilibrium of a continuous space market. Weltwirtschaftliches Archiv, 71:73–89.

Bradley, H., Brown, G., and Gravis, A. (1977). Design and implementation of large-scale primal transshipment algorithms. Management Science, 24:1–33.

Cooper, L. (1967). Solutions of generalized locational equilibrium models. Journal of Regional Science, 7:1–18.

Dantzig, G. (1963). Linear Programming and Extensions. Princeton University Press, Princeton, New Jersey.

Frank, M. and Wolfe, P. (1956). An algorithm for quadratic programming. Naval Research Logistics Quarterly, 3:95–110.

Isard, W., Liossatos, P., Kanemoto, Y., and Kaniss, P. C. (1979). Spatial Dynamics and Optimal Space–Time Development. North-Holland Publishing Co., Amsterdam.

Kallio, M., Pór, A., and Soismaa, M. (1979). A Fortran Code for the Transshipment Problem. WP-79-26. International Institute for Applied Systems Analysis, Laxenburg, Austria.

Koopmans, T. C. and Beckmann, M. (1957). Assignment problems and the location of economic activities. Econometrica, 25:53–76.

Los, M. (1978). Simultaneous optimization of land use and transportation. Regional Science and Urban Economics, 8:21–42.

Lundqvist, L. and Karlqvist, A. (1972). A contact model for spatial allocation. Regional Studies, 7.

Mills, E. D. (1972). Studies in the Structure of the Urban Economy. Johns Hopkins University Press, Baltimore, Maryland.

Nijkamp, P. and Paelinck, J. H. (1975). Operational Theory and Method in Regional Economics. Saxon House, Farnborough, England.

Puu, T. (1981). A model of interdependent continuous space markets for labor, capital and consumer goods. Regional Science and Urban Economics. (Forthcoming.)

Snickars, F. (1977). Convexity and duality properties of a quadratic interregional location model. Regional Science and Urban Economics, 7.

Weibull, J. W. (1976). An axiomatic approach to the measurement of accessibility. Regional Science and Urban Economics, 6.

Regional Development Modeling: Theory and Practice
M. Albegov, A.E. Andersson and F. Snickars (editors)
North-Holland Publishing Company
© IIASA, 1982

Chapter 11

GOALS OF ADAPTIVITY AND ROBUSTNESS IN APPLIED REGIONAL AND URBAN PLANNING MODELS*

Lars Lundqvist
Research Group for Urban and Regional Planning, Department of Mathematics, Royal Institute of Technology, Stockholm (Sweden)

11.1. Examples of adaptivity goals in planning practice

11.1.1. Why adaptivity?

A common feature of most urban and regional planning situations is their genuine multiple criteria character. The reasons for this multiplicity of goals can be traced to a number of sources: urban and regional systems constitute complex socioeconomic units which affect many competing components of welfare or quality of life; urban and regional systems are made up of social and geographical groups whose interests are often in conflict; urban and regional systems are coupled to infrastructures with long technical and economic life spans, which implies that planning has to deal with conflicts and tradeoffs between short-term and long-term goals.

A second important aspect of many urban and regional planning situations can be described as genuine uncertainty. The main reason for the existence of broad and qualitative uncertainty, which is difficult to express in ordinary decision theory, is the long planning periods resulting from durable elements within the infrastructure. Thus, urban and regional planning has to consider the conditions likely to prevail in the distant future, which will be affected by a number of factors: shifting behavioral, technical, and economic relations within the systems resulting from, for example, changing values and new technical options; uncertainty concerning the availability of resources; uncertainty about environmental developments, i.e., national and international constraints on urban and regional systems; and uncertainty concerning political goals and welfare criteria, i.e., the relative weights given to various components of composite social objectives.

The existence of multiple criteria in urban and regional planning stresses the need

─────────────
* This research was supported by the Swedish Council for Building Research and the Bank of Sweden Tercentenary Foundation.

for tools that are capable of dealing with conflicts. In most cases, the conflicting objectives stem from differences in interests (political parties, social groups) or responsibilities (administrative levels). Thus, it is not easy to combine these criteria into a single objective by any sort of weighting procedure, nor are they well suited for compromise solution through any interactive process.

A third approach to the treatment of multiple criteria concentrates on the set of efficient solutions. An efficient solution is characterized by the Pareto property: a solution is efficient if it is feasible and not dominated by any other feasible solution. Thus, the efficient solution can only be improved in terms of one criterion at the expense of one or several other criteria. The set of efficient solutions (or the efficiency frontier) may be generated without any a priori knowledge of the relative priorities given to the different criteria by decision-makers. This is an attractive feature in situations where many goals (or evaluation indicators) have been expressed but where genuine uncertainties or unwillingness to reveal explicit preferences prevent the formulation of a single welfare objective.

Summarizing, an efficiency frontier approach, which illustrates the possible courses of action over a certain period of time, seems to be most suitable for tackling problems that include multiple criteria and genuine uncertainty. This approach focuses on the set of available opportunities. In a multistage planning problem, the options available in later periods are conditional on the actions taken in the past. Hence, the concept of adaptivity, or preservation of available opportunities, is a natural generalization of static efficiency frontier analysis to a multistage planning situation.

We have already given preliminary reasons for the strong emphasis given to adaptivity in this chapter. The arguments concerned with uncertainty have been related to uncertainty about future preferences, although a similar line of reasoning can be carried through in terms of resources or system constraints. In the next sections we illustrate the growing interest in adaptivity or freedom of action using statements taken from official Swedish studies or planning documents. The concepts of adaptivity, flexibility, and robustness are defined and some results discussed. Finally, we consider the application of optimization models in regional and urban planning situations where adaptivity is one goal of primary interest.

11.1.2. Energy–economic adaptivity in a multiregional situation

Energy policy. The energy program adopted by the Swedish Parliament in 1975 contained four main goals: reduction in the growth of energy consumption through energy conservation measures; adoption of an active oil policy, i.e., reducing the risk of total dependence on oil by substituting other fuels for oil and increasing oil-supply security; provision for the projected increase in electricity demand through hydropower and nuclear power; and participation in international efforts to use energy more efficiently.

The role of future freedom of action was stressed very clearly:

Planning must aim at creating a broad space for action at future decision points. Our ideas of society at the beginning of the next century, e.g., the volume and composition of production, the nature of housing and living conditions, are in many ways unclear. For future generations not to become the prisoners of an energy system of our creation, our present energy policy must aim at preserving a high degree of freedom of action . . . . We are urged to take decisions that to some extent will be binding in the future. However, these decisions should be designed to keep the maximum number of options open.

A similar expression of the need for adaptivity is found in the Government's 1979 energy proposal:

The energy policy should be designed to preserve freedom of action. The energy system should therefore be able to adapt to new techniques and changing societal conditions.

Economic policy. The main goals of economic policy in Sweden are full employment, price stability, economic growth, equitable income distribution, and regional balance. Various sources of uncertainty were listed in the 1978 long-term economic survey: international developments, the relations between the elements of the economic system (e.g., prices and demand, wages and prices, and input–output interactions), and forecasts of exogenous variables:

Against this background we must be prepared to reconsider our economic policy when new information is available about internal and external developments . . . . The claims for adaptivity in economic policy are very strong . . . .

Similar arguments were given in the analysis of longer-term planning:

The uncertainties in judgments about the future increase as we look further into the future . . . . Economic policy should be designed to preserve a maximum degree of freedom of action within the framework of a purposeful long-term strategy.

The balance of payments plays a special role in discussions of economic policy because it imposes an important constraint on freedom of action.

Regional policy. During the 1970s there was a great deal of interest in the structural properties of local labor markets. The aim was to create robust settlement systems with differentiated labor markets which should be less vulnerable to changes in the structure of production and employment than conventional systems.

Since the mid-1960s, regional and employment policy in Sweden has been based partly on programs proposed in each county. In this way it is hoped to provide more equitable opportunities in terms of employment, services, and environmental protection in the regions. These programs are based on projections of expected "spontaneous" developments over a time horizon of about 10 years.

In the 1980 guidelines for county planning, the Government states that the planners should "judge the effects of structural changes on local and regional levels and make it possible to take action on various development alternatives over a 20-year period". In later documents the county boards were recommended to treat uncertainties explicitly and to plan for future freedom of action or flexibility.

11.1.3. Location–transportation adaptivity in urban situations

Lundqvist (1978a) has reviewed the planning goals of Swedish cities in the mid-1970s and his main findings are summarized in Table 11.1.

Since then the emphasis on adaptivity, flexibility, and future freedom of action has become even more pronounced. This is illustrated by a few recent examples:

The Public Transport Board recommends that studies of long-term freedom of action shall be pursued as part of regional planning in the future.
(Public Transport Board, Stockholm, 1978)

Future regional planning should focus on long-term development . . . . The relations between long-term considerations and short-term decisions should be studied in order to clarify the nature of strategic choices in urban development.
(Proposed regional plan, Stockholm, 1979)

The municipalities should be prepared to meet different future development alternatives. They should, when possible, preserve their freedom of action and avoid being bound to a detailed plan at an early stage of development.
(Proposed regional plan, Malmö region, 1979)

The municipality program will cover a 10-year planning period. However, the need for freedom of action should be studied over a longer perspective.
(Uppsala municipality program, 1979)

11.2. On the meaning and measurement of adaptivity

11.2.1. Meaning·

It is obvious from most of the quotations given above that adaptivity, flexibility, and freedom of action are used as general concepts without any high degree of precision. For example, an analysis of various interpretations of freedom of action in Swedish energy policy reveals some interesting differences (Leckius, 1978): freedom of action is seen variously as a means of keeping open the option to implement various energy technologies; a means of keeping open the option to implement various energy-supply structures; a means of counteracting a sudden energy-supply crisis; or as a

Table 11.1. Aims in urban development planning in Swedish cities in the mid-1970s.

City	Criteria				
	Efficiency	Distribution	Access	Environment	Adaptivity
Stockholm (1976)	Good conditions for the production of goods and services	Equal opportunities to reach and choose job, services, housing, and recreation	Good access between dwellings and jobs, dwellings and services, and dwellings and recreation areas	Ecological basis for planning; conserving land for agriculture	Range of long-term developments made possible by short-term allocation
Göteborg (1972)	Differentiated labor market and high productivity	Free choice of types of housing; equal living conditions for all residential areas	Good access so that the region behaves as a uniform labor and housing market	Protection of environment; conservation of natural resources	
Malmö (1972)	Improved conditions for efficient production, industrial location, and provision of social services	Equal opportunities to choose freely among jobs, residences, etc.	Good access from residential areas to work places, services, and recreation areas	Protection of environmental and cultural values	
Uppsala (1972)	Efficient location of residential areas; stimulatation of increased rate of employment	Equal quality of residential areas; freedom of choice among jobs, housing, etc.	Good access regardless of the choice of residence and job	Good facilities for cultural and physical recreation	Possibility of adapting residential areas according to new circumstances
Linköping (1973)	Good conditions for efficient production and supply of services	Differentiated social structure in all communities; freedom of choice regarding housing	Good access to jobs, services, and recreation areas, regardless of the mode of transportation	Environmental protection	
Helsingborg (1975)	Sizing of residential areas with regard to efficient provision of services and public transportation	Equal and differentiated supply of housing; opportunities to choose freely among offered alternatives	Access from dwellings to jobs, services, and recreation areas by public transportation	Adaptation to present agricultural use; environmental conservation	Flexibility of timing development stages; possibility of adapting to future conditions

means of adapting the energy system to future supply–demand conditions without seriously affecting economic and social objectives.

More precise definitions of the concepts of flexibility, adaptivity, and robustness are given below. These definitions are consistent with standard interpretations given in the planning literature; see, for example, Strangert (1974).

Flexibility is a property which expresses the ability of a plan to meet static uncertainty over a certain planning period. This assumes a planning situation in which several uncertain alternatives are possible and in which there is no indication that the uncertainty may change over time or that it can be diminished by the planners. Flexibility-creating measures will make the plan less sensitive to the outcome of uncertain elements during the planning period. This often encourages a compromise between actions that would have been optimal with regard to each outcome taken separately. Examples of flexibility-creating measures could be investment in structures designed for multiple uses or in heating systems that can use a number of fuels.

Adaptivity is a corresponding property which expresses the ability of a plan to meet dynamic uncertainty or uncertainty that may be resolved over a certain planning period. Adaptivity-creating measures can be regarded as actions that prepare the way for the various decisions that could be made at future stages on the basis of the information then available. Thus, an adaptive plan is a strategy giving decisions for the first period of time and alternative courses of action which may be followed in the future depending on the information available at the time of the decision. Examples of adaptivity-creating measures include parallel research and development projects or preparations for the modification of buildings and other infrastructure.

It is obvious that both flexibility-creating and adaptivity-creating measures involve costs greater than those of an a posteriori optimal action program. These "regrets" have to be weighed against the advantages gained by the ability to deal with uncertain outcomes.

Robustness is taken to be a property of the first decision in a multistage planning situation. A robust initial choice is a first-period decision that keeps open as many good subsequent options as possible. A good option is a near-optimal (or near-efficient) solution corresponding to one of the possible information states at the next decision point. Thus, robustness as defined here is a certain type of adaptivity in which all possible second-stage conditions are considered without any attempt to estimate their probabilities. This makes robustness a desirable property in planning situations in which there is genuine uncertainty about future conditions.

This discussion concentrates on cases of dynamic uncertainty in which adaptivity or robustness are especially relevant. We will also pay special attention to the problem of genuine uncertainty.*

---

* For methods useful under less broad types of uncertainty, see, for example, Werczberger (1980).

11.2.2. Measurement

A few attempts to measure adaptivity and robustness have been described in the literature. The different approaches used are outlined below.

Koopmans (1964) discussed the choice between sets of action programs. Initially a set Z of action programs is available. If all these programs contain the same first-stage decision, then the choice between action programs can be postponed. If not, the set Z is partitioned into sets of opportunities $(X \subset Z)$ according to the decision made during the first period. A choice between these subsets has to be made immediately, while subsequent choices can be postponed. Koopmans proposed a number of postulates or axioms reflecting a preference for the postponement of choices.

Kreps (1979) has recently presented a representation theorem for choices among sets of opportunities. Let Z be a finite set of choices and let X be the set of nonempty subsets of Z. Let z and x be generic elements of Z and X, respectively. Two expressions of the desire for freedom of action are formulated as axioms to be fulfilled by the preference relation $\gtrsim$ on opportunity sets X; these are $x \supseteq x'$ implies $x \gtrsim x'$; and $x \sim x \cup x'$ implies for all x'' that $x \cup x'' \sim x \cup x' \cup x''$. These are reasonable properties in cases where a large opportunity set is preferred to a smaller one. The second axiom states that if the adaptivity gained by joining x' to x is negligible, then adding x' to $x \cup x''$ (which is larger than x) should also produce a negligible improvement in adaptivity.

When Z is finite, a complete and transitive preference relation $\gtrsim$ on X satisfies both axioms if and only if there exists a finite set S and a function $U : Z \times S \to R$ such that

$$v(x) = \sum_{s} [\max_{z \in x} U(z, s)]$$

represents the binary relation $\gtrsim$. S can be regarded as a set of states representing the possible outcomes of uncertain second-period preferences. Kreps also proves a more general ordinal version of the representation theorem and generalizes the result to infinite Z.

Gabor (1969) argued for a planning policy that does not unduly restrict the freedom of future planners. This extreme case of maximum freedom of action for the next generation of planners is illustrated by dynamic systems in which freedom of action is measured in terms of phase volume or action radius. These pure freedom-of-choice measures may be traded off against conventional objectives.

Gupta and Rosenhead (1968) studied stability and robustness in sequential investment problems. The robustness of an initial decision was defined in terms of the number of "good" second-period solutions (for possible external conditions) that remain open after the first period. One simple measure is the ratio of the number of options remaining open after the first period to the total number of good second-period solutions (end states) considered. (The end-states may be weighted subjectively according to their "closeness" to the "best" possible solution.) This measure was used in a two-period plant location problem with uncertainty about future demand (number of plants to be built). Lundqvist (1975) suggested the use of a similar measure of adaptivity in an integrated urban location–transportation problem. A more recent application is summarized in Section 11.4.

Strangert (1974) presented a general method for evaluating action programs and strategies. Both optimizing and satisficing decision-making behavior were considered, as well as cases of risk and genuine uncertainty. The method provides rules for the evaluation of

(a) Action programs

(b) Conditional decisions (may be terminal or nonterminal). I_j represents possible information states

(c) Nonterminal options

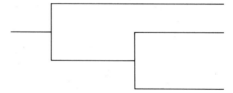

By combining these definitions, arbitrary decision trees

can be generated and assessed using the corresponding evaluation rules. The value of a conditional decision at a certain stage may be seen as a measure of the freedom of action at that point and may indicate how adaptivity-creating measures can improve the freedom of action. Robustness, as defined here, is the value of a simple nonterminal option where there is genuine uncertainty about the second-period information states and only second-period effects are counted.

The measures of adaptivity outlined above provide a starting point for the application of this concept in urban and regional planning models. This is especially important in countries such as Sweden, where great emphasis has been placed on adaptivity and robustness in the planning process.

11.3. Methodological approaches and adaptivity

There are many methodological approaches which could be used to take the need for adaptivity and robustness into account; some of the possible techniques are outlined below.

Scenario-writing is widely used to highlight structural differences in long-term development options. The scenarios are generally produced by nonformal methods because of the difficulties involved in using observations from today's society for projecting situations with qualitatively different social conditions. However, in some cases behavioral models have been used to improve the internal consistency of various scenarios, e.g., the Energy Policy Project of the Ford Foundation and other energy–economic investigations. In a study of the future of Sweden, two extreme scenarios were suggested, according to which the Swedish energy system in 2005 will be almost totally based on either nuclear power or soft technologies. These extreme scenarios were used in a discussion of the dependence of future freedom of action on short-term policy decisions.

Deterministic simulation models can easily be integrated into a decision-tree evaluation approach (Strangert, 1974). In a similar way, descriptions of macro-economic systems or multisectoral economic growth models can be used to evaluate the adaptivity or robustness of various courses of action. A multisectoral economic growth model was used by Bergman and Mäler (1979) to study the robustness of short-term energy policies with respect to uncertainty in future oil price developments.

The next section contains some examples in which optimization models are used to study adaptivity or robustness. Dynamic programming is suitable for solving multi-stage decision problems of moderate size; some care in model construction is needed when using linear or integer programming models to search for adaptive strategies. However, in the two-stage decision problems considered here the use of optimization models is generally straightforward (even in the risk case; see, for example, El Agizy, 1967; Midler, 1970).

11.4. Examples of urban and regional planning models used to study the adaptivity or robustness of structural developments

This section gives three examples of how urban and regional planning models can be used to define possible resource allocations, taking into account genuine uncertainties and multiple objectives. Options of this type defined for various periods of time could be used as a starting point for a qualitative assessment of the adaptivity or robustness of different policy decisions. In the cases discussed below, robustness properties will be illustrated with regard to both uncertainty concerning future objectives (Sections 11.4.1 and 11.4.2) and uncertainty concerning external conditions in a more general sense (Section 11.4.3). A robust initial decision may be identified by considering the stability of end-state solutions to possible changes of criteria or external conditions during the second period.

11.4.1. A multiregional energy–economic input–output model

A multiregional model based on the estimated interregional input–output balance of Sweden in 1975 (Snickars, 1979) has been constructed in order to study the relations between regional development, employment, and energy use. The input–output system provided regional technical coefficients, aggregate coefficients of interregional trade (intraregional supply of intermediate goods and final demand, and interregional distribution of domestic imports), and regional coefficients of foreign trade. Other sources of Swedish data, especially the 1975 long-term economic survey, national statistics, and the 1968 household budget survey, were used to calculate consumption, capital, and investment coefficients. Special efforts were made to obtain regional capital, labor, and energy use coefficients. The model can be described mathematically as follows:

$$\underset{x \geqslant 0, i \geqslant 0, h \geqslant 0}{\text{maximize}} \quad v_1 f x^c + v_2 l x + v_3 u P x$$

subject to

$$x_r^t + \mathbf{D}_r^t[m] \, x_r^t \geqslant \mathbf{D}_r[\alpha] \, \mathbf{A}_r^t x_r^t + \mathbf{D}_r[\beta] \, [q_r^t x_r^{ct} + \mathbf{B} i_r^t] + g_r^t h^t + T_r^t$$

$$T_r^t = \sum_{s \neq r} \mathbf{D}_{rs}[\omega] \, \{ \mathbf{D}_s[1 - \alpha] \, \mathbf{A}_s^t x_s^t + \mathbf{D}_s[1 - \beta] \, (q_s^t x_s^{ct} + \mathbf{B} i_s^t) \}$$

$$x_r^{ct} \geqslant \gamma_r \cdot w_r^t x_r^t$$

$$\mathbf{D}_r^t[c] x_r^t \leqslant \mathbf{D}_r[(1 - \delta)^t] c_r^0 + \mathbf{D}_r[(1 - \delta)^{t-1}] \, i_r^1 + \ldots + i_r^t$$

$$\underline{l}_r^t \leqslant l_r^t x_r^t \leqslant \overline{\overline{l}}_r^t$$

$$\mathbf{P}^t x^t \leqslant \overline{\overline{p}}^t$$

$$-m^t x^t + h^t \geqslant \overline{z}^t$$

$$\overline{c}_t \leqslant (1 - \delta)^t c^0 + (1 - \delta)^{t-1} i^1 + \ldots + 1 i^t$$

where:

x, i, h are the production (x^c = consumption), investment, and total foreign export variables

$\mathbf{P}, l, c$ are the energy (several types), labor, and capital coefficients

m, g are the import and export coefficients

$\mathbf{A}, q$ are the input–output and consumption coefficients

α, β, ω are the intraregional supply (α, β) and interregional trade (ω) shares

$\mathbf{D}[s]$ is a diagonal matrix with diagonal elements s

T is the interregional trade demand

f is the discounting factor

w, γ are the value-added share and minimum consumption rate

$\mathbf{B}, c^0, \delta$ are the investment goods structure, the initial capital stock, and the capital depreciation

$\underline{l}, \overline{\overline{l}}, \overline{\overline{p}}, \overline{c}, \overline{z}$ are upper (double overbar) and lower (single overbar) bounds on labor, energy use, total capital stock, and balance of payments

v_1, v_2, v_3 are tradeoff coefficients

u are the weights attached to various types of energy

The constraints represent commodity and capital balances, limits on the use of labor and energy, a balance-of-payments restriction, and a lower bound on total capital formation. A lower bound on consumption in each region is also assumed. A more complete discussion of the model is given in Lundqvist (1980).

The model was used to analyze the Swedish economy, which for the purposes of the study was divided into nine sectors and eight regions. The period studied (1975–1990) was divided into three five-year periods. Table 11.2 summarizes three calculations based on different objectives: maximizing consumption, maximizing employment, and minimizing use of electricity.

It should be noted that the first of these objectives (maximizing total consumption) is expressed in terms of projected regional consumption patterns, and includes certain shares which express minimum levels of public consumption. The model often indicates that it is necessary to have surplus production in the public sector to achieve a feasible regional solution. Table 11.2 gives the levels of consumption both excluding and including this surplus public-sector production.

We will concentrate on the stability of the results given in Table 11.2. First, it is clear that the three objectives considered are in conflict; this is especially true for maximization of consumption and minimization of electricity use. Maximizing employment can be regarded as a compromise solution. However, it should be emphasized that all three solutions satisfy the same set of constraints, i.e., they are all located on the efficiency frontier of the regional energy–economic problem.

Table 11.2 also contains projected 1980 and 1990 production and consumption volumes for two regions (Stockholm and Northern Sweden). It is clear that the choice of objective has a greater influence on production and consumption in Northern Sweden than in Stockholm. This is true for production, private consumption, and the provision of public services. For example, private-sector production in Stockholm in 1990 varies by 18% depending on the objective chosen; the corresponding figure for Northern Sweden is 24%. The widest variations are observed in the construction sector and the manufacturing industry. Thus, the economic structure of the Stockholm region seems to be more robust with regard to changes in economic and energy policies than that of Northern Sweden.

One method of increasing the adaptivity of the economy to future changes in policy is to prepare selective transportation or employment support systems that might counteract the undesirable effects of any policy change. The model may be used to simulate the impacts of such systems on the regional economy, e.g., by tracing their effects on technology and patterns of interregional trade.

11.4.2. A long-term planning location–network design model

We shall now turn to intraregional planning, and consider some applications of a system of interrelated location–transportation models for the Stockholm region. The politically determined program behind the 1978 regional plan contained one point that directly referred to adaptivity or robustness: "The plan report should outline the long-term options that are kept open by the short-term regional plan".

During 1978, the Public Transport Board of the Stockholm County Council used an integrated land-use–network-design model to study the long-term consequences of two

Table 11.2. Summary of results from the Swedish multiregional input–output model.

Economic variable	Objective					
	Maximize average consumption (1975–1990)		Maximize average employment (1975–1990)		Minimize average electricity use (1975–1990)	
	1980	1990	1980	1990	1980	1990
Stockholm						
Employment	839.5	854.4	854.0	886.6	839.5	854.4
Consumption[a]	26,959/17,388 (32,394)	31,964/19,977 (36,854)	27,375/17,657 (32,970)	32,700/20,437 (38,521)	26,976/17,399 (32,304)	29,808/18,630 (38,202)
Production						
Process industry	2,124	2,572	2,184	2,579	2,133	2,178
Manufacturing industry	16,966	23,547	17,087	23,627	16,628	19,436
Construction	6,409	4,935	7,194	4,956	7,954	2,993
Transportation	5,367	6,943	5,321	7,041	5,164	6,344
Housing services	3,100	3,356	3,148	3,434	3,102	3,130
Private services	14,961	18,055	14,860	18,160	14,398	15,825
Public services	15,006	16,877	15,313	18,084	14,905	19,572
Total	71,107	85,596	72,275	87,176	71,344	77,888

Northern Sweden

Employment	243.9	245.3	249.5	274.4	232.9	245.3
Consumption[a]	9,474/5,703	9,157/5,329	7,863/4,733	9,317/5,422	7,400/4,455	8,060/4,691
	(9,474)	(10,302)	(8,565)	(11,620)	(7,925)	(10,586)
Production						
Process industry	2,888	3,484	2,953	3,491	2,845	2,956
Manufacturing industry	2,549	3,130	2,542	3,137	2,395	2,350
Construction	3,386	2,006	3,879	1,754	3,568	1,035
Transportation	1,432	1,839	1,411	1,862	1,365	1,681
Housing services	1,023	897	849	913	799	790
Private services	1,852	2,078	1,810	2,130	1,717	1,796
Public services	3,771	4,973	3,832	6,198	3,470	5,895
Total	20,355	22,823	20,765	23,864	19,562	20,283

National averages

Consumption[b]	154.5(162.8)		148.7(165.4)		141.1(159.0)	
(10⁹ Skr per year)						
Employment	4234.3		4396.5		4214.2	
(1000 persons)						
Electricity use	95.6		95.4		90.4	
(TWh per year)						

[a] Given in the form: total consumption with fixed share of public consumption/total private consumption (total consumption including "excess" public consumption).

[b] Given in the form: total average consumption with fixed share of public consumption (total consumption including "excess" public consumption).

proposed plans (A and B). A detailed account of the evaluation has been given by Lundqvist (1978b); this discussion will concentrate on the robustness aspects.

The model allocates dwellings and places of work to geographical areas, and selects a combination of transportation improvements that minimizes a composite objective (contact cost, density cost), subject to constraints on investment and operating costs:

$$\text{minimize}_{x,t} \; \beta_1 \left[\alpha_1 P(x, t) + \alpha_2 C(x, t) + \alpha_3 O(x, t) \right] + \beta_2 D(x, t)$$

subject to

$$Ax = \bar{x} \qquad \text{(activity volumes)}$$
$$B_1 x + B_2 t \leqslant \bar{F} \text{ (resource constraints)}$$
$$l \leqslant x \leqslant \bar{\bar{u}} \qquad \text{(initial stock, maximum development)}$$

where

x is the volume of activity in each geographical area (zone)

t is the network structure

α, β are tradeoff coefficients

P is an index representing average travel time (no discounting, no competition)

O is an index based on the number of opportunities available within an area corresponding to a certain travel time (i.e., discounting)

C is similar to O but also takes into account the competition for opportunities from other zones (i.e., discounting and competition)

D expresses the desire for high spatial standards in terms of average development density

Here P, O, and C are contact cost indicators representing various properties of discounting and competition, while D is a simple density cost indicator. We used two combinations of contact cost indicators: one pure average contact cost indicator (P); and one average composite opportunity cost indicator based on O for social contacts (dwelling–dwelling) and business contacts (workplace–workplace), and based on C for travel between home and work.

Figure 11.1 summarizes the results of a number of calculations (varying β) over both the short term (1975–1990) and the long term (1975–2005). The effects of two proposed short-term plans on long-term options are also illustrated. All of these calculations were carried out assuming 12 zones, 2 activities (with the demand split into six equally sized "bundles" for each activity), 25 improvements in transportation, and 2 resource categories.

It may be concluded that the proposed short-term plans are far from efficient, indicating that certain objectives and constraints not included in the model are important in practice. In addition, both of the proposed alternatives (A and B) are biased toward the use of space at the expense of accessibility. The differences between the plans in terms of future freedom of action are small but not negligible. Figure 11.2 gives a clearer illustration of how the choice of a robust short-run development pattern

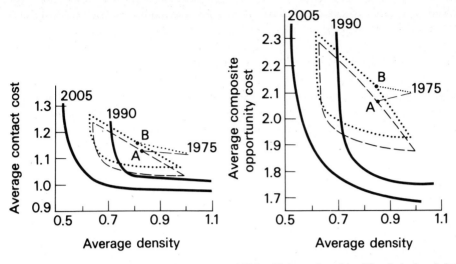

Figure 11.1. Short-term (1990) and long-term (2005) efficiency frontiers. The dashed and dotted lines show the conditional efficiency frontiers in 2005 that would result if development between 1975 and 1990 followed plan A or plan B, respectively. Source: Lundqvist (1978b). (Reproduced with kind permission of the publishers).

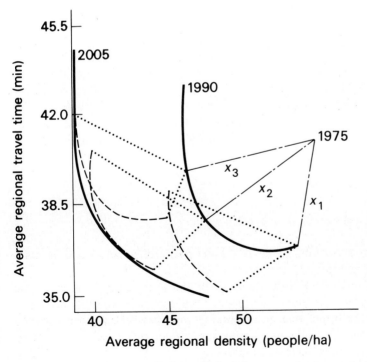

Figure 11.2. The robustness of three efficient short-term developments. The dashed lines show options that are kept open by the short-term choices.

can lead to greater adaptivity in the future. The balanced alternative obviously preserves more freedom of action than the highly dispersed or very compact land-use patterns.

The Public Transport Board finally concluded that:

Alternative A and even more alternative B will restrict the chances of achieving a regional structure which reduces traveling time (compared to 1975) in the period after 1990 . . . . This loss of long-term freedom of action should be considered against the background of a possible future shortage of energy.

One way of measuring the loss of long-term freedom of action (1975–2005) brought about by a given short-term development x (1975–1990) is to calculate the percentage reduction in the range of action in each indicator dimension:

$$L_I^{1990}(x) = \frac{|I_{1975}^{2005}(x^0) - I_{1990}^{2005}(x)|}{|I_{1975}^{2005}(x^0) - I_{1975}^{1975}(x^0)|} \times 100$$

where $I_{t_1}^{t_2}(x)$ is the minimum value of indicator I at time t_2 starting from urban structure x at time t_1.

The loss in adaptivity produced by a given plan x for 1990 may then be expressed as:

$$A^{1990}(x) = \max_I L_I^{1990}(x)$$

For the three short-run policies (x_1, x_2, x_3) in Figure 11.2, $A^{1990}(x)$ is 38.8, 20.0, and 52.0, respectively. The most robust decision obviously also minimizes the loss in adaptivity (which in this case is a measure of the maximum regret of a second-stage decision as compared to an a posteriori optimal action over the whole planning period). If this simple measure is to be plausible, the efficiency frontiers must be convex. The method may be modified by introducing weights reflecting the priority given to various indicators.

11.4.3. Transportation systems and residential location

Our final examples are taken from studies of the interrelationships between the location of places of residence and transportation systems.

Four different schemes have been proposed for the future public transportation system in the northeastern corridor of Stockholm. The issue has been under discussion for some years and the future role of a railway system has been especially controversial. The model used to study the possible land-use consequences of these four transportation systems may be outlined as follows:

$$\min_b \alpha b k(a, \beta, \mathbf{C}) + (1 - \alpha) b \mathbf{S} b$$

subject to

$$eb = 1$$
$$bf_l \leqslant v_l/B$$
$$\bar{b} \leqslant b \leqslant \bar{\bar{b}}$$

where

e is the unit vector

b is the proportion of the population living in the various zones

k is the average time taken to travel from each residential zone to (the closest) workplace areas containing at least a given proportion β of all workplaces

a is the location of workplaces

C is a matrix of traveling times

S is a diagonal matrix of inverted zonal development space or local capacity

f_l are loading factors denoting the proportion of residents traveling through certain cuts l (one or more links) defined around the CBD

v_l is the maximum transport capacity of cut l (one or more transport links) defined around the CBD

B is the total population

$\bar{b}, \bar{\bar{b}}$ are the lower and upper bounds on the proportion of the total population in the zones

The model was calibrated (by adjusting α and β) using the regional structure for 1990 predicted by the regional plan. The location of places of work and other characteristics for the year 2005 were then entered and the matrix of traveling times for each public transportation system was used to calculate the resulting pattern of land use in 2005. These traveling times also affect the split between modes of transportation and hence have a twofold influence on the average traveling times that determine the final population distribution. The northeastern corridor was divided into 41 zones and the rest of the region was represented by 10 zones.

When evaluating the results of the model, the Public Transport Board paid special attention to the robustness of the allocations in view of the uncertainty surrounding the choice of transportation system. Each zone in the northeastern corridor was assessed on the basis of the stability of its attractivity. A zone was described as robust if it showed the same tendency for population increase or decrease in the reference plan and in the three other transportation options. Using this definition, 7 out of 34 zones were shown *not* to be robust. Table 11.3 shows the effects produced on population distribution by different public transportation systems in the two most carefully analyzed municipalities. The population ranges were obtained by using three different methods for estimating zonal development spaces (**S**).

These results are reasonably stable with respect to the variation in development spaces (**S**) and emphasize that the transportation systems affect the attractivity of municipalities in a nonuniform way. It is possible that the municipalities may rank the transportation systems differently. Alternative 3 produces a high average attractivity but is unstable to changes in local capacities and produces nonuniform effects.

Table 11.3. Effects of different public transportation systems on the populations of two munici-palities, measured by the difference from the reference system (alternative 1).[a]

Municipality	Transportation system		
	2	3	4
Täby	1300–1500	500–1300	− 700–400
Vaxhdm	1400–1700	1900–3400	900–1900

[a] The difference ranges were obtained by using three different methods to estimate zonal develop-ment spaces (**S**).

Alternative 2 is more robust to changes in local capacities, more uniform in terms of population effects, but produces a slightly lower average attractivity. In later work, the costs of the transportation systems and their effects on the environment and levels of services were studied in terms of their projected impacts on land use.

The residential allocation model discussed above has also been used to produce con-sistent long-term population distributions on the basis of various assumptions about the level of central city car traffic, residential densities, and private consumption. The scenarios generated represent a number of long-term options that are kept open by the regional plan for 1990.

11.5. Final remarks

This chapter has dealt exclusively with the issues of adaptivity and robustness in applied regional and urban modeling. It is felt that modeling could be useful in gener-ating consistent long-range development alternatives which take a number of impor-tant interdependencies into account. The flexible use of optimization models seems to be important in generating consistent scenarios and in providing planners and decision-makers with information on the limits to freedom of action and the long-term binding effects imposed by short-term decisions.

References

Bergman, L. and Mäler, K.-G. (1979). Energy policy and economic vulnerability. In W. Bergström and B. Rydén (Editors), Strategic Choices in Swedish Policy-making (in Swedish). SNS, Stock-holm.

El Agizy, M. (1967). Two-stage programming under uncertainty with discrete distribution function. Operations Research, 15: 55–70.

Gabor, D. (1969). Open-ended planning. In E. Jantsch (Editor), Perspectives of Planning. OECD, Paris.

Gupta, S. K. and Rosenhead, J. (1968). Robustness in sequential investment decisions. Manage-ment Science, 15: B18–29.

Koopmans, T. C. (1964). On flexibility of future preference. In M. W. Shelley and G. L. Bryan (Editors), Human Judgements and Optimality. Wiley Publishing Co., New York.

Kreps, D. M. (1979). A representation theorem for "preference for flexibility". Econometrica, 47: 565–577.

Leckius, I. (1978). Adaptive planning for future options – concepts and applications in the energy field (in Swedish). Economic Research Institute, Stockholm School of Economics, Stockholm.

Lundqvist, L. (1975). Transportation analysis and activity location in land use planning – with applications to the Stockholm region. In A. Karlqvist, L. Lundqvist, and F. Snickars (Editors), Dynamic Allocation of Urban Space. Saxon House, Farnborough, Hampshire.

Lundqvist, L. (1978a). Planning for freedom of action. In A. Karlqvist, L. Lundqvist, F. Snickars, and J. W. Weibull (Editors), Spatial Interaction Theory and Planning Models. North-Holland Publishing Co., Amsterdam.

Lundqvist, L. (1978b). Urban planning of locational structures with due regard to user behaviour. Environment and Planning A, 10: 1413–1429.

Lundqvist, L. (1980). A dynamic multi-regional input–output model for analyzing regional development, employment and energy use. Paper presented at the European Congress of the Regional Science Association, held at Munich. TRITA-MAT-1980-20, Department of Mathematics, Royal Institute of Technology, Stockholm.

Midler, J. L. (1970). Investment in network expansion under uncertainty. Transportation Research: 4: 267–280.

Snickars, F. (1979). Construction of interregional input–output tables by efficient information adding. In C. P. A. Bartels and R. H. Ketellapper (Editors), Exploratory and Explanatory Statistical Analysis of Spatial Data. Martinus Nijhoff, The Hague.

Strangert, P. (1974). Information, Uncertainty and Adaptive Planning. National Defense Research Institute, Stockholm.

Werczberger, E. (1980). The versality model in decision-making under uncertainty, with regard to goals and constraints. In P. Nijkamp and J. Spronk (Editors), Multicriteria Analysis: Practical Methods. Gower Publications, London.

Regional Development Modeling: Theory and Practice
M. Albegov, A.E. Andersson and F. Snickars (editors)
North-Holland Publishing Company
© IIASA, 1982

Chapter 12

A NONDIFFERENTIABLE OPTIMIZATION TECHNIQUE FOR USE IN MULTIOBJECTIVE REGIONAL PLANNING

Alexander Umnov
International Institute for Applied Systems Analysis, Laxenburg (Austria)

12.1. Introduction

The main purpose of the research carried out in the **IIASA** Regional Development Task is to provide decision-makers with sufficient information to formulate development policies for a given region. This information is usually in the form of scenarios that describe a number of development paths for the region under a given set of objectives and constraints. However, it should be emphasized that the information obtained from the mathematical models is meant to supplement the decision-maker's regional analysis, not replace it (Albegov, 1978).

This supplementary information usually consists of the solutions of some single-objective problems. Typical problems are: to find the optimal point of a given model for each of the given objectives; to find the optimal point for a given linear combination of objectives; and to maximize the coefficient of consistency for a given set of objectives, i.e., to find the point at which the relative differences in the values of the objectives (from their individual optimal values) are minimized.

All of these problems can be formulated and solved in a standard way without particular difficulty. However, the situation is complicated by the fact that the regional development models have to be linked to form a system. Thus, each of the models will contain a set of parameters describing the "environment" of the model in addition to a set of endogenous variables describing the state of the model under given external conditions. The linkage problem is described more fully in Umnov (1979).

The main methodological difference between parameters and variables in this regional planning system is that the values of the parameters are chosen by the decision-makers, but the optimal values of the variables are defined by the solution of mathematical programs.

In this situation, it seems reasonable that the decision-makers should have some knowledge of the way in which the values of the parameters chosen may influence the solution of the programming problems. The dependence of programming solutions on parameter values is considered in more detail in the following sections.

12.2. The coefficient of consistency

A nondifferentiable optimization method was adopted to analyze the role of model-linkage parameters within a system of interrelated models. This technique is based on the concept of a coefficient of consistency, which will be defined later in this section. However, it is first necessary to formulate a general programming frame-work.

Let x be a vector of decision variables and u a vector of parameters such that $u \in E^L$. Assume that there are n variables and L parameters. Then a single-objective problem may be stated as follows:

$$\underset{x \in E^n}{\text{maximize}} \, f_k(x, u)$$

subject to $\hspace{8cm}$ (12.1)

$$y_s(x, u) \geqslant 0 \qquad s = [1, m]$$

where $\{f_k(x, u), \; k = [1, N]\}$ represents a given set of objective functions and $y_s(x, u)$ represents a set of constraints.

Given the above, it is also possible to find the optimal solution of the model for a given linear combination of objectives:

$$\underset{x \in E^n}{\text{maximize}} \sum_{k=1}^{N} \lambda_k f_k(x, u) \qquad \lambda_k \geqslant 0$$

subject to $\hspace{8cm}$ (12.2)

$$y_s(x, u) \geqslant 0 \qquad s = [1, m]$$

The multiobjective problem may also be stated in terms of maximization of the coefficient of consistency (which will be denoted by μ); the procedure is outlined below.

Assume that the minimum value $f_k^{\min}$ of each f_k is known and finite. Let $f_k^*(u)$ be the optimal value of the single-objective problem (12.1). Then the problem of maximizing the coefficient of consistency may be formulated as:

$$\underset{\mu, x}{\text{maximize}} \, \mu$$

subject to $\hspace{8cm}$ (12.3)

$$f_k(x, u) \geqslant \mu f_k^*(u) \qquad k = [1, N]$$

$$y_s(x, u) \geqslant 0 \qquad s = [1, m]$$

$$u \in \Omega \subset E^L$$

where Ω is the set of parameter values chosen.

The main aim of this analysis is to solve problems (12.1), (12.2), and

$(12.3)^{\dagger}$ with respect to $\boldsymbol{u}$. Since the same difficulties arise in the solution of all three problems, a full description will be given only for problem (12.3), which is the most complex. The final aim is to solve the problem:

$$\underset{\boldsymbol{u}}{\text{maximize }} \mu^*(\boldsymbol{u})$$

subject to   (12.4)

$$\boldsymbol{u} \in \Omega \subset \mathrm{E}^L$$

where $\mu^*(\boldsymbol{u})$ is the solution of (12.3) for some fixed $\boldsymbol{u}$.

The set Ω of parameter values chosen is not assumed to have any special structure. In standard sensitivity tests, it would have a very simple, polyhedral structure because slight changes in the values of selected parameters are common in such tests. However, in the type of model linkage application considered here the set Ω is likely to be considerably more complicated.

It is clear that problem (12.4) cannot be solved using standard mathematical programming methods because it is impossible to obtain an explicit form for the functions $f_k^*(\boldsymbol{u})$ and $\mu^*(\boldsymbol{u})$ except in some trivial cases. In addition, it is necessary to overcome the following difficulties: functions $f_k^*(\boldsymbol{u})$ and $\mu^*(\boldsymbol{u})$ are not defined for all points in Ω – in other words, problems (12.1) and (12.3) may not have a feasible solution for any $\boldsymbol{u}$ in Ω; and functions $f_k^*(\boldsymbol{u})$ and $\mu^*(\boldsymbol{u})$ are nondifferentiable even if $f_k(\boldsymbol{x}, \boldsymbol{u})$ and all $y_s(\boldsymbol{x}, \boldsymbol{u})$ are differentiable.

It seems that nondifferentiable optimization is the only technique that could be used to solve problem (12.4). One such method is described in the following section.

12.3. A nondifferentiable solution technique

The basic idea is to use a "smooth image" of $f_k^*(\boldsymbol{u})$ and $\mu^*(\boldsymbol{u})$ to solve the problem of nondifferentiability. We will use images which it is impossible to evaluate numerically and so we must also use numerical algorithms which do not require the functions involved to be given in explicit form. Smooth images should be defined not only for points close to $\mu^*(\boldsymbol{u})$ and $f_k^*(\boldsymbol{u})$, but also for all points in the parameter space E^L. In this case it is not necessary to consider whether problem (12.1) has a feasible solution for all $\boldsymbol{u}$.

The approach considered here involves use of the "smooth exterior point" version of the Sequential Unconstrained Minimization Technique (SUMT) described by Fiacco and McCormick (1968). Other techniques are available, but the analysis presented

---

† It is assumed that problem (12.3) satisfies the constraint $f_k^*(\boldsymbol{u}) > 0$. However, it is not essential from a theoretical point of view to include this assumption; from a practical point of view it may be simpler to use

$$f_k(\boldsymbol{x}, \boldsymbol{u}) \geqslant \mu f_k^*(\boldsymbol{u}) + (1 - \mu) f_k^{\min}(\boldsymbol{u})$$

where $f_k^{\min}(\boldsymbol{u})$ is the minimum attainable value of $f_k(\boldsymbol{x}, \boldsymbol{u})$.

here deals exclusively with SUMT. The solution process involves a sequence of uncon-strained minimizations of some auxiliary function associated with the mathematical programming problem to be solved.

12.3.1. Solving a subproblem by means of SUMT

We shall now solve problem (12.1) using SUMT. Let the auxiliary function be

$$\epsilon_k(x, u) = -f_k(x, u) + \sum_{s=1}^{m} P[T, y_s(x, u)] \tag{12.5}$$

where T is a positive fixed parameter defining the penalty for violation of constraints and the penalty function $P(T, \alpha)$ satisfies the conditions

$$\lim_{T \to +0} P(T, \alpha) = \begin{cases} 0, & \alpha > 0 \\ +\infty, & \alpha < 0 \end{cases} \tag{12.6}$$

The main property of SUMT is that

$$\lim_{T \to +0} \epsilon_k[\bar{x}^k(u), u] = -f_k^*(u)$$

where

$$\bar{x}(u) = \underset{x}{\operatorname{argmin}} \epsilon(x, u)$$

that is

$$\min_x \epsilon(x, u) = \epsilon[\bar{x}(u), u]$$

If this is so, we can use

$$\bar{f}_k(u) = -\epsilon_k[\bar{x}^k(u), u]$$

as the image of $f_k^*(u)$. The closeness is provided by the main property of SUMT defined above.

Secondly, the implicit function theorem can be applied to the minimum of (12.5), i.e., to

$$\nabla_x \epsilon_k(x, u) = 0 \tag{12.7}$$

Equation (12.6) implicitly defines $\bar{x}^k(u)$, and hence $\bar{f}_k(u)$. If $f_k(x, u), y_s(x, u)$, and $P(T, \alpha)$ are sufficiently smooth, $\bar{x}^k(u)$ is differentiable.

Since problem (12.5) is unconstrained, $\bar{x}^k(u)$ and $\bar{f}_k(u)$ will exist for any $u \in E^L$. The fact that problem (12.1) may not have a feasible solution for all u is then un-important.

Finally, any other necessary quantities can be evaluated from (12.5) and (12.7). For example, the partial derivatives $\partial \bar{x} / \partial u$ can be calculated by differentiation of (12.7) with respect to all components of u.

12.3.2. Solving the general problem using SUMT

Now let us return to problem (12.4). This can be solved using the above version of SUMT, in which $f_k^*(u)$ has been changed to $\bar{f}_k(u)$. The auxiliary function for problem (12.4) is

$$E(x, u, \mu) = -\mu + \sum_{s=1}^{m} P[T, y_s(x, u)] + \sum_{k=1}^{N} P[T, R_k(x, u, \mu)] \tag{12.8}$$

where

$$R_k = f_k(x, u) - \mu \bar{f}_k(u)$$

To complete the general description it is necessary to show how (12.8) can be minimized without using the explicit form of $\bar{f}_k(u)$. Consider, for example, the first partial derivatives. Formally, we have

$$\frac{\partial E}{\partial x_i} = \sum_{s=1}^{m} \frac{\partial P}{\partial y_s} \cdot \frac{\partial y_s}{\partial x_i} + \sum_{k=1}^{N} \frac{\partial P}{\partial R_k} \cdot \frac{\partial f_k}{\partial x_i} \qquad i = [1, n]$$

$$\frac{\partial E}{\partial \mu} = -\left[1 + \sum_{k=1}^{N} \frac{\partial P}{\partial R_k} \bar{f}_k(u)\right]$$

$$\frac{\partial E}{\partial u_j} = \sum_{s=1}^{m} \frac{\partial P}{\partial y_s} \cdot \frac{\partial y_s}{\partial u_j} + \sum_{k=1}^{N} \frac{\partial P}{\partial R_k} \left(\frac{\partial f_k}{\partial k_j} - \mu \frac{\partial \bar{f}_k}{\partial u_j}\right) \qquad j = [1, L]$$

In order to use these formulae, $\bar{f}_k(u)$ and $\partial \bar{f}_k / \partial u_j$ must be known for some point u. It is possible to find $\bar{f}_k(u)$ by minimizing (12.5) at u, but it is more difficult to evaluate the following expression for $\partial \bar{f}_k / \partial u_j$:

$$\frac{\partial \epsilon_k}{\partial u_j} + \sum_{i=1}^{n} \frac{\partial \epsilon_k}{\partial x_i} \frac{\partial \bar{x}_i}{\partial u_j}$$

However, using (12.7) it can be shown that

$$\frac{\partial \bar{f}_k}{\partial u_j} = -\frac{\partial \epsilon_k}{\partial u_j}$$

12.3.3. Modifications for practical implementation

At every iteration in the process of minimizing (12.8) it is necessary to find $\bar{x}^k(u)$ for all $k = [1, N]$; substitution of $\bar{x}$ for x in the last equation above gives:

$$\frac{\partial \bar{f}_k}{\partial u_j} = -\frac{\partial f_k}{\partial u_j} - \sum_{s=1}^{m} \frac{\partial P}{\partial y_s} \cdot \frac{\partial y_s}{\partial u_j} \qquad \text{if } x = \bar{x}$$

Thus, the general scheme is as outlined above. From a practical point of view, however, there are several reasons why the process should be divided into two stages. First, we consider u to be a parameter and solve problem (12.8) with respect to μ and x only. Then we alter u in an attempt to produce a better value of μ. This two-stage procedure can be useful in two ways. First, it is possible that decision-makers might be interested in values of μ at points other than the optimum. Second, the model is often supplied with software which enables it to be solved at fixed parameter values. It therefore seems sensible to use this software, especially when n (the number of variables) is much greater than L (the number of parameters) in the model.

This two-stage process can also be carried out using SUMT, as follows. First, we choose some initial point within E^L and minimize (12.8) with respect to x and μ. Naturally, all problems (12.1) must be solved for the given value of u. Values for $\hat{x}(u)$ and $\hat{\mu}(u)$ are found as $\operatorname*{argmin}_{x,\mu} E(x, u, \mu)$. Second, we must try to find a new point u which gives a better value of μ. In other words, it is necessary to minimize

$$\hat{\Phi}(u) = E[\hat{\mu}(u), u, \hat{x}(u)] \tag{12.9}$$

with respect to $u \in \Omega \subset E^L$.

The functions $\hat{\mu}(u)$ and $\hat{x}(u)$ are defined implicitly by the equations

$$\frac{\partial E}{\partial x_i} = \sum_{s=1}^{m} \frac{\partial P}{\partial y_s} \cdot \frac{\partial y_s}{\partial x_i} + \sum_{k=1}^{N} \frac{\partial P}{\partial R_k} \cdot \frac{\partial f_k}{\partial x_i} = 0 \qquad i = [1, n] \tag{12.10}$$

$$\frac{\partial E}{\partial \mu} = -\left[1 + \sum_{k=1}^{N} \frac{\partial P}{\partial R_k} \bar{f}_k(u)\right] = 0 \tag{12.11}$$

Analogously, we have

$$\hat{\Phi}'_{u_j} = \frac{\partial E}{\partial \mu} \cdot \frac{\partial \hat{\mu}}{\partial u_j} + \sum_{i=1}^{n} \frac{\partial E}{\partial x_i} \cdot \frac{\partial \hat{x}_i}{\partial u_j} + \frac{\partial E}{\partial u_j} \qquad j = [1, L]$$

Using (12.10) and (12.11) this leads to

$$\hat{\Phi}_{u_j} = \left. \frac{\partial E}{\partial u_j} \right|_{\substack{x = \hat{x}(u) \\ \mu = \hat{\mu}(u)}}$$

The formulae for the second-order partial derivatives will be more complex:

$$\hat{\Phi}''_{u_j u_l} = \frac{\partial^2 E}{\partial u_j \partial u_l} + \frac{\partial^2 E}{\partial u_j \partial \mu} \cdot \frac{\partial \hat{\mu}}{\partial u_l} + \sum_{i=1}^{n} \frac{\partial^2 E}{\partial x_i \partial u_j} \cdot \frac{\partial x_i}{\partial u_l}$$

In this case, it is necessary to find the values of $\partial \hat{\mu} / \partial u_l$ and $\partial x_i / \partial u_l$. These can be deduced from the following system of linear equations:

$$\sum_{j=1}^{N} \frac{\partial^2 E}{\partial x_i \partial x_j} \frac{\partial \hat{x}_i}{\partial u_l} + \frac{\partial^2 E}{\partial x_i \partial \mu} \cdot \frac{\partial \hat{\mu}}{\partial u_l} = -\frac{\partial^2 E}{\partial x_i \partial u_l} \qquad i = [1, n]$$

$$\sum_{j=1}^{n} \frac{\partial^2 E}{\partial \mu \partial x_i} \frac{\partial \hat{x}_j}{\partial u_l} + \frac{\partial^2 E}{\partial \mu^2} \frac{\partial \hat{\mu}}{\partial u_l} = -\frac{\partial^2 E}{\partial \mu \partial u_l} \qquad l = [1, L]$$

These equations are produced by differentiation of (12.10) and (12.11) with respect to all components of u.

After obtaining $\partial\hat{\mu}/\partial u_l$ and $\partial\hat{x}_i/\partial u_l$, a procedure such as the Newton–Raphson scheme is used to minimize (12.9).

12.4. Conclusions

This chapter describes a theoretical method for solving multiattribute planning problems using nondifferentiable optimization techniques. A practical method of solution, based on the principle of maximizing the coefficient of consistency, has been developed, and used to link models for agriculture and water resources development.

This scheme was originally developed for the Generalized Regional Agriculture Model (GRAM); see Albegov et al. (1981) for further details. More recently, however, this methodology has been used for water-resource planning in other systems; the practical details will be documented elsewhere.

References

Albegov, M. (1978). The strategy of future regional economic growth. In M. Albegov (Editor), Proceedings of a Task Force Meeting on Regional Development, April 19–21, 1977. CP-78-1. International Institute for Applied Systems Analysis, Laxenburg, Austria.

Albegov, M., Kacprzyk, J., Orchard-Hays, W., Owsinski, J., and Straszak, A. (1981). Generalized Regional Agriculture Model (Application to a Region in Poland). International Institute for Applied Systems Analysis, Laxenburg, Austria.

Fiacco, A. V. and McCormick, G. P. (1968). Nonlinear Programming: SUMT. Wiley Publishing Co., New York.

Umnov, A. (1979). An Approach to Distributed Modeling. WP-79-48. International Institute for Applied Systems Analysis, Laxenburg, Austria.

PART V

STRUCTURAL ANALYSIS OF REGIONAL DEVELOPMENT

Regional Development Modeling: Theory and Practice
M. Albegov, A.E. Andersson and F. Snickars (editors)
North-Holland Publishing Company
© IIASA, 1982

Chapter 13

A REGIONAL STUDY OF THE DISTRIBUTION OF VINTAGES AND PROFITS OF INDUSTRIAL ESTABLISHMENTS: A STOCHASTIC TRANSITION MODEL

Börje Johansson*
Department of Economics, University of Gothenburg, Gothenburg (Sweden)

Ingvar Holmberg*
Demographic Research Institute, University of Gothenburg, Gothenburg (Sweden)

13.1. Introduction

This chapter is based on the work of a research project aimed at developing methods for the analysis of industrial structure and change.** These methods, which include programming approaches and input–output methods, have been designed to analyze the rigidities and vintage effects caused by the creation of fixed capital and other durables at fixed locations.

There have been difficulties associated with empirically verifying the vintage assumption and also in applying it in different situations. The fundamental reasons for this are that production units may be composed of equipment of different vintages and that renewal and ageing processes are closely linked. To cope with this problem we shall use a stochastic model that describes the transition of these units between vintage classes. This may be compared with the way in which Brownian processes are modeled in physics.

Structural change at the regional level is analyzed with the help of transition matrices. By analyzing steady-state scenarios for each region, it is possible to reach conclusions not only about structural change but also about the evolution of the process of structural change itself.

* The authors wish to thank Åke E. Andersson, Håkan Persson, and Henrik Lutzen for valuable comments at different stages of the work.
** The project was the joint work of Börje Johansson, Ingvar Holmberg, Ulf Strömqvist, and Bertil Marksjö. See also Johansson and Strömqvist (1980).

13.2. The vintage approach

The vintage approach to production analysis is based on the assumption that capital equipment embodies the technology of its date of construction. The most elaborate version of the vintage approach is the putty–clay model, in which the amount of labor associated with a unit of equipment is fixed by the design of the equipment. This model has been used to describe technologies in which factor combinations are variable ex ante (before investment takes place) and fixed ex post (when capital has actually been committed). This type of analysis was pioneered by, among others, Svennilson (1944), Johansen (1959, 1972), Salter (1960), Solow (1960, 1962), and Phelps (1962, 1963). An important feature of the vintage approach is the distinction between the technical, or physical, lifetime of the plant, and its economic lifetime, a distinction between physical depreciation and economic obsolescence. The latter may be empirically observed as a successively decreasing stream of quasirents, here called gross profits. This process reflects the economic ageing of the plant, which may continue until gross profits no longer cover prime costs.

New plants are being constructed all the time, taking advantage of the most recent technological developments to achieve greater efficiency than their predecessors. Competition from new plants is one of the main reasons for the decline in the gross profits of older plants.

The vintage assumption has been difficult to incorporate into empirical analysis. This is largely because a plant usually consists of production equipment of different vintages and new production techniques are often introduced at different times in different parts of the plant.

The vintage approach makes a clear distinction between the physical depreciation of existing capital in an establishment and the economic obsolescence of the establishment itself. We shall therefore consider a set M of indexes, τ, such that

$$M = \{\tau : \tau = 1, 2, \ldots, \eta\}$$

and such that increasing values of τ indicate increasing economic age. The units that we consider are industrial establishments. Each unit is assumed to consist of a number of subsystems which may be of different vintages. Investment bringing new capital equipment into the subsystems, causing reorganization, is assumed to make the establishment more economically viable and hence reduce its τ-value.

When using empirical observations to measure the τ-value of a single unit, we wish to distinguish between the long-run ageing process and the short-run deviations produced by market fluctuations. Furthermore, we also wish to have a model which incorporates the effect of the reinvestment pattern on the τ-value. We have chosen to use a stochastic model to describe and analyze these processes. The model treats the process of ageing and renewal as a transition process, i.e., as a process of structural change.

13.3. Profit share and economic age

The production of an establishment (production unit) over a given period (e.g., a year) may be described using the following variables:

$F = B + W$, where F denotes value added, B denotes gross profits, and W denotes wages

$W = wS$, where w denotes wage level and S denotes number of employees

$b = B/F$, where b denotes gross profit share

For each industrial sector we shall use the gross profit share to determine profit share classes, ξ_τ, in the following way:

$$\xi_\tau = \{b : (1.1 - 0.1\tau) \geqslant b > (1 - 0.1\tau)\} \quad \text{for all } \tau \in M \tag{13.1}$$

where each ξ may be regarded as a 10% interval on a profit share scale.
The age specifications may then be described as a mapping T, such that

$$T(b) = \tau \quad \text{if } b \in \xi_\tau \tag{13.2}$$

It is possible to interpret τ as a vintage classification representing some kind of average of the vintages of the subsystems in an establishment. However, it should be emphasized that τ really expresses the relative age of an establishment within a sector, since τ is affected by both internal and external factors. Capital formation in the unit constitutes the internal component and the combination of market development and capital formation in competing units constitutes the external component. Figure 13.1 illustrates how internal and external processes are assumed to affect the vintage classification τ.

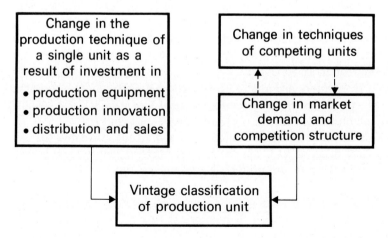

Figure 13.1. The internal and external factors affecting the vintage classification of a production unit.

13.4. Short-term and long-term changes in the vintage classification

At time t, the gross profits $B(t)$ of an establishment may be specified as

$$B(t) = C(t) + N(t)$$

where $N(t)$ denotes net profits and $C(t)$ the sum of fixed costs, such as overheads, repair and maintenance costs, and capital depreciation. Note that $C(t) > B(t)$ implies negative net profits. It is also obvious that a large value of $b(t)$ does not necessarily imply that net profitability is high.

The average establishment in the Swedish manufacturing industry is characterized by the following values (Johansson and Strömqvist, 1980): $b = 0.4$, $C/F = 0.24$, and $N/F = 0.16$. The fixed costs component $c = C/F$ is a reflection of past investment; it therefore represents sunk costs. The value of c may differ between sectors and between units in the same sector; $c(t)$ may also vary with economic age, since depreciation generally decreases over time while repair and maintenance costs increase. Renewal of an establishment also affects $c(t)$. It would be important to include all these aspects in a complete study of profitability, but it is not necessary in this study, which concentrates on the identification and measurement of structural change.

The value of $b(t)$ is influenced both by the long-run ageing/renewal process and by short-run market fluctuations. To describe these two factors we shall express $b(t)$ as follows:

$$b(t) = 1 - v(t)$$

$$v(t) = w(t)S(t)/q(t, t_0)F(t_0) \quad t \geqslant t_0$$

where $q(t, t_0)$ denotes the effect on value added of price changes between times t_0 and t. If $w(t)/q(t, t_0)$ is constant, or decreases over time, ageing will be caused simply by "physical depreciation", in that labor productivity, F/S, decreases over time in each unit. Economic obsolescence will therefore emerge as the result of a long-run increase in $w(t)/q(t, t_0)$. Assuming continued technical progress, capital formation in existing and new units will counteract this trend by decreasing the labor input requirements, S/F, in these units. However, this type of capital formation will actually reinforce the upward trend in $w(t)/q(t, t_0)$.

Short-run market fluctuations give rise to variations in b if $w(t)$ and $q(t, t_0)$ vary in any unsynchronized way around the long-run trend, and if $S(t)$ and $F(t)$ for each unit vary with different amplitudes. In general, $S(t)$ varies much less than $F(t)$.

Let $(\hat{w}, \hat{q})$ denote a long-run wage and price trend and let the associated changes in the gross profit share of a unit over time be $\hat{b}$, given that the unit is not renewed. Referring to (13.2) we may then express the τ-value mapping as follows:

$$T[b(t)] = T[\hat{b}(t)] + \epsilon(t) \tag{13.3}$$

where $\epsilon(t)$ describes the effect of market fluctuations at time t. For a given trend $(\hat{w}, \hat{q})$, we may determine time intervals, Δt, such that $T[\hat{b}(t)] = \tau$ and $T[\hat{b}(t + \Delta t)] = \tau + 1$. This represents the trend-related transition of a unit between two vintage classifications.

Table 13.1. Short-term variation of b-values for Swedish industry, 1968–1977.[a]

Vintage classification of establishments in year t	Proportion of establishments that change one class or less between year t and $(t + 1)$	Proportion of value added in establishments that change one class or less between year t and $(t + 1)$
1	94	98
2	76	90
3	70	88
4	70	82
5	70	84
6	71	80
7	71	80
8	69	75
9	63	67
10	51	45
11	43	43
12	52	43
13	55	33

[a] τ is related to b by eqn. (13.1): $\tau = 1$ if $b > 0.9$, $\tau = 13$ if $-0.2 > b$.

Observation of the values of b for groups of establishments will provide reliable information about the economic age of each group only if $\epsilon(t)$ is small relative to $T[\hat{b}(t)]$. If the expected value of $\epsilon(t)$ is small and if $\epsilon(t)$ also exhibits a small variance, $b(t)$ will remain almost constant over the short term. As can be seen from Table 13.1, there is not much change when the value of τ is low; the variation increases as the value of τ increases. This is exactly what we should expect given the arguments outlined above.

13.5. Transition arrays and transition probabilities

Consider an algebraic scheme $\mathbf{Y}(t) = \{y_{ij}(t)\}$, such that $i, j = 1, 2, \ldots, n + 1$. Here y_{ij} is a vector of components describing the properties of establishments that are characterized by age index $i \in M^e$ in year t and index $j \in M^o$ in year $t + 1$, where $M^e = \{M, e\}$, $M^o = \{M, o\}$, and e and o denote the entry and exit, respectively, of an establishment to and from the market.

An element, y_{ij}, of the transition array $\mathbf{Y}$ may be expressed as follows:

$$y_{ij} = [a_{ij}, l_{ij}, \omega_{ij}, \pi_{ij}, \ldots] \tag{13.4}$$

where a_{ij} is the proportion of all establishments characterized by (i, j)
            l_{ij} is the proportion of all jobs characterized by (i, j)
            ω_{ij} is the proportion of the wage sum characterized by (i, j)
            π_{ij} is the proportion of value added characterized by (i, j)

Let x_{ij} be any of the four variables specified in (13.4). The following marginal distributions may then be defined:

$$x_i(t) = \sum_{j \in M^o} x_{ij}(t) \quad i \in M$$

$$x_i(t+1) = \sum_{j \in M^e} x_{ji}(t) \quad i \in M$$

Now consider eqn. (13.3), which is assumed to refer to a single unit, and consider a set of establishments, each of which satisfies $T[b(t)] = \tau$, $\tau \in M$, in year t. Each of these establishments will have a specific (individual) long-run transition path and, in addition, an individual ϵ-term in year $t + 1$. This suggests the formulation

$$h_{ij}(t) = x_{ij}(t)/x_i(t) \quad j \neq i$$
$$h_{ii}(t) = [x_{ii}(t) + x_{ie}(t) - x_{io}(t)]/x_i(t) \tag{13.5}$$

where h_{ij} denotes the probability of transition from class i to class j. Note that h_{ii} also incorporates the probabilities that new units will enter the market in class i, and that existing units will leave the market from class i.

From (13.5), is is clear that $x_{ij}(t) = x_i(t-1)h_{ij}(t)$. This is an important relationship since we shall use the h_{ij} to characterize the process of structural change. We therefore introduce transition matrices $\mathbf{H} = \{h_{ij}\}$ and form a system $\{\mathbf{H}^a, \mathbf{H}^l, \mathbf{H}^\omega, \mathbf{H}^\pi\}$, where the indexes a, l, ω, and π refer to the notation used in (13.4). Since each of the four matrices is assumed to be nonnegative, their incidence matrices will be identical by construction. In other words: $a_{ij} = 0 \Rightarrow l_{ij} = \omega_{ij} = \pi_{ij} = 0$, and $a_{ij} > 0 \Rightarrow l_{ij}, \omega_{ij}, \pi_{ij} > 0$. This follows from (13.4). We may therefore unambiguously use $\mathbf{H}$ to denote any of the four matrices listed above.

13.6. Ergodicity and steady-state solutions

Consider a sequence $\{\mathbf{H}(t) : t = 0, 1, \ldots\}$ of transition matrices for an industrial sector in a specific region. Each of these matrices gives a momentary picture of the structural change taking place at time t, and the differences between the matrices reflect the evolution of the structural change process over a longer period of time. The structural change at time t may be characterized by the steady-state solution of the matrix $\mathbf{H}(t)$; the evolution of structural change may be characterized by the corresponding change in the steady-state solutions. In this section we shall outline the theoretical background to this approach, and in subsequent sections we shall illustrate and extend the method.

Let $T = \{0, 1, \ldots T_o\}$ be a sequence of years. The structural change taking place between 0 and T_o is characterized by the steady-state solution of the average matrix

$$\mathbf{H} = \frac{1}{T_o + 1} [\mathbf{H}(0) + \ldots + \mathbf{H}(T_o)] \tag{13.6}$$

To ensure that the method implied by (13.6) is feasible, we shall make two assumptions. Let $t = 0$ denote the initial time (year) and let $x(0) = [x_1(0), x_2(0), \ldots, x_n(0)]$ be a normalized initial distribution. The first assumption is then in three parts:

1(a). $x_i(0) \geqslant 0$ for $i \in M$, and $\sum_{i \in M} x_i(0) = 1$ (13.7)

1(b). $h_{ij}(t) \geqslant 0$, and $\sum_{j \in M} h_{ij}(t) > 0$ (13.8)

1(c). The matrix H defined in (13.6) is assumed to represent the transition process during the period

$$T = \{0, 1, \ldots, T_o\} \quad T_o \geqslant 0$$ (13.9)

Condition (13.7) is a natural convention, since $x(0)$ represents a frequency distribution that has been normalized over the index set M. Condition (13.8) states that transition probabilities must be nonnegative. In addition, (13.8) assumes that $x_{ii} + x_{ie} \geqslant x_{io}$, which imposes a constraint on the exit process.[†] Condition (13.9) forms the basis of the method described in (13.6). Using this assumption, it is possible to compare the structural changes taking place in a number of subperiods within T.

In order to ensure the existence of steady-state solutions we also have to impose constraints on the matrix H. Our second assumption is based on observation rather than theory:

2. H, as defined in (13.6), is a primitive matrix. This means that there is a $k > 0$ such that

$$H^k > 0$$ (13.10)

Given assumptions (13.7)–(13.10), three important properties may be deduced.

Property 1. Assumptions (13.7)–(13.10) imply that there exists a maximum eigenvalue λ of H such that λ is real and strictly positive. The left and right eigenvectors associated with λ are unique up to constant multiples. This result is a special case of the Perron–Frobenius Theorem; for a proof, see Seneta (1973, pp. 1–6).

Property 2.[††] Assumptions (13.7)–(13.10) ensure that the matrix H possesses strong ergodicity. This means that the normalized rows of H^k tend to the same value as the normalized left eigenvector of H as $k \to \infty$.

Let us now introduce the normalized version of $x(t)$:

[†] Note that since h_{ii} is constructed as the difference between positive probabilities, the constraint on the frequency of exit from each class i is not based on logical arguments but on empirical observations; see Johansson and Strömqvist (1980).

[††] Property 2 is examined in more detail in Appendix 13A, and a more general formulation is given in Section 13.8. A proof is given in Seneta (1973, pp. 73–75).

$$z(t) = [z_1(t), z_2(t), \ldots, z_n(t)]$$

$$z_i(t) = x_i(t)/\sum x_i(t) \tag{13.11}$$

The transition process associated with $\mathbf{H}$ may be described as

$$x(t) = x(t-1)\mathbf{H} = x(0)\mathbf{H}^t \tag{13.12}$$

A steady-state solution to this process is denoted by $\overset{*}{x}$ and can be characterized as follows:

$$\left. \begin{array}{l} z(t)\mathbf{H} \to \lambda\overset{*}{x} \\ z(t) \to \overset{*}{x} \end{array} \right\} \quad \text{as } t \to \infty \tag{13.13}$$

$$\sum \overset{*}{x}_i = 1$$

Property 3. Assume that $x(0)$ and $\mathbf{H}$ satisfy (13.7)–(13.10). Process (13.12) will then converge so that $z(t) \to \overset{*}{x}$ and $z(t)\mathbf{H} \to \lambda\overset{*}{x}$ as $t \to \infty$, where λ is the maximum eigenvalue of $\mathbf{H}$ and where $\overset{*}{x}$ is the normalized left eigenvector associated with λ. Property 3 is proved in Appendix 13A.

The solution described in (13.13) corresponds to a steady-state scenario for the system $\{\mathbf{H}^a, \mathbf{H}^l, \mathbf{H}^\omega, \mathbf{H}^\pi\}$. A scenario of this type may be written

$$\overset{*}{\chi} = [\overset{*}{a}, \overset{*}{l}, \overset{*}{\omega}, \overset{*}{\pi}] \tag{13.14}$$

Each solution of the type (13.14) is associated with four positive eigenvalues $\lambda_a, \lambda_l, \lambda_\omega$, and λ_π, such that: λ_a is the eigenvalue of $\mathbf{H}^a$, where $(\lambda_a - 1)$ describes the annual net increase in the number of establishments; λ_l is the eigenvalue of $\mathbf{H}^l$, where $(\lambda_l - 1)$ describes the annual net increase in the number of workers employed as a *direct* effect of the market exit and entry of production units; λ_ω is the eigenvalue of $\mathbf{H}^\omega$, where $(\lambda_\omega - 1)$ describes the annual net increase in the wage sum caused by the market exit and entry of production units; and λ_π is the eigenvalue of $\mathbf{H}^\pi$, where $(\lambda_\pi - 1)$ describes the annual net increase in the value added caused by the market exit and entry of production units.

13.7. Structural change scenarios for six regions

Consider the transition array $\mathbf{Y}$ specified in (13.4). Let the indexes of y_{ij} be constrained to the set M, which means that production units cannot enter or leave the market. Let $\mathbf{A} = \{a_{ij}: i, j \in M\}$ be a "truncated" matrix of this type; the changes in this matrix between 1968 and 1976 have been examined for the Swedish manufacturing industry. The pairwise correlation between the matrices $\mathbf{A}(t)$, $\mathbf{A}(t+1), \ldots,$ $\mathbf{A}(t+8)$ varies between 0.96 and 0.99. Analogous comparisons for the matrices $\{l_{ij}\}$, $\{\omega_{ij}\}$, and $\{\pi_{ij}\}$ also reveal a high degree of correlation (Johansson and Strömqvist, 1980). This means that when production units cannot enter or leave the market, the first four components of each "cell vector" y_{ij} remain almost constant. Hence, any significant alteration in the process of structural change at the national level must, to a large extent, be explained by market exit and entry processes.

We shall now illustrate the use of scenarios of the type specified in (13.14). We divide Sweden into six regions and compare three different scenarios for each region. The matrices (**H**) associated with each scenario have been estimated for three periods: I = 1968–1974, II = 1974–1977, and III = 1968–1977; the corresponding scenarios are denoted by $\overset{*}{\chi}_I$, $\overset{*}{\chi}_{II}$, and $\overset{*}{\chi}_{III}$. Each of these scenarios describes the long-run effects of the transition pattern observed over the corresponding period. A difference between $\overset{*}{\chi}_I$ and $\overset{*}{\chi}_{II}$ will then reveal a difference in the process of structural change in these two periods.

Our empirical results may be illustrated by focusing on two regions in which the composition of the industrial sectors is quite different. The two regions, Mid-Sweden and Eastern Sweden, also have different patterns of structural change, implying different market exit and entry processes.

Vectors $\overset{*}{\pi}_I$ and $\overset{*}{\pi}_{II}$ are associated with scenarios $\overset{*}{\chi}_I$ and $\overset{*}{\chi}_{II}$, respectively. These two vectors may be used to construct Salter-type curves (Salter, 1960). Each curve shows how the cost of labor (wage sum) increases as a percentage of value added as we gradually incorporate establishments with lower and lower gross profit shares. In this way we obtain a marginal cost curve which may be interpreted as a quasisupply curve.[†] The quasisupply curves of Mid-Sweden are shown in Figure 13.2(a) and those of Eastern Sweden are plotted in Figure 13.2(b). The broken line in each figure indicates

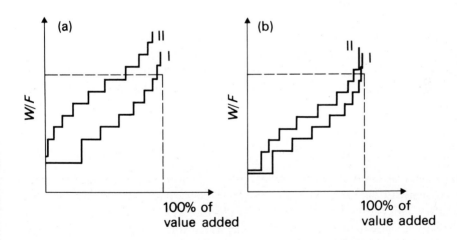

Figure 13.2. Marginal cost curves, which may be interpreted as quasisupply curves, for (a) Mid-Sweden and (b) Eastern Sweden. Curves I cover the period 1968–1974, curves II the period 1974–1977, and the broken lines represent zero gross profits.

[†]The prefix "quasi" is used to indicate that the curve describes cost conditions rather than actual supply behavior.

Table 13.2. Percentage changes in the productivity and wage share of six regions, produced as a direct effect of market exit and entry processes.

Scenario[a]	Eastern Sweden	Southeastern Sweden	Southern Sweden	Western Sweden	Mid-Sweden	Northern Sweden
Productivity						
I	+ 3.3	+ 1.4	+ 0.4	+ 0.6	+ 0.6	+ 0.1
II	+ 0.7	+ 0.7	+ 0.5	+ 0.6	+ 0.2	+ 0.1
III	+ 1.7	+ 0.9	+ 0.5	+ 0.7	+ 0.5	− 0.1
Wage share						
I	− 3.0	− 1.0	+ 1.4	− 0.1	− 0.3	+ 0.0
II	− 0.5	− 0.5	− 0.1	− 0.2	+ 0.1	+ 0.0
III	− 1.3	− 0.7	+ 0.6	− 0.2	− 0.3	+ 0.2

[a]Scenario I is based on the transition pattern over the period 1968–1974; II on the pattern over the period 1974–1977; and III on the average pattern over the period 1968–1977.

zero gross profits. When the wage sum of an establishment crosses this line the establishment makes negative gross profits, i.e., a loss. It may therefore be argued that if this is a permanent situation the establishment should be closed down.

Quasisupply curves constructed in this way have one particular advantage; they are normalized in such a way that regions may be compared in a straightforward manner. For example, it is obvious from the figures that Eastern Sweden has been affected much less than Mid-Sweden by the post-1974 alteration in the structural change process. Comparisons of this type may be used as a basis for analyzing the distribution of investment funds between regions.

The overall effects of market exit and entry may be calculated from the eigenvalues defined at the end of Section 13.6. Of special interest is the effect of exit and entry on the annual change in productivity, $\lambda(\pi, l)$, and wage share, $\lambda(\omega, \pi)$. These are defined as

$$\lambda(\pi, l) = (\lambda_\pi - \lambda_l)/\lambda_l$$

$$\lambda(\omega, \pi) = (\lambda_\omega - \lambda_\pi)/\lambda_\pi$$

and are listed (for the three scenarios) in Table 13.2. Note that the average change in labor productivity between 1968 and 1977 lies between 2% and 3% (for industry as a whole). This means that the exit/entry process has a significant effect on total productivity, in spite of the fact that the value added corresponding to units entering the market is only about 1% of the total value added in industry as a whole.

13.8. Changes in sectoral composition

The sectoral composition of industry in Sweden varies quite widely across the country. The two regions discussed earlier demonstrate this point clearly: industry in Mid-Sweden is largely composed of sectors with high capital coefficients (e.g., the paper and pulp industry, the iron and steel industry); these sectors represent a much smaller proportion of the industry in Eastern Sweden.

It is relatively easy to deduce the transition patterns of each industrial sector at the national level; it is more difficult at the regional level because information about individual establishments may not always be available. We shall now consider two indirect methods of obtaining information about changes in the sectoral structure of regional industry. Both methods are discussed in terms of the distribution of value added. First, we should note that it is possible, for each region r and sector s, to estimate the vector[†]

$$\pi^{rs}(t) = [\pi_1^{rs}(t), \pi_2^{rs}(t), \ldots, \pi_n^{rs}(t)] \quad \text{for each year } t$$

Let $\mathbf{H}_r$ be a transition matrix (in terms of value added) for the manufacturing industry in region r, and let $\mathbf{H}_j$ be a nation-wide transition matrix for sector $j \in J$, where $J = \{1, 2, \ldots, J_0\}$ is an index set of sectors. The steady-state solutions associated with these matrices are $\overset{*}{\pi}{}^r$ and $\overset{*}{\pi}{}^j$, respectively. Let $\gamma^r(t) = [\gamma^{r1}(t), \gamma^{r2}(t), \ldots, \gamma^{rJ_0}(t)]$ be a vector describing changes in the sectoral composition of industry in region r, where $\gamma^r(t)$ is normalized such that $\Sigma_j \gamma^{rj}(t) = 1$ for all $t \geqslant 0$. We may now form the following hypothetical scenarios:

$$\underline{\overset{*}{\pi}}{}^r = \sum_{j \in J} \gamma^{rj}(0) \overset{*}{\pi}{}^j$$

(13.15)

$$\underline{\underline{\overset{*}{\pi}}}{}^r = \sum_{j \in J} \overset{*}{\gamma}{}^{rj} \overset{*}{\pi}{}^j$$

where $\overset{*}{\gamma}{}^r$ is the sectoral composition to which region r is converging.[††] We may use (13.15) to form two vector differences that indicate the extent to which the steady-state solution depends on the sectoral composition of industry in region r:

$$\Delta \underline{\overset{*}{\pi}}{}^r = \overset{*}{\pi}{}^r - \underline{\overset{*}{\pi}}{}^r$$

$$\Delta \underline{\underline{\overset{*}{\pi}}}{}^r = \overset{*}{\pi}{}^r - \underline{\underline{\overset{*}{\pi}}}{}^r$$

It is possible to use an alternative method if we construct the regional matrices

$$\tilde{\mathbf{H}}_r = \sum_{j \in J} \gamma^{jr}(0) \mathbf{H}_j$$

(13.16)

$$\tilde{\tilde{\mathbf{H}}}_r(t) = \sum_{j \in J} \gamma^{jr}(t) \mathbf{H}_j$$

[†] Aggregate data about total value added in entering and exiting establishments are also available for each region.
[††] Whether $\gamma^r(t)$ converges or not depends partly on the method used to estimate the function.

The steady-state solutions of $\tilde{\mathbf{H}}_r$ and $\tilde{\tilde{\mathbf{H}}}_r(t)$ are $\tilde{\pi}^r$ and $\tilde{\tilde{\pi}}^r$, respectively. We then obtain the differences $\Delta\tilde{\pi}^r = \overset{*}{\pi}{}^r - \tilde{\pi}^r$ and $\Delta\tilde{\tilde{\pi}}^r = \overset{*}{\pi}{}^r - \tilde{\tilde{\pi}}^r$. The existence of $\tilde{\pi}^r$ follows from Property 3 in Section 13.6 and the existence of $\tilde{\tilde{\pi}}^r$ follows from Property 4 below.[†]
Property 4. Assume that $\mathbf{H}_j$ and $\pi^r(0)$ satisfy the conditions of Property 3. The system then converges to $\tilde{\tilde{\pi}}^r$ if, for each j, $\gamma^{jr}(t)$ converges to $\overset{*}{\gamma}{}^{jr} \geqslant 0$.

13.9. Conclusions

Structural change may be described by changes in an array of elements $\mathbf{Y}(t)$ of the type specified in (13.4); the array should also include information about gross profits, number of employees, etc., for each position (i, j). Structural change may, for example, be represented by the path $\{l(t), \omega(t), \pi(t)\}$. It has been argued that a vintage classification based on b-values may be complemented by specifying the transition pattern of the class to which the b-value belongs. In this way we have obtained a method for classifying plants that may also be used when the sectors are aggregated. The same approach also makes it possible to analyze the evolution of the process of structural change itself. This technique should prove very useful in studies of regional development in the future.

References

Johansen, L. (1959). Substitution versus fixed production coefficients in the theory of economic growth: a synthesis. Econometrica, 27: 157–177.

Johansen, L. (1972). Production Functions. North-Holland Publishing Co., Amsterdam.

Johansson, B. and Strömqvist, U. (1980). Vinster och sysselsättning i svensk ekonomi (Profits and employment in Swedish industry). SIND, Vol. 2, Utredning Håu Statens Industriverk, Liber Förlag, Stockholm.

Phelps, E. S. (1962). The new view of investment: a neoclassical analysis. Quarterly Journal of Economics, 83: 548–567.

Phelps, E. S. (1963). Substitution, fixed proportions, growth and distribution. International Economic Review, 4: 265–288.

Salter, W. E. G. (1960). Productivity and Technical Change. Cambridge University Press, London.

Seneta, E. (1973). Non-Negative Matrices. George Allen and Unwin, London.

Solow, R. M. (1960). Investment and technical progress. In K. J. Arrow, S. Karlin, and P. Suppes (Editors), Mathematical Methods in the Social Sciences 1959. Stanford University Press, Stanford, California. ·

Solow, R. M. (1962). Substitution and fixed proportions in the theory of capital. Review of Economic Studies, 29: 207–218.

Svennilson, I. (1944). Industriarbetets växande avkastning i belysning av svenska erfarenheter (Growing industrial productivity in view of Swedish experiences). In Studier i Ekonomi och Historia tillágnade Eli F. Heckscher (Studies in Economics and History in Honor of Eli F. Heckscher). Almqvist and Wicksell, Uppsala, pp. 238–252.

[†]An even less restricted condition showing how the proposition may be proved is given in Appendix 13A.

Appendix 13A. Mathematical background

This appendix explains in more detail the mathematical basis of Properties 2, 3, and 4. Equations (13.7)–(13.10) assume that H is a nonnegative primitive matrix and that the initial distribution $x(0)$ satisfies $\Sigma x_i(0) = 1$, and $x_i(0) \geqslant 0$. From this, it follows that:

$$\left. \begin{array}{l} z(t)H \;\to\; \lambda \overset{*}{x} \\[6pt] z(t) \;\to\; \overset{*}{x} \\[6pt] H^k \lambda^{-k} \to\; R \end{array} \right\} \quad \text{as } t \to \infty \tag{13.17}$$

where $\Sigma z_i(t) = 1$ as specified in (13.11) and $\Sigma \overset{*}{x}_i = 1$ as specified in (13.13). Further, λ is the maximum eigenvalue of H, and R is a matrix satisfying the following conditions: R is the vector product of $\overset{*}{u}$ and $\overset{*}{x}$ so that $R = \overset{*}{u}\overset{*}{x}$, where $\overset{*}{u}$ is a column vector and $\overset{*}{x}$ a row vector; $\overset{*}{x}$ is the positive left Perron–Frobenius eigenvector of H, normalized such that $\Sigma \overset{*}{x}_i = 1$; and $\overset{*}{u}$ is the positive right eigenvector of H, normalized such that the scalar product $\overset{**}{xu} = 1$.

Restating the last part of (13.17) (proved in Seneta, 1973, pp. 7, 8):

$$H^k \lambda^{-k} \to R \text{ as } t \to \infty \tag{13.18}$$

This leads to

$$x(0)H^k \lambda^{-k} \to x(0)R \text{ as } t \to \infty \tag{13.19}$$

which gives

$$\lim_{t \to \infty} x(t)\lambda^{-t} = x(0)R = \overset{*}{q} \tag{13.20}$$

From the properties of R we have

$$\overset{*}{q}/(\overset{**}{qu}) = \overset{*}{x} \tag{13.21}$$

which implies that $\overset{*}{q}_i/\overset{*}{q}_j = \overset{*}{x}_i/\overset{*}{x}_j$ for all (i, j). Since $\overset{*}{q}H = \lambda \overset{*}{q}$, (13.18)–(13.21) imply that the conditions of (13.17) are fulfilled. It is now possible to select any row $(r_1, r_2, \ldots, r_n) = r$ and normalize it so that $\Sigma r_i = 1$. Then, according to (13.20) and (13.21), $\overset{*}{x} = r$, as stated in Property 3.

Now consider the matrix $\overset{\approx}{H}(t) = \overset{\approx}{H}_r(t)$ in (13.16). It may be shown (Seneta, 1973, pp. 73–75) that the conditions of (13.17) are fulfilled if $\overset{\approx}{H}(t)$ has no row composed only of zeros, and $\overset{\approx}{H}(t) \to \overset{*}{\overset{\frown}{H}}$ as $t \to \infty$, where $\overset{\frown}{H}$ is primitive. Note that this assumes that $x(0) \geqslant 0$ and $\Sigma x_i(0) = 1$.

Finally, consider $\overset{\approx}{H}(t) = \Sigma_j \gamma^{jr}(t)H_j$. If every $\gamma^{jr}(t) \to \overset{*}{\gamma}^{jr}$ as $t \to \infty$, $\overset{\approx}{H}(t)$ will converge to $\overset{*}{\overset{\frown}{H}}$, since we have assumed that every H_j is primitive.

Regional Development Modeling: Theory and Practice
M. Albegov, A.E. Andersson and F. Snickars (editors)
North-Holland Publishing Company
© IIASA, 1982

Chapter 14

A LINEAR ACTIVITY ANALYSIS OF LABOR ALLOCATION AND TRAINING SCHEMES

Paolo Caravani, Agostino La Bella, and Alberto Paoluzzi
Istituto Analisi dei Sistemi e Informatica, National Research Council, Rome (Italy)

14.1. Introduction

Much of the current debate over efficient allocation of scarce resources involves a revised discussion of social sources of profit (Commoner, 1975; Slessner, 1978; Knox, 1979). Inefficient allocation often results in so-called external diseconomies such as ecological damage, transportation congestion, communication costs, and the like. As these factors are coming to play a dominant role in profit formation and growth, it seems appropriate that their effects should be explicitly introduced into economic analysis. In this chapter we address the problem of inefficient allocation of labor to the vacancies occurring in a regional labor market.

As has been observed by several investigators (Doeringer and Piore, 1971; Gordon, 1972; Valkenburg and Vissers, 1978), there often exist structural imbalances between labor demand and supply at a local level, even when there is an excess labor supply at the national level. This has led to the introduction of several theories, some reinforcing each other, others overlapping, and often conflicting.

Neoclassical theory postulates an automatic saturation of the labor market sustained by wage-rate adjustment to the value of marginal labor product. Thus, excess supply is met by lower wages, excess demand by higher wages. However, a closer look into the modern industrial infrastructure soon reveals the existence of jobs where qualification requirements are of greater importance than wage levels, i.e., lack of skill cannot always be traded off against lower wages on the management side. A similar rigidity is observable on the supply side. Relatively high wages in conjunction with unsatisfactory working conditions (instability, environment, shift work, etc.) may actually depress the labor supply or simply fail to attract labor.

Recognizing this feature, some authors introduced the conceptual device of a "dual labor market". This comprises a primary market linked to large, capital-intensive industries providing high wages, job stability, and excellent working conditions, and a secondary market with all the opposite features. Within each sector, the laws of marginal theory still hold, but communication and information flow between sectors — if present — is one-directional, i.e., from primary to secondary (Doeringer and Piore, 1971; Gordon, 1972; Valkenburg and Vissers, 1978; Edwards et al., 1975). In order for this theory to be workable at a regional level, it is necessary to assume spatial homogeneity within each market. While this may be true for certain advanced industrial cities, where large high-wage industries have moved out of the city centers and small labor-intensive industries have remained in the center, it appears questionable to postulate homogeneity in an integrated interregional analysis. Different regions may have very different development patterns and the distinction between primary and secondary markets may lose much of its explanatory value when projected onto a spatial dimension.

An extension of the dual labor market concept was recently introduced by Freiburghaus (1978), Piore (1975), and Osterman (1975) in the form of a multi-segmented labor market. Such a finely articulated description may well be warranted on a microscale for sociological assessment of behavioral complexity. From the standpoint of economic analysis, however, we see no reason to pursue this methodological trend beyond an attempt to recover within each segment of the labor market the validity of marginal theory. Being unwilling to enter the dispute here by pledging allegiance to a new theory or a new definition on this sociologically overworked subject, we will rather assume, along with Malinvaud (1978), a Keynesian framework of downward sticking wages and the existence of equilibrium under partial employment. Our purpose in this chapter is to propose a method for assessing the economic "disvalue" arising from the imperfection of labor market mechanisms in combining supply and demand at a regional level.

Imbalances in local labor markets take the form of unsatisfactory matching of skills within the same region as well as the existence of redundant supply and unsatisfied demand among the regions. It is assumed that each of these imbalances can be assigned a social disvalue in terms of retraining and/or commuting–migrating costs. This assumption permits the problem of allocating labor to vacancies in the short-run to be formulated as the minimization of the economic disvalue of market imperfection. This minimization problem will be translated into the standard format of a linear programming problem. The economic interpretation of the associated dual problem permits the social cost–benefit* of creating a job in a given region and skill to be evaluated, as well as the marginal value–disvalue of a new unemployed person entering the labor market. It is hoped that the results of this kind of analysis will offer valuable information for an active regional policy on both the supply and the demand side.

14.2. A typology of unemployment

The main feature of our approach is to regard unemployment in terms of two fundamental dimensions, skill and space. Within the same region labor demand and supply

---

* Note that the term "cost–benefit" is used here in a more technical sense than in traditional cost–benefit analysis.

may not match because of skill discrepancies, although some of these discrepancies could be removed by allowing supply in one region to meet demand in another. This viewpoint requires, of course, a rather precise definition of a labor market region. Several definitions have been proposed in this regard; see for instance Slessner (1978) and Booth and Hyman (1978). For our purposes it will be sufficient to introduce the following "closure" criterion: labor market regions are to be self-contained with respect to home–work commuting. On this basis it is possible to categorize unemployment under the five headings suggested by Gleave and Palmer (1979) and freely rephrased here:

1. Frictional: unemployed workers who could be employed in the same labor market (region) and occupation (skill) because sufficient vacancies exist.
2. Spatial–structural: unemployed workers who could find a job in the same occupation but in a different labor market.
3. Occupational–structural: unemployed workers who could gain employment within their region if they could be retrained and learn another skill.
4. Spatial–occupational–structural: unemployed workers who would need to change both region and occupation to obtain employment.
5. Demand-deficient: persons unemployed because of an excess of supply over the total number of vacancies.

In Gleave and Palmer (1979) a partition algorithm is proposed to classify the unemployed by region and skill in the above five categories. This procedure, however, suffers from two major shortcomings. First, the solution obtained is dependent on the order in which the single classes are defined. Second, the proportional criterion used in the assignment of vacancies to unemployed persons seems devoid of theoretical justification.

A theoretical foundation for a classification procedure should be sought under the assumption of a rational allocation of unemployed persons to vacancies. Let us assume throughout that the labor market comprises N regions and M skills. Assume also that a social utility function*

$$U = U(x, c, g, w) \tag{14.1}$$

is assigned, where:

$x = \{x_{ir}^{js}\}$ is the vector of reallocated unemployed persons (x_{ir}^{js} is the number of unemployed persons having skill i in region r reallocated to skill j in region s)

$c = \{c_{rs}\}$ is a cost vector associated with movement of labor (from region r to s)

---

* The term "social utility function" will be freely used throughout this chapter as a surrogate for "objective function", "performance index", "optimality criterion", etc. Although it will be occasionally abbreviated as "utility function", it must be stressed that no relationship between utility and individual preferences is investigated in the present context.

$g = \{g_{ij}\}$ is a cost vector associated with retraining (from skill i to j)
$w = \{w_{ir}\}$ is the vector of wage distribution over skills and regions

with $i, j \in \{1, 2, \ldots, M\}$, and $r, s \in \{1, 2, \ldots, N\}$.

A rational reallocation of unemployed persons to vacancies requires — in principle — the solution of the following optimization problem:

$$\max_{x} U(x, c, g, w) \tag{14.2}$$

subject to

$$\sum_{j} \sum_{s} x_{ir}^{js} \leqslant D_{ir} \qquad (\forall i, r) \tag{14.3}$$

$$\sum_{i} \sum_{r} x_{ir}^{js} = V_{js} \qquad (\forall j, s) \tag{14.4}$$

$$x_{ir}^{js} \geqslant 0$$

where

D_{ir} is unemployment in sector i and region r
V_{js} is vacancies in sector j and region s

Note that constraints (14.3) and (14.4) imply that

$$\sum_{i} \sum_{r} D_{ir} \geqslant \sum_{j} \sum_{s} V_{js}$$

i.e., that total unemployment, over all regions, exceeds the total number of vacancies. Assuming that conditions for a solution of problem (14.2)–(14.4) exist, let $^*x_{ir}^{js}$ be the optimal solution of the problem. Then, the spatial distribution of unemployment according to the five categories mentioned above results in the following definitions:

1. Frictional: $\{^*x_{ir}^{ir}\}$
2. Spatial–structural: $\{^*x_{ir}^{is}\}$, where $s \neq r$
3. Occupational–structural: $\{^*x_{ir}^{jr}\}$, where $j \neq i$
4. Spatial–occupational–structural: $\{^*x_{ir}^{js}\}$, where $j \neq i$, and $s \neq r$
5. Demand-deficient: $\{e_{ir}\}$, where $e_{ir} = D_{ir} - \Sigma_j \Sigma_s {}^*x_{ir}^{js}$.

14.3. Optimal labor assignment and workplace location

The classification of unemployment requires a precise definition of the utility function. We will assume that social utility is measured by the increase in GNP made possible by the reallocation of the labor force, after deduction of relocation costs. Therefore, the utility function will be assumed to be

$$U = \Delta Q - C$$

where ΔQ is the total production increase and C the total relocation cost.

Assuming that Q has differentiable dependence on capital K, labor l, and human capital H, then

$$Q = Q(K, l, H)$$

where labor is regarded as a vector indexed by skill and region.

Over the short term we may assume constant capital. Therefore, to a first-order approximation

$$\Delta Q = \sum_{js} (\partial Q/\partial l_{js})\Delta l_{js} + (\partial Q/\partial H)\Delta H \qquad (14.5)$$

Letting

$$(\partial Q/\partial l_{js}) \triangleq \beta w_{js}$$

be the prevailing wage for skill j in region s, then

$$\Delta l_{js} = \sum_{ir} x_{ir}^{js}$$

The increase in human capital can be measured by the increase in potential wage rates; that is

$$\Delta H = \sum_{ijrs} (w_{js} - w_{ir})x_{ir}^{js}$$

Assuming constant marginal productivity of human capital

$$\partial Q/\partial H = \alpha$$

we obtain

$$\Delta Q = \sum_{ijrs} [(\beta + \alpha)w_{js} - \alpha w_{ir}]x_{ir}^{js}$$

The cost term C in the utility function is the sum of the costs of movement and retraining

$$C = \sum_{ijrs} (c_{rs} + t_{ij})x_{ir}^{js} = \sum_{ijrs} [\beta w_{js} + \alpha(w_{js} - w_{ir})]x_{ir}^{js}$$

where c_{rs} and t_{ij} are the prevailing unit costs for moving from r to s and for retraining from i to j.

In compact vector notation, the utility function takes the linear form

$$U = \gamma^T x$$

where all cost components have been arranged in the vector

$$\gamma = \{\gamma_{ir}^{js}\}; \qquad \gamma_{ir}^{js} \triangleq (\beta + \alpha)w_{js} - \alpha w_{ir} - c_{rs} - t_{ij} \tag{14.6}$$

Note that this notation requires the components of vectors γ and x to be arranged in the same order. To avoid ambiguity, we will state that component x_{ir}^{js} occupies the position[†]

$$k \triangleq 1 + (s-1) + (j-1)N + (r-1)NM + (i-1)NM^2 \tag{14.7}$$

in vector x (and similarly for the components of γ).

On these premises it is possible to formulate the optimization problem

$$\max \gamma^T x \tag{14.8}$$

subject to

$$Ax \leqslant D \tag{14.9}$$

$$Bx = V \tag{14.10}$$

$$x \geqslant 0$$

where A is an $(NM) \times (NM)^2$ matrix with the following structure (induced by (14.7))

$$A \triangleq \begin{bmatrix} 1^T 0^T \dots 0^T \\ 0^T 1^T \dots 0^T \\ \cdots \quad \cdots \\ 0^T 0^T \dots 1^T \end{bmatrix}; \qquad \begin{aligned} 1^T &= \{111 \dots 1\} \sim 1 \times NM \\ 0^T &= \{000 \dots 0\} \sim 1 \times NM \end{aligned}$$

B is an $(NM) \times (NM)^2$ matrix

$$B \triangleq [I|I| \dots |I]; \qquad I \sim (NM) \times (NM) \text{ identity matrix}$$

$$D = \{D_{ir}\}; \qquad V = \{V_{js}\}$$

and component D_{ir} occupies the position

$$k = 1 + (r-1) + (i-1)N = r + (i-1)N$$

in vector D (and likewise for V with r and i replaced by s and j, respectively).

It should also be noted that this is a linear programming problem. As matrix $C = [A|B]$ is unimodular and vectors V and D are integer-valued, any basic solution is also integer.

The solution x^* of problem (14.8)–(14.10) yields the optimum assignment of unemployed workers to vacancies.

[†] This is a standard way of storing multidimensional arrays in a one-dimensional vector.

14.4. Shadow prices of vacancies and unemployment

In this section we will attempt to assign a "market value" to the inefficient allocation of labor. It is clear from the foregoing discussion that different distributions of vacancies and unemployed persons result in different values for the utility function.

Inefficient allocations in the labor market yield a suboptimal value of the utility function. Therefore, it is reasonable to define the "inefficiency" of the labor market as the difference between the current and the optimal value of the utility function. However, we have so far been unable to disaggregate inefficiency in terms of its most significant components. Two major contributory factors are, on the supply side, the unproductiveness of a potential employee who has not been able to gain employment; and on the demand side, for a given production level, the higher (lower) relocation–retraining costs resulting from inappropriate (appropriate) vacancy assignment.

A formal definition of these concepts is now required. Let $y_{ir}^{(1)}$ be the contribution to inefficiency resulting from one unemployed person in region r having skill i, and $y_{js}^{(2)}$ be the contribution resulting from one unfilled vacancy in region s requiring skill j. While it is reasonable to assume $y_{ir}^{(1)} \geq 0$, no restriction will be placed on the sign of $y_{js}^{(2)}$. In this framework it is meaningless to consider the case in which a component of the reallocation cost γ exceeds the value of the corresponding component of the labor market inefficiency, otherwise there would be no reason for reallocation to take place. Therefore, the following constraint results:

$$y_{ir}^{(1)} + y_{js}^{(2)} \geq \gamma_{ir}^{js} \qquad (\forall i, j, r, s)$$

or, in matrix notation

$$A^T y^{(1)} + B^T y^{(2)} \geq \gamma \qquad (14.11)$$

Inefficiency will be at a minimum when the total

$$L = D^T y^{(1)} + V^T y^{(2)} \qquad (14.12)$$

is minimized on the set (14.11).

As might have been expected, problem (14.11)/(14.12) is the dual formulation of problem (14.8)–(14.10). Since the solution of the primal problem is bounded, the solution of this problem will also be bounded. The optimal solution y^* of problem (14.11)/(14.12) yields the shadow prices of unemployment and vacancies:

- $^*y_{ir}^{(1)}$ can be interpreted as the shadow cost of one unemployed person in region r having skill i
- $^*y_{js}^{(2)}$ can be interpreted as the shadow price of one vacancy in region s requiring skill j

The solution of the dual problem permits the cost vector γ to be disaggregated into a part $\{y_{ir}^{(1)}\}$ associated with unemployment (supply side) and a part $\{y_{js}^{(2)}\}$ associated with vacancies (demand side).

Furthermore, this allows a four-fold classification of labor market inefficiency to be established by means of the following comparative indices:

1. $\quad \bar{y}_r^{(1)} = \sum_i {}^*y_{ir}^{(1)} D_{ir} \Big/ \sum_i D_{ir}$

(Average inefficiency of labor market in region r, measured on the supply side.)

2. $\quad \bar{y}_i^{(1)} = \sum_r {}^*y_{ir}^{(1)} D_{ir} \Big/ \sum_r D_{ir}$

(Average inefficiency of labor market with respect to skill i, measured on the supply side.)

3. $\quad \bar{y}_r^{(2)} = \sum_i {}^*y_{ir}^{(2)} V_{ir} \Big/ \sum_i V_{ir}$

(Average inefficiency of labor market in region r, measured on the demand side.)

4. $\quad \bar{y}_i^{(2)} = \sum_r {}^*y_{ir}^{(2)} V_{ir} \Big/ \sum_r V_{ir}$

(Average inefficiency of labor market with respect to skill i, measured on the demand side.)

14.5. Optimal workplace location policy and training schemes

Assume that the reallocation of the labor force has been completed according to the scheme (14.8)–(14.10). It is obvious that, in each region, there will remain a number of unemployed persons, given by

$$\sum_i e_{ir} = \sum_i \left(D_{ir} - \sum_j \sum_s {}^*x_{ir}^{js} \right) \qquad (r = 1, 2, \ldots, N)$$

If a number W of new vacancies were subsequently made available, how should a policy-maker distribute them over all skills and over space to minimize inefficiency? Retaining the assumption of excess labor supply, let

$$W \leqslant \sum_i \sum_r e_{ir}$$

The unknown distribution of vacancies W_{ir} will have to satisfy the constraints

$$\sum_i W_{ir} \leqslant \sum_i e_{ir} \qquad (r = 1, 2, \ldots, N)$$

$$\sum_{ir} W_{ir} = W$$

$$W_{ir} \geqslant 0 \qquad (\forall i, r)$$

The optimality criterion we suggest will again be based on the concept of inefficiency. Using the inefficiency measure introduced in Section 14.4, we define an objective function

$$I = \sum_{ir} y_{ir}^{(2)} W_{ir}$$

Minimizing this index is equivalent to using the newly created vacancies W_{ir} to reduce as far as possible labor market inefficiency.

On the supply side an optimal training scheme under conditions of excess labor supply can do no better than promote educational outputs in accordance with the skill-ranking arising from the coefficients $^*y_{ir}^{(1)}$, i.e., in each region r, efforts should be directed toward retraining from skill i to skill j whenever $^*y_{ir}^{(1)} > {}^*y_{jr}^{(1)}$.

14.6. Conclusions and suggested extensions

The analysis carried out has been based on an essentially static approach. Our main concern was to provide a computational tool to interpret data on regional unemployment and to assist the decision-maker in devising regional policies. This justifies the assumptions of exogenous costs and wage levels. Within the framework of our definitions and basic assumptions, three conclusions can be drawn:

1. The notion of labor market inefficiency can be given a quantitative basis.
2. Inefficiency can be decomposed into a supply component and a demand component.
3. Regional and occupational indices can be defined to draw an "inefficiency map" of a regional market system.

In the next stage of this work, our method will be applied to analyze data for the Tuscany region in Italy. Over the longer term, we feel that a more comprehensive analysis would be valuable. As some authors (see, for instance, Malinvaud, 1978) have pointed out, it is unlikely that meaningful results could be obtained outside a general equilibrium framework. To our knowledge such an approach has never been attempted on a regional scale. Quite apart from the problem of assembling a data set sufficiently comprehensive and complete to test a dynamic theory of unemployment, some of the assumptions contained in the work of Malinvaud (1978) (for instance, the market compatibility assumption) would have to be discussed in a regional context. The introduction of space into existing theories of unemployment may prove to be more than a trivial extension of those theories (see Caravani and La Bella, 1980).

References

Booth, D. and Hyman, G. (1978). A Time Series Analysis of Population and Employment Change. Paper presented at the Regional Science Association meeting, British Section, 11th Annual Conference, September 1978.

Caravani, P. and La Bella, A. (1980). Labor–production equilibrium in a multi-sectoral multi-regional system. Papers of the Regional Science Association, 47.

Commoner, B. (1975). Energy, Environment and Economics. An Essay in Energy: The Policy Issues. University of Chicago Press, Chicago, Illinois.

Doeringer, P. B. and Piore, M. (1971). Internal Labor Markets Manpower Analysis. D. C. Heath, Lexington, Massachusetts.

Edwards, R. C., Reich, M., and Gordon, D. M. (Editors) (1975). Labor Market Segmentation. D. C. Heath, Lexington, Massachusetts.

Freiburghaus, D. (1978). Arbeitsmarktsegmentation: Wissenschaftliche Modeerscheinung oder arbeitsmarkttheoretische Revolution? (Labor Market Segmentation: Modification or Revolution in Labor Market Theory?) Papers of the International Institute of Management, Berlin.

Gleave, D. and Palmer, D. (1979). Spatial Variations in Unemployment Problems: A Typology. Paper presented at the Regional Science Association meeting, held at University College, London.

Gordon, D. M. (1972). Theories of Poverty and Underemployment. D. C. Heath, Lexington, Massachusetts.

Knox, F. (1979). Labour Supply in Economic Development. Saxon House, Farnborough, Hampshire.

Malinvaud, E. (1978). The Theory of Unemployment Reconsidered. Basil Blackwell, Oxford.

Osterman, P. (1975). An empirical study of labor market segmentation. Industrial and Labour Relations Review, 28: 508–528.

Piore, M. (1975). Notes for a theory on labor market stratification. In R. C. Edwards (Editor), Labor Market Segmentation. D. C. Heath, Lexington, Massachusetts.

Slessner, M. (1978). Energy in the Economy. Macmillan, London.

Valkenburg, F. C. and Vissers, A. M. C. (1978). Theorie van de Dubbele Arbeidsmarkt (Theory of the Dual Labor Market). University of Tilburg, The Netherlands.

Regional Development Modeling: Theory and Practice
M. Albegov, A.E. Andersson and F. Snickars (editors)
North-Holland Publishing Company
© IIASA, 1982

Chapter 15

SYSTEMS OF SERVICE PROVISION UNDER CONDITIONS OF CHANGE: PROBLEMS OF MODELING

John B. Parr*
Department of Social and Economic Research, University of Glasgow, Glasgow (UK)

15.1. Introduction

The broad aim of this chapter is to suggest the outlines of a framework for modeling the way in which systems of service provision adjust to change. A system of service provision is assumed to consist of a set of facilities concerned with the supply of a given range of services. Such a system may involve publicly provided services (e.g., health care or education) or privately provided services (e.g., commercial retailing). Moreover, each type of service-provision system can exist either at the regional (inter-urban) scale or at the intraurban (intrametropolitan) scale. There are thus a number of quite distinct emphases or perspectives, as indicated by the four cells in Table 15.1. Each cell has traditionally been dominated by a particular discipline of the social sciences. Thus, the supply of privately provided services at the regional scale (I) has tended to be dominated by economics and economic geography, as exemplified in the work of von Böventer (1962) and Berry (1967). In contrast, the supply of privately provided services at the intraurban scale (II) has mainly been the concern of urban planning (Carol, 1965; Chapin, 1965). Cell (III) refers to the supply of publicly provided services at the intraurban scale, and this has been largely dominated by operations research (Toregas et al., 1971; Helly, 1975). Cell (IV), referring to the supply of publicly provided services at the regional scale, has not really been dominated by any discipline, but work in sociology (Christenson, 1976) and political science (Kochen and Deutsch, 1969) appears to have been the most prominent. It should be stressed that no cell in Table 15.1 is the exclusive domain of a particular discipline, and that no discipline (of those mentioned) has restricted its attention to a particular cell. Furthermore, work in this general area has been undertaken by a number of other disciplines not mentioned in Table 15.1.

In the following discussion of systems of service provision, an attempt will be

* The author wishes to thank the Carnegie Trust for the Universities of Scotland for its support in the preparation of this chapter.

Table 15.1. The disciplines used to study the provision of services at two different scales.

Scale of analysis	Provision of services	
	Private	Public
Regional	(I) Economics	(IV) Sociology
	Economic geography	Political science
Intraurban	(II) Urban planning	(III) Operations research

made to keep the perspective as general as possible, without concentrating on any particular scale of service provision. The analysis will also be concerned with multielement systems. Thus, rather than dealing with an individual service (such as secondary education or durable consumer goods), this study will examine sets of related services (such as primary, secondary, and tertiary education, or convenience goods, nondurable goods, durable goods, and specialized personal services). The usual framework for dealing with multielement systems of service provision is to be found in that branch of location theory known as central place theory. This was developed by Christaller (1933/1966) and Lösch (1940/1954) and extended by Tinbergen (1964), Beckmann (1968), and Nourse (1979), among others. While central place theory is generally used to analyze the market provision of services at a regional scale, it can readily be adapted to deal with other perspectives. Much of central place theory has been conceived in static terms and this has seriously reduced its applicability, particularly with respect to the framing of policy. Since a major concern of this chapter is the question of change, it will be necessary to go beyond the conventional models of central place theory and recast them in comparative-static or multistatic terms. Before doing so, however, it may be useful to review the basic version of the static model of service provision, since the proposed modifications and extensions are derived from it.

15.2. The underlying static model

The basic element of virtually every central place model is the spatial configuration of supply for an individual service. This basic element varies slightly, according to whether a privately provided service or a publicly provided service is being considered. Lösch (1940/1954) outlined the relevant analysis for a privately provided service. He assumed a homogeneous plain, which was inhabited by consumers with identical demands. These consumers bore the cost of transport, so that the distance from a given consumer to the closest point of supply would affect the real price (i.e., the f.o.b. price plus transport cost) facing this consumer and thus the quantity of the service that he would demand. It was further assumed that there was free entry into production, and that economies of scale existed, as implied by a falling average cost curve. Lösch derived the equilibrium price and output levels, as well as the equilibrium service area radius, each of these three elements being identical for every individual supplier. This spatial equilibrium configuration resulted in the maximization of the number of independent suppliers, subject to the requirement that each made normal profits. And while this structure was in equilibrium, it did not necessarily represent a social optimum (Denike and Parr, 1970; Beckmann, 1972).

For publicly provided services, the analysis is slightly different: the notion of equilibrium is replaced by the concept of social optimality, and this was the focus of the analysis outlined by Schultz (1970). Following Lösch, a homogeneous plain was assumed, throughout which service requirements were identical. Three components were considered for each service. The first component involved service cost, the sum of capital costs and operating costs. The average cost curve was assumed to be U-shaped. The second component was social benefit. This is the monetary value of the total number of service units utilized. It was assumed that the service units utilized per capita would decline in a negative exponential manner with distance from a facility, and that the social benefit per service unit utilized would decline as the utilization level increased. The third component was travel cost. This represented the cost in terms of time, inconvenience, and money of traveling between consumer residences and facilities, and it was assumed to be proportional to the straight-line distance. With these three components, it was possible to specify the objective function of the model — the maximizing of net social benefit (the difference between the social benefit and the sum of service cost and travel cost) throughout the area under consideration. The optimal size of the service area for an individual facility was determined, together with the optimal output of this facility and the market-clearing price, in cases where prices were charged.

Within a system of service provision, whether private or public, there is likely to be more than one service involved. Each individual service element has its own equilibrium (or social-optimum) configuration, as determined by the size, frequency, and spacing of service facilities or supply points, and these configurations are likely to differ among the various service elements. It is the aggregation of the individual configurations that yields the overall system of service provision. A number of aggregation procedures exists, and which is chosen depends on the desired nature of the overall system of service provision. Nevertheless, aggregation usually involves placing certain restrictions on the number of possible sizes of service area. One type of aggregation, utilized by Lösch (1940/1954, pp. 130–132 of the 1954 translation) and also by Schultz (1970), restricts the number of service area sizes in such a way that if a location contains a service facility with a service area of a given size, that location will also contain all other service facilities with identical or smaller service area sizes. This aggregation procedure has important implications for the spatial structure of the system, and gives rise to a successively inclusive hierarchy in which the number of levels of the hierarchy corresponds to the number of service area sizes and thus to the number of sets of services. Within such a hierarchy, a facility center of level m (where $m = 1, 2, \ldots, N$) supplies a set of services of level m (i.e., a characteristic set of services not supplied by centers of lower levels), in addition to sets 1 through $(m - 1)$, which are characteristic of levels 1 through $(m - 1)$, respectively. It is possible to have only a single service associated with each level of the hierarchy, but it is often more convenient to think in terms of sets of services.

Table 15.2. A successively inclusive hierarchy.

Set of services supplied	Level of center				
	1	2	3	4	5
1	x	x	x	x	x
2		x	x	x	x
3			x	x	x
4				x	x
5					x

This hierarchical structure is illustrated in Table 15.2, where it can be seen that a facility center of level 3, for example, supplies level-3 services, in addition to level-2 and level-1 services. Looking ahead for a moment to Figure 15.3(a), we observe the spatial structure of a system of service provision which contains this successively inclusive hierarchy. It can be seen that the single center of level 3 not only has a level-3 service area to which level-3 services are supplied, but also a level-2 service area and a level-1 service area to which are supplied level-2 services and level-1 services, respectively. Such a highly ordered structure implies some interdependence in the supply of (or demand for) the various service elements, although in the analyses of both Lösch and Schultz this interdependence was absent in the sense that neither the demand (or social benefit) curves nor the cost curves of the various service elements were related to each other.

In the real world this successively inclusive hierarchical structure is a common feature of service provision and appears to stem from interdependence among the various service elements. In the case of privately provided services, these interdependencies are based on the advantages of agglomeration on the revenue side (resulting from the possibility of multiple-purpose visits by consumers) and, at higher levels of the hierarchy, on the cost side, involving urbanization and urban-complex economies (Parr, 1978). In the case of publicly provided services, there are also advantages of agglomeration on the cost side (e.g., shared heating and laboratory facilities in hospitals) and the benefit side (e.g., reductions in both the money and time spent in traveling by the consumer). However, it must be noted that the successively inclusive hierarchy is not always present in a multielement system of service provision. In cases where the various elements are relatively independent of one another, either on the cost side or on the revenue (or benefit) side, the need for such a highly ordered structure would only be justified on the grounds of general social advantage, e.g., reduced public-utility expenditures or a more efficient transport system.

15.3. Mechanisms of adjustment to changed conditions

After a system of service provision has been established, it is reasonable to expect that eventually the underlying conditions will change and that the system will be required to operate in a fundamentally different setting. There are many possible changes; these may include new demographic structures, different levels of regional economic activity, alterations in consumer demand, the development of new technologies in production, transport, and marketing, and so on. While these changed conditions may influence the entire system, they more usually affect a particular part of it. Since changes of this nature typically involve a disruption of some previously existing equilibrium (or previously planned social optimum), the system must adjust (or be made to adjust) to these changed conditions. However, the system cannot be shut down and then replaced by a new one that reflects these different conditions; adjustment must take place within the locational constraints of the existing system. Two simple examples of adjustment mechanisms are now considered, one involving a privately provided service and the other a publicly provided service.

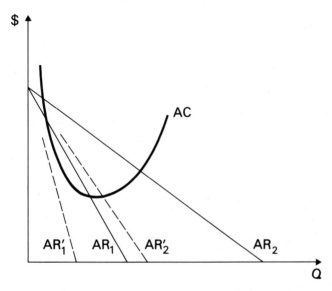

Figure 15.1. Cost and demand curves for an individual privately provided service, showing conditions under which the service will be transferred from level 1 to level 2 or higher.

Figure 15.1 indicates the demand and supply conditions for an individual privately provided service. For the sake of convenience all the demand curves are shown as straight lines and the relative positioning of these curves is only approximate. The curve AR_1 indicates the quantities Q that are demanded at various f.o.b. prices from an individual supplier located in a center of level 1. This curve summarizes the demand of consumers located within the level-1 market area, as well as within the level-1 center. It can also be viewed as the revenue received by the supplier per unit of output at different levels of production Q. The curve AC represents the cost (including "normal profit") per unit of output at different levels of production Q. Curve AR_2

shows the demand conditions facing an individual supplier at a level-2 center, in the absence of supply from level-1 centers. The fact that curves AC and AR_1 intersect indicates that supply of the service can be commercially undertaken from level-1 centers, so that the curve AR_2 can be disregarded for the moment. The service can thus be said to be characteristic of level 1, i.e., part of the level-1 set of services. The service will, of course, be supplied from centers of all higher levels, since we are assuming the existence of a successively inclusive hierarchy. Suppose, now, that tastes change and demand for the service in question decreases. This is reflected in the new demand curve AR_1' to the left of AR_1. Curve AR_2 is similarly replaced by curve AR_2'. Since AR_1' neither intersects nor is tangential to AC, the service cannot now be commercially supplied from level-1 centers. However, AR_2' does intersect AC, indicating that the service can be supplied, on a more centralized or consolidated basis, from the smaller number of centers of level 2 or higher. Thus, the service that was initially characteristic of level 1 (part of the level-1 set of services) now becomes characteristic of level 2 (part of the level-2 set of services).

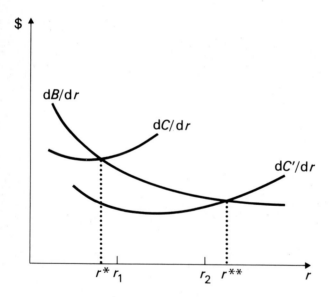

Figure 15.2. Marginal cost and marginal benefit curves for an individual publicly provided service, showing conditions under which the service will be transferred from level 1 to level 2 or higher.

While profitability is the most important factor for privately provided services adjusting to changed circumstances, the level of net social benefit is the relevant measure for publicly provided services. A slightly different mode of analysis is employed to demonstrate the adjustment to changed conditions in this second example. The curve dC/dr in Figure 15.2 is the marginal cost curve (with respect to service area radius r) for an individual service. The curve dB/dr represents the corresponding marginal social benefit curve (with respect to service area radius r) for the same service. The intersection of the two curves at r^* indicates that this is the service area radius which maximizes net social benefit. In the multielement system, where all services are arranged in a successively inclusive hierarchy, the service area radius that approximates r^* most closely is r_1, the radius of a service area for level-1 services. The service in question is thus supplied from centers of level 1 or higher; i.e., it is characteristic of level 1. It is now assumed that there is a major technological change on the cost side, as represented by the new marginal cost curve dC'/dr. The intersection of dB/dr with dC'/dr indicates that net social benefit is now maximized with a service area radius of r^{**}. This is close to r_2, the radius of a service area for level-2 services. The service is now best supplied from centers of level 2 or higher and, as in the previous example, a service that was originally characteristic of level 1 becomes characteristic of level 2.

15.4. Adjustment to change and modifications to the hierarchy

The two examples given in Section 15.3 were concerned with individual services, and the adjustment mechanisms merely lead to the reallocation of an individual service to a different level of the hierarchy, and thus to a redefinition of the various sets of services. These kinds of adjustment occur with great frequency in systems of service provision, particularly systems involving privately provided services. However, adjustments can also be of a more radical nature. If, instead of a single service, all (or a substantial number) of the services of a given level were subject to the type of change discussed above, the adjustment process could well result in a modification in the structure of the hierarchy. Such modifications in the structure of the hierarchy do not generally involve the appearance and disappearance of facility centers, but rather changes in their importance within the hierarchy (Parr, 1981). In dealing with models that involve modifications to the structure of the hierarchy, attention will be confined to those adjustments that preserve the successively inclusive nature of the hierarchy. One reason for this restriction is that the successively inclusive hierarchy is frequently present in systems of service provision. Also, as we have seen, this structure does have a theoretical rationale, which is based on the advantages of agglomeration on both the revenue (or benefit) side and the cost side. Furthermore, and on purely pragmatic grounds, without this restriction the range of possible adjustments becomes enormous and thus not amenable to analysis.

Before discussing the various modifications to the hierarchical structure, a number of points must be made. First, the system of service provision prior to the change is always represented by Figure 15.3(a). This is a three-level structure with a successively inclusive hierarchy. The distribution of supply points is triangular, with hexagonal service areas. While other service area shapes are possible, the hexagonal service area structure does have a particular theoretical significance (Parr, 1980). Second, it is assumed that the existing or initial system is close to some kind of equilibrium (or social optimum), that it is then affected by new conditions, and that thereafter it

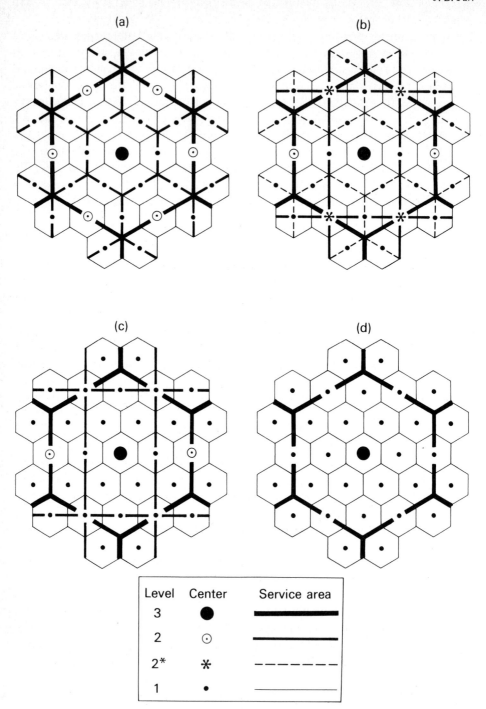

Figure 15.3. Modifications of the hierarchical structure: (a) initial structure; (b) formation of a new level; (c) concentration within a level; (d) disappearance of a level.

emerges close to a new equilibrium (or new social optimum). This is then a two-stage, comparative-static model in which the precise nature of the transition between the two stages or end-points is not specified. Third, in order to be able to compare the various types of modification, the process of adjustment is always assumed to involve level-2 services, so that attention is focused on level 2 of the hierarchy. With these considerations in mind, we now investigate the manner in which adjustments to change may produce modifications in the structure of the hierarchy.

15.4.1. Formation of a new level of the hierarchy

Assume that, as a result of reduced population density, there is a decrease in the equilibrium (or optimum) size of the service area for *certain* of the services provided at level 2. Thus, it is necessary to supply a subset of level-2 services from fewer centers, although more than would be the case if the services were supplied only from centers of level 3. One possible adjustment to this situation would be for the subset of level-2 services to be supplied from a reduced number of level-2 centers and for the remaining former level-2 centers to be downgraded to form a new level of the hierarchy. Those facility centers that continue to provide all the former level-2 services are still referred to as level-2 centers. However, the former level-2 centers that offer a reduced set of former level-2 services are now termed level-2^* centers, this reduced set of services being termed level-2^* services. The centers offering all the former level-2 services now have larger level-2 service areas, and as can be seen in Figure 15.3(b), the hexagonal level-2 service areas have been replaced by larger rectangular service areas. The centers offering level-2^* services have level-2^* service areas corresponding to those of the former level-2 service areas.

15.4.2. Concentration within a level of the hierarchy

This second kind of structural modification is caused by increased economies of scale in the provision of *all* level-2 services, leading to an increase in the equilibrium (or optimum) size of the service area for each service. As a consequence, all level-2 services need to be supplied from fewer facility centers, although more than would be the case if the services were supplied only from centers of level 3. During the adjustment process some of the former level-2 centers lose their level-2 services and are thus downgraded to level-1 centers. From Figure 15.3(c) it can be seen that those level-2 centers that remain in existence have larger level-2 service areas, which are now rectangular rather than hexagonal. This concentration within level 2 of the hierarchy causes a reduction in the number of level-2 centers, and produces a corresponding increase in the number of level-1 centers.

15.4.3. Disappearance of a level of the hierarchy

Consider the case where a reduction in consumer demand (or need) drastically diminishes the requirement for the existing number of facility centers supplying level-2 services, to the extent that the services can only be supplied from the relatively few centers of level 3. Under these circumstances level-2 services are combined with level-3 services and offered as a single set (the level-3 set) from level-3 centers only. Figure 15.3(d) indicates that the former level-2 centers have now become functionally indistinguishable from level-1 centers and are downgraded to level-1 status. Level 2 of the hierarchy thus disappears, although the centers formerly of this level continue to exist in their downgraded status, supplying only level-1 services.

15.4.4. Structural modifications: a comparison

The various kinds of adjustment (each of which involves a degree of centralization) can now be compared. Since the pattern is continuous, the number of centers at each level is presented in terms of level-3 centers; i.e., how many level-1 and level-2 centers are there to every level-3 center? To calculate this, it is necessary to find the number of centers of various types within the level-3 service area boundary. If a center is located on the edge of this boundary, it is counted as one-half of a center. The initial structure (Figure 15.3a) was such that level-2 services were supplied from four centers (one of level 3 and three of level 2). When a new level forms (Figure 15.3b), the centralization process is not very pronounced. Some of the former level-2 services continue to be offered from four facilities (one of level 3, one of level 2, and two of level 2*), with the other former level-2 services being offered from only two centers (one of level 3 and one of level 2). In the case of concentration within a level of the hierarchy (Figure 15.3c), the process of centralization is stronger, since after the adjustment all level-2 services are offered from only two centers (one of level 3 and one of level 2). The process of centralization is the most intense when a level disappears (Figure 15.3d); after the adjustment the former level-2 services are provided by only one center (the single center of level 3).

It is possible that as the need for centralization increases, a system may pass through each of the stages indicated in Figure 15.3. Centralization currently appears to be the dominant trend in systems of service provision, both at the regional level and within the inner districts of metropolitan areas. The underlying reasons for this are numerous and include declining population densities, changing patterns of demand (or need), technological change, and improvements in transport. This process of centralization usually means that there are fewer facilities (sometimes producing on a larger scale), each supplying an enlarged service area. A trend of this type can often raise serious problems connected with the accessibility of facilities and levels of consumption, and it may be necessary to impose additional constraints to ensure that certain criteria of equity are met.

15.5. Future directions

In this chapter a general framework for dealing with the question of change in systems of service provision has been examined. An attempt has been made to demonstrate how the underlying static model of central place theory can be extended to deal with adjustments to change, particularly those that involve a modification in the hierarchical structure (and thus in the spatial structure) of a system of service provision. It is quite clear that a number of improvements and extensions are necessary before this framework can be used in practice. There are four broad areas in which further research can usefully be undertaken. First, there is a need to move away from the simplifying assumption of a uniform population surface and to introduce realistic patterns of population distribution and demand (or need). Second, more attention should be paid to the difficult problem of interdependence, which in the past has generally been neglected. Not only should there be more explicit recognition of the interdependence between the various elements of a single system (e.g., education, health-care provision, or retailing), but the possible interdependence between the different systems of service provision should also be explored. Moreover, the fact that systems of service provision may be associated, through a complicated set of external factors, with other facets of a broader socioeconomic system (e.g., the housing market) should be recognized. Third, the assumption of a successively inclusive hierarchy, although useful, needs to be relaxed, since it is clearly not present in all systems of service provision. This may be due to the various service elements being independent of each other in supply and/or demand terms, or to the fact that the different service elements are administered by different government departments, or even to the nature of historical processes or decisions of the past. Fourth, and very important, the timing and sequencing of market-induced or planned adjustments to change (a set of factors wholly neglected in this brief summary) should be fully integrated into any analysis concerned with modifications in the spatial structure of systems of service provision.

References

Beckmann, M. J. (1968). Location Theory. Random House, New York.

Beckmann, M. J. (1972). Equilibrium versus optimum spacing of firms, and patterns of market areas. In R. Funck (Editor), Recent Developments in Regional Science. Karlsruhe Papers in Regional Science. Volume 1. Pion Press, London.

Berry, B. J. L. (1967). Geography of Market Centers and Retail Distribution. Prentice-Hall, Englewood Cliffs, New Jersey.

von Böventer, E. (1962). Theorie des räumlichen Gleichgewichts (Theory of Spatial Equilibrium). J. C. B. Mohr, Tübingen.

Carol, H. (1965). Der Standort des "Shopping Center" (The Location of Shopping Centers). Plan, 22: 208–214.

Chapin, F. S. (1965). Urban Land Use Planning. University of Illinois Press, Urbana, Illinois.

Christaller, W. (1933/1966). Die zentralen Orte in Süddeutschland (1933). Translated (1966) by C. W. Baskin as Central Places in Southern Germany. Prentice-Hall, Englewood Cliffs, New Jersey.

Christenson, J. A. (1976). Quality of community services: a macro-unidimensional approach with experimental data. Journal of Rural Sociology, 41: 509–525.

Denike, K. G. and Parr, J. B. (1970). Production in space, spatial competition and restricted entry. Journal of Regional Science, 10: 49–63.

Helly, W. (1975). Urban System Models. Academic Press, New York.

Kochen, M. and Deutsch, K. W. (1969). Toward a rational theory of decentralization: some implications of a mathematical approach. American Political Science Review, 63: 734–739.

Lösch, A. (1940/1954). Die räumliche Ordnung der Wirtschaft (1940). Translated (1954) by W. Woglom and W. F. Stolper as The Economics of Location. Yale University Press, New Haven, Connecticut.

Nourse, H. O. (1979). The economics of central place theory: a more general approach. In R. Funck and J. B. Parr (Editors), The Analysis of Regional Structure: Essays in Honour of August Lösch. Karlsruhe Papers in Regional Science. Volume 2. Pion Press, London.

Parr, J. B. (1978). Models of the central place system: a more general approach. Urban Studies, 15: 35–49.

Parr, J. B. (1980). Health care facility planning: some developmental considerations. Socio-Economic Planning Sciences, 14: 121–127.

Parr, J. B. (1981). Temporal change in a central place system. Environment and Planning A, 13: 97–118.

Schultz, G. P. (1970). The logic of health care facility planning. Socio-Economic Planning Sciences, 4: 363–393.

Tinbergen, J. (1964). Sur un modèle de la dispersion géographique de l'activité économique (On a model for the geographic dispersion of economic activity). Revue d'Economie Politique, 74: 30–44.

Toregas, C., Swain, R., and ReVelle, C. (1971). Location of emergency service facilities. Operations Research, 19: 1363–1373.

Regional Development Modeling: Theory and Practice
M. Albegov, A.E. Andersson and F. Snickars (editors)
North-Holland Publishing Company
© IIASA, 1982

Chapter 16

CONTINUOUS FLOW MODELING IN REGIONAL SCIENCE

Tönu Puu
Department of Economics, University of Umeå, Umeå (Sweden)

16.1. Introduction

The continuous approach to regional modeling is not embodied in any one model (Beckmann and Puu, 1981). It is more a set of models developed for various purposes, or, equivalently, a self-contained philosophy of model building. The continuous approach involves many mathematical concepts such as existence, uniqueness, dynamic adjustments, etc.; these will not be discussed in this chapter but are dealt with elsewhere. Similarly, the continuous approach has many applications, for example, to product specialization, road-investment planning, natural-resource extraction, and water-supply management: again, these various applications will not be discussed in any detail. This chapter will, in fact, be limited to a brief description of the fundamental tools of continuous flow modeling, illustrated by the use of the method in a single application.

It seems reasonable to begin by outlining the reasons for using the continuous approach (see Beckmann, 1952, 1953; Isard and Liossatos, 1979). Discrete spatial analysis generally starts with the division of the region under analysis into a number of subregions as shown in Figure 16.1. The various relationships between the subregions are then studied in terms of matrices $[a_{ij}]$, which may represent commuting trips, migration, or any other flow that can be specified for a pair of subregions.

This approach does not require the geometric information given in Figure 16.1; the whole picture could be torn into pieces, figuratively speaking, and the shapes of the various subregions forgotten without affecting the analysis. If this were done, it would be impossible to reproduce the original sizes, shapes, and positions of the subregions, but this is not important when using the discrete approach. It seems a shame to discard this geometric information at the very beginning of the analysis, especially when it can be shown that this is not actually necessary. The continuous model employs techniques adopted from physics and vector analysis to use geometric information that would otherwise be discarded.

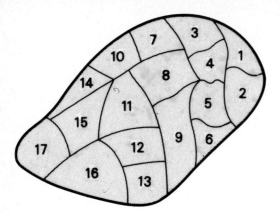

Figure 16.1. Division of the region under analysis into subregions.

However, there are a number of objections to the continuous approach: first, that any empirical data available must refer to discrete subdivisions, and second, that the design of the computer algorithms would make discretization necessary. These objections can be countered by arguing that the case is exactly analogous to the situation in physics. No physicist has yet been able to measure the pressure of a fluid at all points in a continuous flow. These points are not only infinite in number, but, as they represent a continuum, such experiments could not even be arranged in *any* sequence. This is due to the fact that a continuum is nondenumerable. Hence, no continuous model can ever be verified, even by an infinite number of experiments. Despite this, however, continuous flow modeling has been of considerable use in hydrodynamics. It may be argued that spatial economic phenomena are sufficiently similar to physical phenomena to suggest the application of techniques long found useful in analyzing one set of phenomena to the analysis of the other.

It should be stressed, however, that the results which may be obtained through this approach should be regarded as complementing rather than replacing the information obtained using discrete methods.

16.2. Elements of continuous regional modeling

Both discrete and continuous models deal with the same basic types of information: densities and flows. A discrete model is composed of a set of nodes connected by a set of arcs, and considers the densities at the nodes (population, average income, etc.), and the flows along the arcs (commodities, commuters, messages, etc.). In continuous models, however, the densities vary continuously from one location to another, and the flows do the same, changing in both magnitude and direction.

There are two main principles to be observed in continuous modeling:

1. It is necessary to observe a continuity equation, which regulates the change of the inflow and outflow at local sinks and sources. This will be explained in more detail in Section 16.3.
2. Transportation facilities are not represented by a graph or network. Instead, each location is associated with a fixed price representing the cost of movement across it. If transportation facilities are good, the cost is low; if they are bad, the cost is high. The cost of transportation between any pair of locations may then be obtained as a path integral of this local transportation cost function; the optimum routing problem can be solved as a well-defined variational problem using an appropriate Euler equation. To simplify the problem it is assumed that the local cost function is isotropic, i.e., independent of direction. This represents the main departure from real networks, in which only a limited number of transit directions are possible at a given point. However, this approximation is certainly no worse than many of the assumptions that are necessary in the physical sciences.

The second principle is borrowed from geometrical optics and corresponds to Fermat's principle, while the first is derived from hydrodynamics and corresponds to the continuity equation for an incompressible fluid.

16.3. Some mathematical theorems

It is necessary to define a number of mathematical concepts to make the explanation more precise.

16.3.1. Vector field

A vector field is a mapping

$$\phi = [\phi_1(x,y), \phi_2(x,y)]$$

This implies that a vector ϕ is associated with each point (x,y) in the region. The unit vector

$$\phi/|\phi| = (\cos \theta, \sin \theta)$$

represents the direction of the flow, and the euclidian norm

$$|\phi| = (\phi_1^2 + \phi_2^2)^{1/2}$$

represents the intensity (or magnitude) of the flow. Thus, the concept of a vector field and the idea of a continuous flow are equivalent.

The left-hand side of Figure 16.2 represents a vector field. The direction of the arrows represents the direction of flow and the length of each arrow indicates the magnitude of the flow. The lines of continuous flow are illustrated on the right-hand side of Figure 16.2.

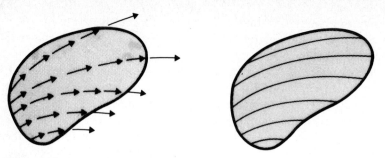

Figure 16.2. A vector field and associated lines of flow.

16.3.2. Gradient of a potential function

The gradient of a potential function is also an important concept. Viewing the potential $p(x,y)$ as a surface in three-dimensional x,y,p-space, the gradient of p is defined as

$$\text{grad } p = (p_x, p_y)$$

where p_x and p_y denote partial derivatives. Grad p is a vector in the direction of steepest ascent of the $p(x,y)$ surface, with a magnitude equal to the rate of increase in this direction.

16.3.3. Divergence

Divergence can formally be defined as

$$\text{div } \phi = \partial\phi_1/\partial x + \partial\phi_2/\partial y$$

The sum of the partial derivatives of the first component of flow with respect to the first space coordinate and of the second component of flow with respect to the second space coordinate may seem an odd choice for an operator. However, the reasons behind this choice will immediately become apparent on studying Gauss' divergence theorem.

16.3.4. Gauss' divergence theorem

This is probably the most important tool in continuous flow modeling. It may be stated as follows:

$$\iint_S \text{div } \phi \, dxdy = \oint_{\partial S} (\phi)_n \, ds \tag{16.1}$$

i.e., the double integral of the divergence of a vector field over a bounded region S equals the line integral of the normal component of the field taken along the boundary

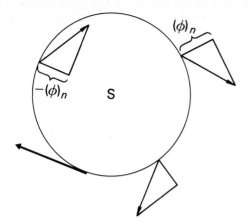

Figure 16.3. Flow vectors and normal projections on the boundary of region S.

∂S. This normal component is denoted $(\phi)_n$, and is, of course, a scalar. Figure 16.3 illustrates the relationship between the bounded region S, the flow vectors on the boundary, and the normal projections $(\phi)_n$.

The right-hand side of eqn. (16.1) thus gives the net outflow from region S. If S is shrunk to a point, then the left-hand side of eqn. (16.1) is the divergence at this point, and the right-hand side is the net outflow from the same point. It is therefore possible to interpret divergence as the net addition to the flow at any point.

The divergence theorem may also be seen as a natural extension of simple integral calculus. If $f(x)$ is defined as $f(x) = dF(x)/dx$, then the fundamental theorem of calculus states that $\int_a^b f(x)dx = F(b) - F(a)$. Thus, the definite integral of a derivative depends on the boundary values of the primitive function. In the divergence theorem the integral is taken over an area, not an interval, and the boundary is a curve rather than a pair of points. The two theorems are equivalent in all other respects and the divergence is, indeed, a type of derivative.

With these preliminaries in mind, it is now possible to restate the two fundamental principles of continuous flow modeling in mathematical form.

16.3.5. The divergence law

Let k, l, and m be the local levels of use of capital, labor, and land services and $f(k, l, m)$ be a production function. If q denotes local consumption then, from the above discussion, excess supply $f(k, l, m) - q$ enters the flow at this point if the divergence is positive, or is withdrawn from it if the divergence is negative. Formally

$$f(k, l, m) - q = \text{div } \phi \tag{16.2}$$

This is also known as the continuity principle.

16.3.6. The gradient law

Let the local cost of transportation be $h(x, y)$. Then

$$h \, \phi/|\phi| = \text{grad } p \tag{16.3}$$

where p is product price. This equation indicates that flows move in the direction of steepest product price increase and that, in this direction, the product price increases according to the transportation cost. (The latter is obvious as (16.3) implies $|\text{grad } p| = h$.)

Intuitively, this gradient law makes good economic sense. It can also be derived formally as the solution to the following optimization problem.

Assume that the local quantity of goods transported is $|\phi|$, and the cost per unit is h; the local transportation cost will therefore be $h|\phi|$. The total cost of transportation in region S is then given by

$$\iint_S h|\phi| \, dxdy$$

Assume that this cost is minimized subject to constraint (16.2). A Lagrangian multiplier p is associated with the constraint and the following expression is obtained for the transportation cost:

$$\iint_S [h|\phi| + p(f - q - \text{div } \phi)] \, dxdy \tag{16.4}$$

The Euler equation which makes (16.4) optimal with respect to the choice of the flow field ϕ may be shown to be eqn. (16.3), i.e., the gradient law.

16.4. An application to production planning

The use of the continuous flow method is now illustrated by applying it to a production planning problem. Assume the problem is to maximize

$$\iint_S U(q_1, q_2, \ldots, q_n, x, y) \, dxdy \tag{16.5}$$

where $q_1, q_2, \ldots, q_n$ are different consumer goods, including housing and services but not transportation. The explicit inclusion of space coordinates in the local utility function U makes it possible to use various weighting coefficients, for example, to reflect different population densities.

Production can be represented by a set of production functions $f^i(k_i, l_i, m_i)$, $i = 1, 2, \ldots, n$. The space coordinates are not included in the functions, which means that the same production opportunities are available everywhere. Subtracting local consumption from local production as in eqn. (16.2) leads to

$$f^i(k_i, l_i, m_i) - q_i = \text{div } \phi_i \qquad (i = 1, 2, \ldots, n) \tag{16.6}$$

Assume that transportation, which is the only kind of activity not represented in the production functions, uses κ_i units of capital and λ_i units of labor for each unit of flow of commodity i. These "fixed" coefficients should be functions of the space coordinates, reflecting the differences in local transportation facilities.

Given the total amounts of labor (L) and capital (K) available, the following constraints hold:

$$\int \int_S \sum_{i=1}^{n} (k_i + \kappa_i |\phi_i|) \, dxdy = K \tag{16.7}$$

$$\int \int_S \sum_{i=1}^{n} (l_i + \lambda_i |\phi_i|) \, dxdy = L \tag{16.8}$$

Unlike capital and labor, the amount of land available is fixed in space; the remaining constraint is therefore local rather than integral in form:

$$\sum_{i=1}^{n} m_i = m \tag{16.9}$$

To summarize, the problem assumes given totals of capital and labor that are to be distributed among different locations to be used for different purposes. Land can also be used for a number of purposes at a given location. Production and transportation are limited by the labor and capital available to produce and transport each commodity. The objective is to maximize welfare given these various constraints.

In mathematical terms this involves maximizing (16.5) subject to constraints (16.6)–(16.9). The optimum conditions for production are

$$p_i f_k^i = r \tag{16.10}$$

$$p_i f_l^i = w \tag{16.11}$$

$$p_i f_m^i = g \tag{16.12}$$

The optimum condition for transportation is

$$(r\kappa_i + w\lambda_i) \, (\phi_i / |\phi_i|) = \text{grad } p_i \tag{16.13}$$

The optimum condition for consumption is

$$U_i = p_i \tag{16.14}$$

The p_i, r, w, and g are Lagrangian multipliers; in optimization methods they are generally interpreted as input and output prices. It should also be noted that r and w are Lagrangian multipliers for integral constraints (16.7) and (16.8), and therefore must be constant independent of location.

This is a solution to a planning problem, but it is interesting to note that most of the conditions listed above may also be interpreted in terms of market equilibrium. Viewed in this light, eqns. (16.10)–(16.12) are the conditions under which private firms make maximum profit; they state that the marginal value productivity of capital, labor, and land must everywhere equal its local price. The fact that capital rent and wages are constant can be taken to mean that in the long run, through capital accumulation and population migration, labor and capital will be found at the point where the rewards are greatest.

Equation (16.13) indicates that all commodity flows choose the directions of steepest price increases, and that in these directions prices increase at the same rate as transportation costs. The latter is now equal to $(r\kappa_i + w\lambda_i)$ since transportation costs have been specified in terms of primary input requirements (labor and capital). This condition also makes sense in a competitive market equilibrium: private transporters ship goods in the directions in which the spatial price differentials are greatest and, because of competition among transporters, the prices increase by the transportation costs.

The only condition which may not be fulfilled in a competitive equilibrium is (16.14). It is, however, interesting to note that, by repeated use of Gauss' theorem, it is possible to derive the condition

$$\int\int_S \sum_{i=1}^{n} P_i q_i \, dx dy = rK + wL + \int\int_S g \, dx dy \qquad (16.15)$$

The derivation is quite lengthy, and is therefore not reproduced here. The only assumptions made are that production functions are linearly homogenous, and that any imported goods are financed by the export of other goods so that a trade balance is maintained. The left-hand side of eqn. (16.5) is the total consumption value of all goods, evaluated at local prices; this is equal to the sum of capital rents, wages, and land rents in the region (right-hand side of the equation). This means that an overall "budget constraint" is obeyed. Since everything is evaluated at local prices, it must be possible to design an intraregional income transfer policy which could meet the conditions of eqn. (16.14), while allowing the consumers complete autonomy within the limits set by local budget constraints.

The above example demonstrates quite clearly the equivalence of planning and competitive equilibrium in the continuous flow approach, although it should be noted that this is only one of many possible applications of the continuous flow model to regional planning and equilibrium.

References

Beckmann, M. (1952). A continuous model of transportation. Econometrica, 20: 643–660.

Beckmann, M. (1953). The partial equilibrium of a continuous space market. Weltwirtschaftliches Archiv, 71: 73–89.

Beckmann, M. and Puu, T. (1981). A Continuous Transportation Model. International Institute for Applied Systems Analysis, Laxenburg, Austria. (Forthcoming.)

Isard, W. and Liossatos, P. (1979). Spatial Dynamics and Optimal Space–Time Development. North-Holland Publishing Co., Amsterdam.

Regional Development Modeling: Theory and Practice
M. Albegov, A.E. Andersson and F. Snickars (editors)
North-Holland Publishing Company
© IIASA, 1982

Chapter 17

INTEGRATED MATERIAL–FINANCIAL BALANCES IN REGIONAL ECONOMIC ANALYSIS

Boris Issaev
Central Institute for Economics and Mathematics, USSR Academy of Sciences, Moscow (USSR)

17.1. Introduction

Regional economic analysis deals with the factors that jointly determine the economic and social situation within a region. Any examination of the material and social factors which affect regional economic activity should include an analysis of local material resources, environmental issues, organization and management systems, as well as certain sociodemographic characteristics of the region. The production process should be described in terms of the flow of goods and the circulation of financial resources. In any regional analysis it is also necessary to consider the changing relationships between regions, as reflected by both national and regional indicators. It would be useful to develop a system which integrates all of these factors in some consistent fashion to produce a comprehensive picture of the economic situation within a region.

One approach to this problem is currently being investigated at the Central Institute for Economics and Mathematics of the USSR Academy of Sciences (CIEM). It has been found that work being carried out on modeling integrated material–financial balances contains some elements that could also be useful in multiaspect economic analysis at a regional level.

17.2. Basic elements of material–financial balances

The first step in this approach is to classify the factors affecting regional activity and express their mutual interdependence in matrix form, as shown in Table 17.1. It is clear that some of the interrelationships depicted in Table 17.1 are explicitly or implicitly tied to the flow of monetary resources. This makes it possible to construct a multiaspect model of the regional economy in the form of an integrated system which balances the circulation of goods and financial resources; one such model, the integrated material–financial balance (IMFB) model, is outlined in Table 17.2.

Table 17.1. The factors affecting regional activity and their interrelationships.[a]

	A	B	C	D	E	F
A	AA					
B	BA	BB				
C	CA	CB	CC			
D	DA	DB	DC	DD		
E	EA	EB	EC	ED	EE	
F	FA	FB	FC	FD	FE	FF

[a] A, B, C, D, E, F represent arrays Y_{itr} where i represents the type of commodity and t and r refer to time and region, respectively. A describes material and natural resources; B environmental factors; C the type of management system; D the demographic characteristics; E the flow of goods; and F the flow of money.

This model links elements of vectors A, E, F, C, and D (representing resources, material, financial, institutional, and social aspects of regional analysis, respectively) in one closed system. A reflects the value of national (or regional) wealth (classification a). E is classified by types of products (e_1) and production sectors (e_2). F contains various classes of financial flows ($f_{1,2}$), and C includes a number of institutional sectors and other decision-making centers ($c_{1,2}$). The sociodemographic characteristics of the system D are modeled by grouping the population in various ways and by identifying the industries fulfilling the social needs of the region and the nation (d).

The matrix presented in Table 17.2 may be interpreted as follows. Let an industrial sector produce a set of products, which is represented by a vector in $e_2 e_1$. The value of production (total financial return) is the sum of the elements of this row vector. The use of resources in each industrial sector is given in the column corresponding to that particular sector. All material inputs are entered in $e_1 e_2$ and the national income produced is entered in $f_1 e_2$. The column sums of the elements of $f_1 e_2$ give the structure of national income produced by each industrial sector, the row sums reveal the industrial origin of national income.

The process of redistribution of each element of national income produced is reflected in submatrices $c_1 f_1$, df_1, and $f_2 f_1$. The resulting income flows are divided among enterprises grouped according to their institutional characteristics, population, and financial and credit systems. The total amount of resources that each sector (c_1) uses is derived from the sector's primary resources originating from national income produced ($c_1 f_1$), as well as from resources available from other sectors ($c_1 c_1$), from the population ($c_1 d$), and from financial and credit systems ($c_1 f_2$). The use of these total resources is given by $c_1 c_1$, dc_1, $f_2 c_1$, $c_2 c_1$, ac_1, and rc_1. The first three blocks represent the expenditure of financial resources during the redistribution of national income, rc_1 represents the flow of money out of the system, $c_2 c_1$ and ac_1 show the use of money for consumption and for accumulation. Total expenditure on fixed assets and stocks in category A are reflected by the row sums of blocks ac_1, ad, and af_2. The submatrix $e_1 a$ represents the supply of products of various types necessary to cover these expenditures.

Table 17.2. One type of integrated material–financial balance (IMFB) model.

Category	Description of subcategories	Subcategory	A	E	E	F	C	D	F	C	A	X	A
			a	e_1	e_2	f_1	c_1	d	f_2	c_2	a	r	a
A	Wealth at beginning of period	a									a_0		
	Production												
E	Goods and services	e_1		$e_1 e_1$	$e_1 e_2$					$e_1 c_2$	$e_1 a$	$e_1 r$	
	Sectors	e_2		$e_2 e_1$									
F	National income	f_1			$f_1 e_2$								
	Distribution and redistribution of national income												
C	Sectors	c_1				$c_1 f_1$	$c_1 c_1$	$c_1 d$	$c_1 f_2$				
D	Households	d				df_1	dc_1	dd	df_2				
F	Finance and credit systems	f_2				$f_2 f_1$	$f_2 c_1$	$f_2 d$	$f_2 f_2$				
C	Consumption	c_2					$c_2 c_1$	$c_2 d$					
A	Accumulation	a	a_0				ac_1	ad	af_2				a_t
X	External transactions	r		re_1			rc_1	rd	rf_2				
A	Wealth at end of period	a									a_t		

The external relations of the system are represented by flows re_1 (imports) and e_1r (exports), and rc_1, rd, and rf_2 (debit/credit balance with the rest of world not directly related to the flow of goods). The index r used in the identification of flows classifies foreign subsystems according to some appropriate characteristic.

The IMFB model can be used in regional economic analysis to reveal to what extent the resources necessary to realize certain projects may be provided from internal sources and therefore what proportion of resources must come from outside. It can also indicate how new projects or policies would affect the structure of regional economies, and what effect this would have on the national economy.

These three objectives can be achieved in practice by the construction of two IMFB balances: one for the region and one for interregional flows of goods and financial resources.

If this system is to be used to highlight the financial and economic implications of regional projects, each individual model must contain financial characteristics that are methodologically consistent with the integrated material–financial balance of the region. This means that a special material–financial balance is advisable for each large regional project or program, and the integrated balance of the region should be associated with satellite balances (see Section 17.3.2) designed for microlevel analysis.

17.3. Application of the IMFB procedure

17.3.1. Estonian studies

The laboratory of financial analysis at the CIEM has been involved in regional analyses based on the IMFB model for several years. The work has been carried out for Estonia, and two experimental balances have been constructed: one for 1960 (matrix dimensions 150×150) and one for 1972 (matrix dimensions 250×250). The balances were compiled using data obtained from annual statistical reports of enterprises and from questionnaires designed for the compilation of input–output tables. In an attempt to find new sources of information for systems of this type, the researchers have tried to compile an integrated material–financial balance using data obtained from payments recorded by the State Bank of the Estonian Republic. These payments identify both the payer and recipient and also provide information about the nature of the transaction.

Analysis of the Estonian IMFB model for 1972 led to a number of significant conclusions regarding the overall effect of changes in a regional economy on the national economy. It was also possible to determine factors reflecting the difference between national income produced in the republic and that used for consumption and accumulation. These factors included the types of goods produced by the republic, investments from funds within the republic and from centralized funds, and flows of household income. The analysis also examined the influence of different local economic activities on the elements of the balance of external transactions in both goods and money.

The Estonian IMFB model for 1972 classified enterprises according to their managerial status (republic or all-union), and in this case it was therefore possible to analyze the differences between decision-making at various levels.

Use of the IMFB model has revealed a number of deficiencies in the statistical data available for the Soviet republics, and these deficiencies considerably limit the scope and the depth of economic analyses. It is clear that there is a need for new accounting and statistical schemes especially designed to collect data for regional systems analysis.

17.3.2. Satellite balances

A number of studies aimed at developing "satellite balances" have been carried out at the regional level. The construction sector of a republic was chosen as the subject of the study and data were collected at the enterprise level. The econometric model constructed simulates the decision-making process involved in preparing a material–financial plan for an enterprise. It takes into account the conflicting interests of decision-making centers within the enterprise and the relations of the enterprise with the institutional environment. The model gives special emphasis to the problem of incomplete data, e.g., when establishing production targets.

The satellite balance is obtained by disaggregating the rows and columns referring to a particular sector and aggregating all other elements to a level useful to the decision-makers of the enterprise. The balance is determined entirely by the economic aspects of the enterprise and by its characteristics as a separate economic unit in both technological and managerial terms.

The main problem encountered in developing a satellite balance was in making it consistent with the republic balance. The internal part of the satellite balance had to be flexible enough to reflect the features of the enterprise, while the external part should be as rigid as possible in order to define the role of the enterprise in the economy of the region (or republic).

17.4. Examples of interregional IMFBs

17.4.1. Intraregional tables

Many different IMFBs for the USSR as a whole have been put forward, but only one 45 × 45 matrix has been checked experimentally. This matrix is now used to follow the dynamics of the Soviet economy and to analyze the financial implications of calculations based on a system of optimization models that present the results in the form of input–output tables. Only five production sectors are considered.

The skeleton of a regional IMFB is repeated in Table 17.3. The table differentiates between intraregional and multiregional sectors and households; when applied to the household sector, "intraregional" indicates residents and "multiregional" denotes migrants. With the exception of rows and columns 15 and 16, the matrix reflects

Table 17.3. Detailed intraregional integrated material–financial balance (IMFB).

Description of class[a]	Class[a]	Initial wealth		Production			Distribution and redistribution of national income					Consumption	Accumulation			External relations		Final wealth		Total
		1	2	3	4	5	6	7	8	9	10	11	12	13	14	15	16	17	18	T
Wealth at beginning of period																				
Intraregional	1												1.12							
Multiregional	2													2.13						
Production																				
Goods	3				3.4	3.5						3.11			3.14	3.15				
Sectors: intraregional	4			4.3																
multiregional	5			5.3																
Distribution and redistribution of national income																				
Incomes	6				6.4	6.5														
Sectors: intraregional	7						7.6	7.7	7.8	7.9										
multiregional	8						8.6	8.7	8.8	8.9							8.16			
Households: intraregional	9						9.6	9.7	9.8								9.16			
multiregional	10																10.16			
Consumption	11							11.7	11.8	11.9	11.10									
Accumulation																				
Sectors: intraregional	12				12.4			12.7		12.9										
multiregional	13					13.5			13.8											
Objects	14	14.1	14.2										14.12	14.13				14.17		
External relations																				
Goods	15			15.3													15.16			
Money	16								16.8	16.9						16.15				
Wealth at end of period																				
Intraregional	17												17.12							
Multiregional	18													18.13						
Total	T																			

[a] Classes of accounts.

relationships between subjects that are situated within the region; rows and columns 15 and 16 cover relations in which one party is in the region and the other is not. The significance of any particular relation depends on the economic importance of the appropriate balances. This scheme was tested in Estonia.

The balance allows policies for regional development to be introduced and analyzed in the context of the regional economy. For instance, the decision to construct a large plant managed by the central authorities will affect first block 14.13, in which the value of the plant will be assigned to a given year, and then block 13.8, which represents the investing sector. Both blocks 14.13 and 13.8 shift the analysis through a system of identities to other types of analysis. The supply of material resources necessary for the construction of the plant is examined in 3.14 in terms of the product-mix required. It is also possible to study the origin of these resources, to determine whether they are produced within the region (4.3) or imported (15.3). The source of the funding is again reflected in the IMFB matrix; financing from centralized funds will affect block 8.16. Finally, it is necessary to achieve an overall balance in the intraregional and multiregional relations of the region studied. Rows and columns 15 and 16 will reflect the balance in external relations.

The flow of income from the population is very important to the regional economy. The sources of in-migrants' income lie outside the region (10.16), but this income is spent within the region (11.10). This should be taken into account when planning the supply of goods for household consumption (3.11). In particular, planners must decide whether the goods required should come from intraregional production (4.3, 5.3), or from outside (15.3). This will also be reflected in the balance in external transactions.

17.4.2. Multiregional tables

A simple interregional balance matrix is presented in Table 17.4. This scheme is designed for joint analysis of aggregated regional balances, and emphasizes the part of the national income of each region that goes into interregional exchange. The scheme does not include the relationships between regions (information is unavailable), and the balance for each region shows only the inflows and outflows resulting from interregional redistribution. At the regional level, the equilibrium between social product (or national income) produced and consumed is obtained by including both of these flows. Interregional exchange is reflected in the matrix by rows and columns 10 and 11; equilibrium is obtained by combining these with the regional balances.

17.4.3. Satellite-balance analyses of regional development projects

The main purpose of satellite balances is to describe the financial aspects of the activity of each plant in the region in terms consistent with those of the regional IMFB. This means that standard financial indicators must be used both for individual plants and for the region as a whole.

Table 17.4. Simple example of a multiregional integrated material–financial balance (IMFB).

Description of class[a]	Class[a]	Production						Interregional redistribution					Consumption	Accumulation			Rest of world	Total
		1	2	3	4	5	6	7	8	9	10	11	12	13	14	15	16	T
Production																		
Goods: i	1				1.4	...	1.6						1.12	1.13	...	1.15	1.16	
...	2.				...	...	...						...	...	...	...	...	
m	3				3.4	...	3.6						3.12	3.13	...	3.15	3.16	
Regions: i	4	4.1	...	4.3														
...	5	...	...	...														
r	6	6.1	...	6.3														
Interregional distribution of national income																		
Regions: i	7				7.4													
...	8										...	...						
r	9						9.6				9.10	9.11						
Centralized funds	10				10.4	...											10.16	
Interregional flows	11				11.4	...												
Consumption	12							12.7	...	12.9								
Accumulation																		
Regions: i	13							13.7										
...	14								...									
r	15									15.9								
Rest of world	16	16.1	...	16.3							16.10							
Total	T																	

[a] Classes of accounts.

Table 17.5. An integrated material–financial balance (IMFB) for a regional economic unit.

Description of class[a]	Class[a]	Opening assets 1	Production 2	Production 3	Production 4	National income produced 5	National income produced 6	National income produced 7	Financial resources 8	Financial resources 9	Financial resources 10	Social outlays 11	Reproduction of fixed assets and stocks 12	Reproduction of fixed assets and stocks 13	Reproduction of fixed assets and stocks 14	Other economic units in region 15	Other economic units in region 16	Other economic units in region 17	External units 18	Closing assets 19	Total T
Opening assets	1													1.13	1.14						
Internal balance																					
Production																					
Goods	2													2.13	2.14	2.15			2.18	2.19	
Material inputs	3				3.4							3.11		3.13	3.14						
Economic unit	4		4.2																		
National income produced																					
Payment for labor	5				5.4																
Profit	6				6.4																
Other items	7				7.4																
Financial resources																					
Own resources	8						8.6														
State budget resources	9																9.16				
Credits	10																10.16				
Social outlays	11					11.5	11.6			11.9											
Reproduction of fixed assets and stocks																					
Financing	12	12.1		12.3					12.8	12.9	12.10						12.16		12.18	12.19	
Investment in fixed assets	13											13.11	13.12								
Investment in stocks	14												14.12								
External balance																					
Other economic units in region																					
Sectors	15			15.3																	
Financial and credit systems	16						16.6	16.7													
Households	17					17.5	17.6					17.11									
Economic units outside region	18			18.3			18.6	18.7													
Closing assets	19													19.13	19.14						
Total	T																				

[a] Classes of accounts.

The regional balance must include the following characteristics of the individual plants: intraregional and multiregional sales of goods produced and services offered; intraregional and external sources of finance for current production; intraregional and interregional expenditure on material and technological inputs for production; the main components of national income produced and their uses both within and outside the region; the sources financing investment in fixed assets and stocks both within and outside the region, and the origin of material and technological supplies; and the financing of various types of social expenditure at the plant level (social consumption) and the sources of this finance.

Since decisions referring to separate plants always involve examination of the dynamics of material and financial flows, the influence of projects on the economy of a region over time should be determined by analyzing a series of regional IMFBs for different times t.

A material–financial balance for a regional unit (plant) is shown in Table 17.5. An important feature is the clear distinction between the internal and external components of the balance. This distinction is maintained because of the need to analyze the internal and external aspects of all stages of the project separately.

The internal mechanism dealing with the movement of goods produced and used for productive purposes is covered by blocks 4.2, 3.4, 3.11, 2.13, 2.14, 3.13, and 3.14; the external aspects are reflected in 2.15, 2.18, 15.3, and 18.3. The uses of national income within the plant are represented by 8.6, 11.5, and 11.6; the external components are given by 17.5, 16.6, 17.6, 16.7, 18.6, and 18.7. The internal mechanism treating the material and financial aspects of capital formation is covered by 12.3, 12.8, 2.13, and 2.14; the external relationships are shown in 12.9, 12.10, 9.16, 12.16, and 12.18.

The balance of a regional IMFB may be interpreted as a system of indicators to which all economic calculations – formalized or nonformalized – for separate projects should be tied. This would improve analysis of individual plants at the level of the regional economy and would also link calculations for a number of plants, making it easier to assess their joint effects. Insertion of autonomous models into the matrix would lead to a better understanding of the results of the calculations performed with these models.

17.5. Conclusions

The IMFBs discussed above have been developed and tested at the CIEM to be used as a set of autonomous tools for regional analysis and planning. During the work it became clear that there are a number of areas in which further research would be helpful – these include the gathering of regional statistics, the standardization of economic indicators, and the matching of variables in classes of regional economic models. The principles used to integrate the different aspects of the analysis in IMFBs may be employed as a basis for a new unified system for regional economic analysis.

Regional Development Modeling: Theory and Practice
M. Albegov, A.E. Andersson and F. Snickars (editors)
North-Holland Publishing Company
© IIASA, 1982

Chapter 18

MODELING REGIONAL DEVELOPMENT PROJECTS

Anatoli M. Alekseyev
Institute for Economics and Industrial Engineering, Siberian Branch of the USSR Academy of Sciences, Novosibirsk (USSR)

18.1. Defining the problem

When studying regional development projects, it is important to understand the mechanism of the process to be simulated and its interrelations with external factors in order to analyze alternative ways of managing the resources needed. Such an approach requires an examination of aggregated input–output relationships and an integrated evaluation of possible future states of the system's development.

Regional projects are characterized by five important types of linkage:

1. Transport accessibility linkages: the development of the transport system, upon which the construction of industrial plants depends.
2. Resource linkages: planned deliveries of resources among projects at different stages of the development program.
3. Technological linkages: special resource linkages that cannot be described in quantitative terms (for example, flooding of the water-storage reservoir of a hydroelectric power station may take place only after timber has been removed from the bed).
4. Operational linkages: the conditions for selling or delivering output (these differ from technological linkages in that they extend over the entire period of operation).
5. Integration linkages: joint exploitation of resources (for example, the joint operation of a number of plants).

Various types of mathematical model can be used to analyze the internal and external linkages of a development project. The most acceptable and convenient for these applications are network planning models, which, other things being equal, have a number of advantages – simplicity, clarity, and an ability to represent a sequence of stages, to update information, and to reiterate model runs. This is the reason for the extensive use of network schedules in many fields of planning and management.

At present, network models are frequently used in construction and in production planning, as well as in the design and implementation of regional economic plans. The universal applicability of the PERT (Program Evaluation and Review Techniques) system (Berman, 1964) can be attributed to its simple description of the production process, which is divided into a fixed sequence of phases with a fixed duration for each job.

Since the number of network parameters (job, event, and time allowed) is small and the calculations of job time are simple, the fulfillment of output targets can easily be analyzed and efficiently controlled. In addition, the manager can directly reduce the time allowed for the project by concentrating efforts on important jobs that would otherwise delay completion. Problems such as reallocation of resources, modification of technological solutions, and adjustment of financing are handled directly by the manager of the complex.

To find formal methods for solving such problems, economic characteristics such as output, rate of production, and costs are introduced into network scheduling. As a result, the nature of the network models changes radically. The changeover from analyzing "bottlenecks" to analyzing decision-making processes involves a qualitative complication, i.e., the replacement of a simple simulation schedule by an optimization model.

One of the first network optimization problems was solved by Kelley (1961). If the dependence of the costs of particular jobs on their length (derived from the time allowed by the project) is given, the algorithm proposed by Kelley (1961) allows durations for each job to be selected such that the total cost for all jobs is minimized.

A network problem that is complicated by the introduction of a concave dependence of costs on the length (i.e., duration) of jobs can also be found in the literature (Berman, 1964). However, the problem is defined without considering the fact that not all costs incurred are simultaneous. Therefore the major economic problem of comparing costs over time (i.e., the process of discounting) is neglected. If this factor is taken into account, the definition of the problem, the optimization method used, and the solution itself will all be different.

An interesting method of comparing costs over time was proposed by Briskin (1966). In terms of graph theory, the optimization task is to find a chain of directed arcs from source to sink over which optimum total costs are achieved. An "optimum" combination of variants is selected according to two criteria: cost-minimization, and reduction in time allowed. These optimization goals are mutually inconsistent: on the one hand, the costs of a job may decrease as duration increases; on the other hand, the client may be willing to make an additional payment for any reduction in production time. This additional payment s for a reduced production time may be modeled as an exponential function of time

$$s = c(1 - e^{kt}) + h \tag{18.1}$$

(c and h are selected by approximation) which is very similar to the standard form of discount dependence

$$1/(1+E)^t = e^{-kt}, \text{ where } k = \ln(1+E) \tag{18.2}$$

The network scheduling problem has been widely discussed (see, for example, Sadovski et al., 1965; Petrova, 1969; Perepelitsa, 1970). Models allocating resources over the network considerably increase the potential for optimizing planning decisions at the final stage of project preparation. However, when applying these models to long-term planning, there are serious information difficulties associated with the determination of resource constraints over a long period.

These difficulties are minimized if an integrated task is created to produce resources and to employ them in the implementation of a regional development project. In the formulation of an extended problem, it is assumed that final production volumes and total duration of the project are not rigidly fixed and may undergo optimization. It is quite reasonable to suppose that, for each of the final products specified, the forecasted efficiency values for each year of the planning period and the corresponding values for resources to be employed will be known.

In these circumstances, the task is to produce, during the allotted period, a specified final output using a certain composition of resources in such a way that the greatest overall benefit is achieved. This task is difficult to perform because the parameters to be modeled include economic measures of resources employed and of output produced. This means that it is impossible to find an optimum plan using only one type of model. The solution can be derived by running a two-level system of models iteratively. On the upper level of the system it is advisable to use a network model, in which various time estimates can be generated according to the variant of project realization selected. On the lower level, a group of linear programming models should be used to determine efficiency values for final production and comparative scarcity values for the resources consumed for each fixed schedule of project events.

18.2. Model of an economic development project and procedures for its implementation[*]

The problem of implementing a regional project may be represented by an oriented graph, each node of which corresponds to the completion of a certain stage in the project. The arcs of the graph correspond to jobs needed for completion of each stage. On each arc are given the allowed ranges for durations of jobs and for compositions, amounts, and dynamics of resources consumed and final output produced. The topology of the graph is constructed according to the specific content and feasible sequencing of project stages. The planning period (total time for project implementation) is assumed to be given. For the sake of simplicity, the consumption of resources and the output produced for any given job are taken to be constant.

Given these assumptions, the problem can be formally presented as follows. Suppose an oriented graph G without closed loops is taken to represent the process

---

[*] The model was developed by the present author in collaboration with L. A. Kozlov and V. N. Kriuchkov.

of implementation of the economic project. Nodes i, j $(i, j = 1, 2, \ldots, n)$ denote completion of jobs $(i, j) \in G$ and are described by a time index. The starting time t_{ij} and duration τ_{ij} are given for each job. Resources $Q_{ij}^k(t)$ for $t_{ij} \leqslant t \leqslant t_{ij} + \tau_{ij}$ are assumed to be fixed, where k indicates the particular resource consumed ($k = 1$, $2, \ldots, K_1$) or the particular final product ($k = K_1 + 1, K_1 + 2, \ldots, K$). Also, the values of U_k^t, the price of resource (or final product) k in year t of the planning period ($t = 1, 2, \ldots, T$), are given.

The fulfillment efficiency of job (i, j) begun at time t_{ij} is defined as

$$S_{ij}(t_{ij}) = \sum_{t=t_{ij}}^{t_{ij}+\tau_{ij}-1} \left(\sum_{k=K_1+1}^{K} Q_{ij}^k(t)U_k^t - \sum_{k=1}^{K_1} Q_{ij}^k(t)U_k^t \right) \tag{18.3}$$

If the economic goals of the project are expressed by a specified fixed final output to be produced in a given time, the efficiency of each job will vary according to the total costs, i.e.,

$$S_{ij}(t_{ij}) = \sum_{t=t_{ij}}^{t_{ij}+\tau_{ij}-1} \sum_{k=1}^{K_1} Q_{ij}^k(t)U_k^t \tag{18.4}$$

Using the above notation, the optimization problem is to find a schedule of starting times $\{t_{ij}^*\}$ for every job that achieves maximum overall efficiency:

$$\sum_{(i,j)\in G} S_{ij}(t_{ij}^*) = \max_{\{t_{ij}\}} \sum_{(i,j)\in G} S_{ij}(t_{ij}) \tag{18.5}$$

subject to

$$t_j = \max(t_{ij} + \tau_{ij}) \qquad (i, j) \in G \tag{18.6}$$

which implies that no event can take place before all preceding jobs are completed, and

$$t_{ij} \geqslant t_i \tag{18.7}$$

which expresses the constraint that no job can begin before the preceding event takes place, and

$$t_1 \geqslant 0; \qquad t_n \leqslant T \tag{18.8}$$

where T is the length of the planning period.

This model is run using graph optimization with a number of alternative speci-fications. The variation in goals (output or maximization of efficiency) affects only the number of terms in eqn. (18.3). Equation (18.3) may be further simplified by assuming that final production is fixed and only a single resource is employed in implementing the development project. The efficiency function for each job will then have the form

$$S_{ij}(t_{ij}) = - \sum_{t=t_{ij}}^{t_{ij}+\tau_{ij}-1} Q_{ij}(t)U_t \tag{18.9}$$

If the minus sign in eqn. (18.9) is replaced by a plus, the optimization criterion becomes that of minimizing the inputs

$$\sum_{(i,j)\in G} S_{ij}(t) \to \min \tag{18.10}$$

subject to constraints (18.6)–(18.8).

Provisionally, let us assume that a function $f_{ij}(t_i, t_j)$ of minimum costs for each job can be found within the time interval $[t_i, t_j]$, $(t_j - t_i \geqslant \tau_{ij})$. For this purpose, it suffices to check all feasible job-beginnings within this interval to select the minimum from the related costs thus generated. Then, problem (18.10) can be formulated as

$$Z(t) = \sum_{(i,j)\in G} f_{ij}(t_i, t_j) \to \min \tag{18.11}$$

subject to

$$t_j - t_i \geqslant \tau_{ij} \qquad (i, j) \in G \tag{18.12}$$

$$t_1 \geqslant 0; \qquad t_n \leqslant T \tag{18.13}$$

Problem (18.11)–(18.13) can be solved by the dynamic-programming procedures proposed by Bellman (1960). This approach allows an accurate solution to be obtained but it is only suitable for graphs with a specific structure, namely, those where all program actions are represented by parallel chains of jobs. Because of the limited possibilities for applying this algorithm and its computational inefficiency, a two-stage descent method has been developed which is simpler and more suitable for computer implementation.

The essence of this latter method is as follows. For each event the earliest (t_i^p) and the latest (t_i^n) starting times for all jobs are computed. The finishing time for the whole complex of jobs and the duration of each job are given.

As the given total length T for a number of jobs cannot be shorter than the critical path, the starting time (t_{ij}) for a job can be varied within the range

$$t_i^p \leqslant t_{ij} \leqslant t_j^n - \tau_{ij} \tag{18.14}$$

By proceeding iteratively within this interval, the cost–time dependence $(f_{ij}(t_i, t_j))$ is obtained in matrix form. At the same time, the starting time $\bar{t}_{ij}$ within each interval $[t_i, t_j]$ that corresponds to the minimum cost function $(S_{ij}(t_{ij}))$ is found. Within the technologically infeasible range, where $t_j - t_i < \tau_{ij}$, artificially high-cost jobs are introduced. Owing to this, inadmissible job schedules turn out to be suboptimal.

Suppose such functions are obtained for all jobs. Then, having a fixed time of occurrence for all events except event i, it is possible (by varying the position of the ith vertex) to find a position where the total costs related to this vertex are minimized. This position is then fixed and the next vertex is treated. After analyzing all vertices, another iteration is carried out. The minimum total cost is assumed to be achieved if the schedule of events is the same in two consecutive iterations. Here the condition of technological feasibility of all job lengths must be retained; otherwise costs increase for those schedules which include job lengths shorter than those that are technologically feasible. The optimum schedule procedure is then achieved.

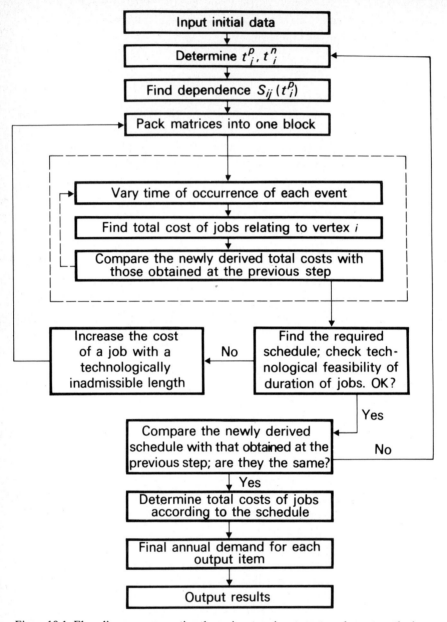

Figure 18.1. Flow diagram representing the major steps in a two-stage descent method.

Major cost-minimization stages for the network are given in Figure 18.1. The preliminary stage involves the determination of an initial approximation, i.e., a schedule of admissible event occurrences. The initial schedule may be arbitrary; however, it can also be derived from the following conditions:

$$t_1 = 0; \qquad t_j = \max_i (t_i + \tau_{ij}) \tag{18.15}$$

The determination of a cost function with exceedingly large values $f_{ij}(t_i, t_j)$ for (t_i, t_j), such that $t_j - t_i \leqslant \tau_{ij}$, fulfills two tasks. First, further calculations are simplified (in that infeasible values of t_i are eliminated). Second, the optimization range is extended, which also facilitates an approximation of the optimum.

Stationary solutions derived from local starting-time optimizations of individual network events, or nodes, are checked once again with respect to feasibility. If the penalties for not keeping to the minimum length prove to be insufficient, the cost function is iteratively increased:

$$f_{ij}(t_i, t_j) = 10 \cdot f_{ij}(t_i, t_j) \tag{18.16}$$

for (t_i, t_j) such that

$$t_j - t_i < \tau_{ij} \tag{18.17}$$

The optimization consists of several series of iterative examinations of schedule events. The times of occurrence of all network events, beginning from the first, are iteratively optimized. The times of occurrence of the other events, excluding the particular event i under analysis, are fixed. From the feasible values $(t_{i_0}^s \leqslant t_{i_0} \leqslant t_i^n)$, the one that achieves minimum total cost is selected:

$$\sum_{(i,j) \in G} f_{ij}(t_i, t_j) \to \min \tag{18.18}$$

Since, in varying t_{i_0}, changes only occur in the lengths and costs of the jobs that "enter" or "leave" event i_0, all jobs not related to event i_0 can be removed from relation (18.14). Moreover, one may simplify (18.11) into the following expression for iteration step λ in the sequence of local minimizations:

$$\sum_{(i, i_0) \in G} f_{ii_0}(t_i^{\lambda+1}, t_{i_0}) \; + \sum_{(i_0, j) \in G} f_{i_0 j}(t_{i_0}, t_j^{\lambda}) \; \to \min \tag{18.19}$$

The first group of terms in (18.19) covers jobs that are inputs of i_0, the second covers outputs of i_0. Since events are analyzed consecutively, all events preceding t_{i_0} are examined at the $(\lambda + 1)$th iteration, subsequent to iteration λ. The value of $\bar{t}_{i_0}$ that minimizes the sum (18.19) is assumed to be locally optimal and is fixed, after which the next event is treated.

After a complete examination of network events, the schedules derived at two successive iterations are compared. If they are identical, the procedure is terminated.

18.3. Coordination of the project with the development plans of subsidiary subsystems

A model of a development project can be used as a device for producing a number of output targets, which then become the goals for each sector engaged in the project.

Consider first the case when the model of a given development project "interacts" with that of a sector whose output serves as input for the project. The operation of the sector is represented by a dynamic model of multicommodity minimum-cost type with continuous variables (Volkonski, 1972). We will begin by defining the following variables:

h is a production unit ($h = 1, 2, \ldots, R_h$)

r is a mode of operation of the unit ($r = 1, 2, \ldots, R_r$)

k are the output items produced by the sector, which serve as input for the development project ($k = 1, 2, \ldots, K$)

t is the year ($t = 0, 1, 2, \ldots, T$), where T is the given total time for implementing the project

a_{ht}^{kr} is the output of k produced by production unit h in year t assuming unit intensity of mode of operation r

c_h^r are the total costs associated with operation of production unit h under mode of operation r at unit intensity

Q_t^k is the planned sectoral output target for item k in year t (this parameter is determined for each fixed schedule of the project by summing inputs of resource k for all jobs in each year)

Z_h^r is the intensity of use of mode of operation r by production unit h

Using this notation, the sectoral problem may be formulated as follows: minimize total costs

$$L(Z) = \sum_{h=1}^{H} \sum_{r=1}^{R_h} c_h^r Z_h^r \tag{18.20}$$

subject to

$$\sum_{h=1}^{H} \sum_{r=1}^{R} a_{ht}^{kr} Z_h^r \geqslant Q_t^k \qquad (t = 0, 1, \ldots, T), \quad (k = 2, 3, \ldots, K) \tag{18.21}$$

which means that the sector must provide the project with the required resources. Additionally,

$$Z_h^r \geqslant 0 \qquad (r = 1, 2, \ldots, R_h), \quad (h = 1, 2, \ldots, H) \tag{18.22}$$

are nonnegativity conditions.

For the sake of simplicity, additional constraints (on scarce resources, production capacity, etc.) have been omitted here; only those conditions (18.21) that are important for considering fundamental problems of coordination between the sectoral model and the development project are retained.

As already shown in Sections 18.1 and 18.2, the values of resources consumed are extremely important in optimizing the courses of economic development projects. These values can be obtained in a number of ways (for example, from price lists, or by consulting experts). For long-term planning, these values may be derived as dual estimates from local systems where the resources are produced. A sector may be taken to be such a subsystem. For example, by solving a problem dual to (18.20)–(18.22), a dynamic system of prices reflecting the production efficiency for item k in year t is built. Denoting these prices by U_t^k, the given dual problem may be formulated as follows. Maximize

$$\sum_{k=1}^{K} \sum_{t=1}^{T} Q_t^k U_t^k \tag{18.23}$$

subject to

$$\sum_{k=1}^{K} \sum_{t=1}^{T} a_{ht}^{kr} U_t^k \leqslant C_h^r \qquad (r = 1, 2, \ldots, R_h), \quad (h = 1, 2, \ldots, H) \tag{18.24}$$

and

$$U_t^k \geqslant 0 \qquad (k = 1, 2, \ldots, K), \quad (t = 0, 1, 2, \ldots, T) \tag{18.25}$$

The prices U_t^k can be interpreted in economic terms as the decrease in value of inputs in a given sector as the planned production target for item k in year t is reduced by one unit.

Assuming this interpretation of the resource values, it is reasonable to introduce them into the model of the development project in the form of a set of indicators, in order to examine various feasible ways of implementing the project (represented as job schedules on the network graph) and to select the most efficient one with respect to sectoral development. Coordination of a sectoral model with the network model of a development project (where the project affects demand for the output of the sector) can be described as an iterative procedure. At each step of this procedure, the data on the vector of demand for output Q_t^k are transferred from the upper-level project model to the sectoral model and efficiency values U_t^k for the production of this output are transmitted from the model of the sector to that of the project. A new optimum schedule for the stages in the development project is found in terms of these values, and a new vector of demand for the output of the sectoral system is determined for this schedule.

In the coordination procedure, however, sharp oscillations in the values of Q_t^k and U_t^k occur, so that the desired results cannot be achieved. To reduce these parameter oscillations, which are detrimental to convergence, the process can be treated by an iterative procedure (Volkonski, 1967). In the present application, the procedure can be applied by adjusting the values U_k^t in two consecutive iterations.[*] To do this, an optimum schedule of the network graph at the $(j + 1)$th step of the iterative process is found not in terms of real values for the sectoral system but in terms of averaged values from all preceding iterations:

---

[*] The scheme is complicated here because the economic project optimization problem is an integer problem.

$$\hat{U}_t^{k(\gamma+1)} = \hat{U}_t^{k\gamma}(1 - \alpha_{\gamma+1}) + U_t^{k(\gamma+1)}\alpha_{\gamma+1} \tag{18.26}$$

subject to

$$\alpha_\gamma \to 0 \text{ if } \gamma \to \infty; \qquad \sum_{\gamma=1}^\infty \alpha_\gamma \to \infty \tag{18.27}$$

where

$\hat{U}_t^{k\gamma}, \hat{U}_t^{k(\gamma+1)}$ are averaged values at iterations γ and $\gamma + 1$, respectively

$\qquad U_t^{k(\gamma+1)}$ are true values derived from the solution of the sectoral problem at iteration $\gamma + 1$

$\qquad\qquad \alpha_{\gamma+1}$ is the weighting of the value from iteration $\gamma + 1$ in the average values

The above coordination procedure has been included in an automated optimization system. The satisfactory operational properties of this system have been confirmed by the solution of many real-world problems.

18.4. Practical use of a two-level system of models for preparing regional development projects

The two-level model was tested for its accuracy in representing the conditions of formation of the Boguchany territorial production complex (TPC).* The possibilities of coordinating the development of the regional construction sector with the levels of project investment were given particular attention.

The model of the construction sector used in this case may be considered as the lower level of the system. It was assembled so as to produce a smooth pattern of construction-sector development over time, taking into account reserve capacity and storage of finished output, as well as possible reallocations of capacity between output for industrial and municipal use. The purpose of this lower-level model was to calculate the industrial and municipal construction output necessary to satisfy the dynamics of demand at the upper (TPC) level, thus reducing total construction costs.

A number of runs of the model enabled us to establish a construction schedule for the industrial units in Boguchany, such that the development and operating costs of the services construction sector were minimized. A full description of the application is given in Alekseyev (1979).

The two-level system of models discussed above was also used to analyze the overall resource characteristics of the economic development project for the zone around the Baikal–Amur railroad. At the upper level of the system, the program of major economic development activities for this region was represented by a network graph with 600 nodes. It reflected the process of formation of nine territorial production complexes and industrial centers in the zone and tasks associated with development of the

---

* For a fuller description of territorial production complex analysis, see Chapter 24.

transportation sector in the area, as well as modeling the resource, operational, and technological linkages of the various development stages.

The lower-level model depicted major intersectoral relationships. The model variables represented imbalances between available and required volumes of resources and output, and the model constraints specified admissible oscillations in annual inputs and outputs. The total imbalances for all components were evaluated in terms of marginal costs for deficient output and resources.

In coordinating the two levels, the total volumes of intermediate and final output, the carrying capacity of the sections of the railroad under construction, and the required volumes of transportation were transmitted to the lower-level model. Dual prices, encouraging changes in the schedule of the project so as to reduce costs, were transferred from the lower to the upper level.

The model runs allowed conclusions to be drawn about the validity of particular relationships reflected in the model and about the time needed for project implementation (Alekseyev et al., 1978).

References

Alekseyev, A. M. (1979). Mnogourovniye Sistemy Planirovaniya Promyshlennogo Proizvodstva (Multilevel Systems for Planning Industrial Production). Nauka, Novosibirsk.

Alekseyev, A. M., Belonogova, A. Ye., Kiselnikov, A. A., Kriuchkov, V. N., Sobolev, Yu. A., and Mitnitskaya, Yu. M. (1978). Voprosy uvyazki planov razvitiya otraslei uchastvuyushchikh v vypolnenii regionalnykh programme (Coordination of the sectoral development plans of regional programs). In A. Alekseyev and T. Tsimdina (Editors), Modelirovanie Vnutrennikh i Vneshnikh Svyazei Otraslevykh Sistem (Modeling Internal and External Linkages for Sectoral Systems). Nauka, Novosibirsk.

Bellman, P. (1960). Dinamitcheskoye Programmirovanie (Dynamic Programming). M. Inostrannaya literatura.

Berman, E. B. (1964). Resource allocation in a PERT network under continuous activity time–cost functions. Management Science, 10(4):734–746.

Briskin, L. E. (1966). A method of unifying multiple-objective functions. Management Science, 12(10):406–416.

Kelley, L. E. (1961). Critical path planning and scheduling on a mathematical basis. Operations Research, 9(3):296–321.

Perepelitsa, V. A. (1970). O dvukh zadachakh iz teorii grafov (Two problems of graph theory). Doklady Akademii Nauk SSSR, Matematicheskaya Seriya, 194(6): 1269–1273.

Petrova, L. T. (1969). Vvedenie v Setevoe Planirovaniya. (An Introduction to Network Planning). Novosibirsk State University, Novosibirsk.

Sadovski, V. N., Kuznetsova, A. M., and Mereinis, D. A. (1965). O programme rascheta setevykh grafikov stroitelstva neskolkikh obiektov s uchetom ogranichenii v obschem raskhode resursov (A computer program for evaluating network models for industrial construction). In Opyt Bumenenia Matematicheskikh Metodov i Vychislitelnoi Tekhniki v Planirovanii Stroitelnogo Proizvodstva (Mathematical Methods and Computational Techniques in Construction Industry Planning): 160. Nauka, Kiev.

Volkonski, V. A. (1967). Model Optimalnogo Planirovaniya v Promyshlennosti (A Model of Optimal Industrial Planning). Nauka, Novosibirsk.

Volkonski, V. A. (1972). Metodicheskie Polozhenia Optimalnogo Otraslevogo Planirovaniya v Promyshlennosti (Methodological Issues of Optimal Industrial Planning). Nauka, Novosibirsk.

PART VI

MODELING REGIONAL DEVELOPMENT PLANNING

Regional Development Modeling: Theory and Practice
M. Albegov, A.E. Andersson and F. Snickars (editors)
North-Holland Publishing Company
© IIASA, 1982

Chapter 19

A MODEL OF REGIONAL DEVELOPMENT AND POLICY-MAKING FOR THE NOTEC REGION OF POLAND

Roman Kulikowski and Lech Krus
Systems Research Institute, Polish Academy of Sciences, Warsaw (Poland)

19.1. Introduction

The Noteć Development Project is concerned primarily with large-scale irrigation of the Upper Noteć region in central-northern Poland. The region specializes mainly in grain production (51.8%); however, potatoes and sugar beet account for 22.1% of the land cultivated. The soil is fertile and all the necessary conditions exist for intensifying agricultural production, despite the fact that the annual precipitation is rather low (450–500 mm/year) in comparison with the rest of the country. Water shortages during the growing period limit agricultural production (and especially that of sugar beet). Moreover, there is concern that future regional growth may be slowed down significantly because of the increasing and competitive demands for water by the urban population, industry, and water transport users, and due to the seasonal shortages in the water supply for agriculture. A large regional water-development project should be launched as soon as possible to prevent losses to the regional economy, primarily to the region's agriculture. The project should be based on detailed analysis of the impact of water supply and demand on integrated regional growth.

Most of the research on this topic is financed by Polish Government Program PR7, which deals with the development and use of water resources. The program was established in 1974 under the general coordination of the Institute of Meteorology and Water Resources (IMWR) with the active participation of the Systems Research Institute (SRI). The main objective of the program is to develop basic scientific, technical, and organizational approaches to water-resource management.

In 1977, the SRI signed an agreement with the International Institute for Applied Systems Analysis (IIASA) to cooperate in the development of a system of computerized models for the Noteć region (Albegov and Kulikowski, 1978a, 1978b). The following models, which use a common data bank, have been built: a regional development model (RDM), a model of water-system development (MWSD), a model of water allocation (MWA), and an agriculture model (AM).

RDM has been built in computerized interactive form (IRDM), thus allowing expected costs and benefits to the region to be estimated under different policy objectives: demographic policy (such as working-age limits); agricultural policy (size of private farms, rents for retired farmers); employment policy; urbanization policy (housing, infrastructure, and transport); industrial and technological policy (size and location of new plants); pricing policy (domestic and foreign market prices); and environmental policy (pollution standards and taxes).

IRDM is linked to MWSD, MWA, and AM as shown in Figure 19.1. In the first and simplest version of IRDM, the regional economy was aggregated into rural R_1 and urban R_2 sectors. A two-level structure was assumed for the management system. The decision-maker R_0 responsible for development of the region divides the regional budget, including subsidies S obtained from the central budget, between sectors R_1 and R_2. The decision-makers responsible for the rural and urban sectors (R_1 and R_2, respectively) divide the subsidies S_1 and S_2 between different spheres of economic activity (capital and labor costs) and aggregate consumption. Within agriculture, private R_p and state R_g sectors are identified. The interactive computerized Policy Evaluation (PE) system enables the decision-makers to evaluate different development policies. In particular, the benefits for the sectors (B_1 and B_2), the upper-level authority (B_0), and for the whole region (B) can be determined.

The broken lines in Figure 19.1 represent information links. MWSD calculates the economic effects of expenditures S_{w0}, S_{w1}, S_{w2} on water-system development and is coordinated with the agriculture model. For given shares of various crops, it derives the optimal irrigated areas, A_{ip} and A_{ig}, for 12 microregions. The agriculture model (AM) receives information about these irrigated areas, investment expenditures S_{kp} and S_{kg}, employment L_p and L_g, and total arable areas A_p and A_g of the private and state sectors. AM calculates the optimal agricultural policies at the microlevel. The parameters of the agricultural production submodel (such as the production efficiency W_1) in IRDM are currently being improved, and it is hoped that new and more-accurate results will be obtained. In addition, the system is linked to MWA, which determines the water-distribution strategies; i.e., it allocates water among particular users over 10-day periods.

The Policy Evaluation (PE) Model (see Figure 19.2) consists of the Regional Development Core Model and several other detailed submodels describing structural changes in agriculture, demographic patterns, the environment, and the urban economy. The core model includes the production and consumption subsystems of rural and urban subregions. Migration is taken into account and the optimal allocation of labor is determined. The PE model allows the consequences of different policies (see Figure 19.2) to be evaluated.

In this chapter we will concentrate on a description of IRDM. The MWSD, MWA, and AM models are described more fully in Gutenbaum et al. (1981).

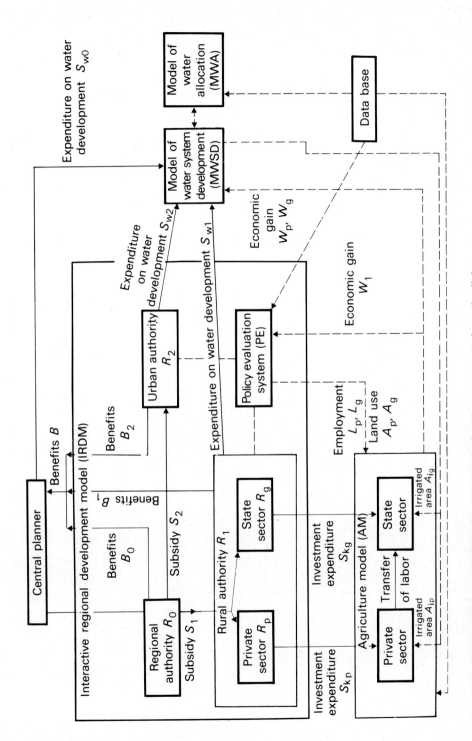

Figure 19.1. Links in the Noteć system of models. The broken lines represent information links.

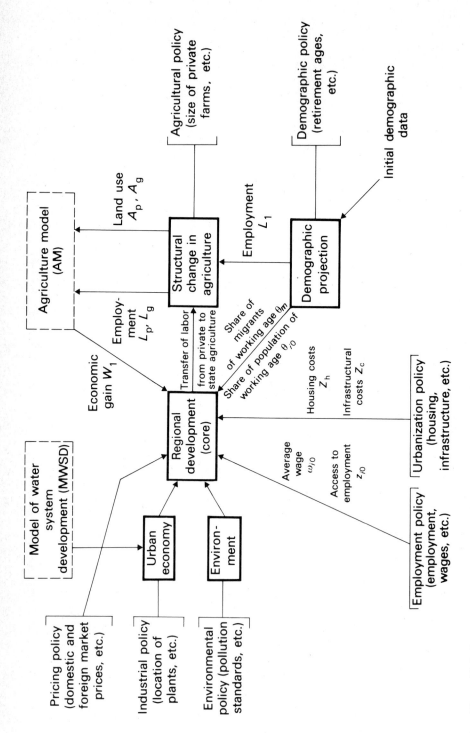

Figure 19.2. Structure of the policy evaluation (PE) model.

19.2. General methodology

It is assumed that the main exogenous factor of regional development is population, predetermined by demographic trends. The regional population participates in production and consumption. In planned economies both activities are controlled by the hierarchically organized planning and management system. The planning of production (in terms of investments and allocation of labor) is carried out by the sectors, which consist of production plants and higher-level planning and decision centers. Sectoral plans based on purely economic efficiency do not generally maximize regional welfare. The regional and higher-level authorities are therefore concerned with allocating and using investments to achieve the desired standard of living, environmental quality, and interregional equity (Zawadzki, 1973). In order to achieve these objectives, two kinds of plans (sectoral and regional) are usually constructed for each planning interval. Discrepancies between them are resolved through a coordination process, which is carried out by a number of committees in which the representatives of sectoral, regional, and higher-level authorities participate.

As shown in Kulikowski (1980), planning and coordination may be studied using some simple concepts of game theory in which the sectors and regions negotiate their planned budgets with a higher-level authority. The game can be implemented in the form of an interactive, computerized system. The main purpose of using such a system is that planners and decision-makers can be aided in evaluating alternative plans and alternative regional and sectoral policies. In particular, when the "characteristic function" of the game is convex, regional policies become stable in the sense that more partners (i.e., producers and consumers) join the regional system. In the case of unstable policies, outflows of labor and capital follow and regional growth declines.

In this section the general concepts of Kulikowski (1980) are presented and extended.

19.2.1. Factor coordination in production and consumption

Consider a production system with output x described by the production function $x = f(x_0, x_1, \ldots, x_m)$, where $x_i (i = 0, 1, \ldots, m)$ are production factors, such as labor, capital, land, etc. Assume that output price p and input prices $\omega_i (i = 0, 1, \ldots, m)$ are given. Generally x, as well as x_i, may vary over time t, so that the production value Y discounted over the given interval $[0, T]$

$$Y = \int_0^T \exp(-\lambda t) px(t) \, dt$$

and input costs (which are generally nonlinear), for example

$$Y_i = \int_0^T \exp(-\lambda_i t) \omega_i [x_i(t)]^{\gamma_i} \, dt \qquad (\gamma_i \neq 1) \tag{19.1}$$

depend on the strategy of input intensities $x_i(t) (i = 1, 2, \ldots, m)$ adopted.

When $x_0(t)$ and the total input cost (Y_i) are given one can try to find an optimum strategy for allocating production factors $x = \hat{x}$, which is the solution of the problem

$$\max_{x \in \Omega} Y(x) = Y(\hat{x})$$

$$\Omega = \left\{ x_i(t): \sum_{i=1}^{m} Y_i(x_i) \leqslant \bar{Y}; \quad x_i(t) \geqslant 0; \quad i = 1, 2, \ldots, m; \quad t \in [0, T] \right\}$$

where $\bar{Y}$ is the given input cost.

To illustrate the procedure, consider the case $m = 1$, and the Cobb–Douglas production function

$$x(t) = A \exp(\mu t) [K(t)]^{\beta_1} [L(t)]^{\beta_0}$$

where A, μ, β_0, and β_1 are known positive constants. In addition, let us assume that the discount rates as well as the production function parameters satisfy a number of economically justified conditions, such as

$$\lambda = \alpha_0 \lambda_0 + \alpha_1 \lambda_1$$

and

$$\alpha_0 + \alpha_1 = 1$$

and condition (19.4), below, where $\alpha_0 = \beta_0/\gamma_0$ and $\alpha_1 = \beta_1/\gamma_1$.

Using the Hölder inequality (see Kulikowski, 1980), the optimum strategy for given $x_0(t) = L(t)$ can easily be derived:

$$\hat{x}_1(t) = \hat{K}(t) = \{(\alpha_1 \omega_0(t)/\alpha_0 \omega_1(t)) [L(t)]^{\gamma_0}\}^{1/\gamma_1} \tag{19.2}$$

$$\hat{Y} = Y(\hat{x}) = W\bar{Y} \tag{19.3}$$

where

$$W = A p(t) \exp(\mu t) [\alpha_0/\omega_0(t)]^{\alpha_0} [\alpha_1/\omega_1(t)]^{\alpha_1} = \text{constant} \tag{19.4}$$

Then

$$Y_0(\hat{x}) = \alpha_0 \bar{Y}; \qquad Y_1(\hat{x}) = \alpha_1 \bar{Y} = (\alpha_1/\alpha_0) Y_0 \tag{19.5}$$

Production is profitable when $W > 1$.

It follows from (19.2) that, for a given technology, a change in the amount of capital $K(t)$ should be coordinated with employment $L(t)$. In the more general case $(m > 1)$, it is possible to show that a change in all the remaining production factors should be coordinated with $L(t)$. This principle, referred to in Kulikowski (1978a, 1978c) as the "factor coordination theorem", can easily be extended to the case of the CES (Constant Elasticity of Substitution) production function or to the case when the production function depends on spatial variables, e.g., the quality of land (Kulikowski, 1978b). However, the most important application of the factor coordination principle is for "agglomeration" economies, which involve scale economies and externalities. When, for example, the number

$$\delta = 1 - (\beta_0/\gamma_0) - (\beta_1/\gamma_1)$$

is equal to zero, one can say that the scale benefits (large β_0 and β_1) compensate for the externalities, expressed by γ_0 and γ_1, and the efficiency index of the economy W does not depend on the size of the population L.

Increasing employment in an agglomeration stimulates an increase in wages (i.e., externalities in employment appear) and γ_0 increases while δ becomes positive. Then, according to the factor coordination principle, W, which can be regarded as the ratio of exogenous product price p to marginal production cost $\bar{p}$, that is

$$W = p(t)/\bar{p}(t)$$

where

$$\bar{p}(t) = A^{-1} \exp(-\mu t) [\omega_0/\alpha_0]^{\alpha_0} [\omega_1/\alpha_1]^{\alpha_1} [L(r)]^{\delta\gamma_0}$$

decreases with an increase in $L(t)$.

When the scale benefits prevail, δ becomes negative and $W(L)$ increases with $L(t)$. In this last case, it pays to increase the scale of production, and the agglomeration exhibits a tendency to increase.

The factor coordination theorem can also be used to find the optimum strategy for allocating consumption of goods and services in such a way that the given utility function attains a maximum. With regard to regional utilities, planners often use broad concepts such as per capita consumption Z_0 within the planning interval or per capita aggregate consumption $(Z_\nu, \nu = 1, 2, \ldots, M)$, which includes education, housing, health, recreational facilities, and other such factors. They believe that if regional budgets for wages and services are planned in a proper way, maximum public satisfaction will follow. In this chapter, the concept of utility as perceived by planners or decision-makers will therefore be used. Thus, the difficult problem of aggregating individual utilities is avoided, as is the question: does the planner's utility correctly reflect individual utilities? An attempt is also made to emphasize the concept of accessibility to employment, services, and recreational amenities, using the approach followed by Kulikowski (1978c). In order to explain this concept and its relation to per capita consumption, consider (19.1) for $i = 0$. This relation can be written as

$$\int_0^T \exp(-\lambda_0 t) \Omega_0 [z_0(t)]^{\gamma_0} \, dt = Z_0 \tag{19.6}$$

where

$$z_0(t) = x_0(t)/P_0; \qquad \Omega_0 = P_0^{\gamma_0} \omega_0 P^{-1}; \qquad Z_0 = Y_0/P$$

and

P is the total population
P_0 is the population of working age
z_0 is the index of accessibility to employment
Ω_0 is the "cost" of providing jobs with a per capita income equal to Z_0

Similarly, one can introduce access indices for services

z_ν = amount of services of category ν/number of people demanding the service

and the cost relations

$$\int_0^T \exp(-\lambda_\nu t) \, \Omega_\nu \, [z_\nu(t)]^{\gamma_\nu} \, dt = Z_\nu \qquad (\nu = 1, 2, \ldots, M) \qquad (19.7)$$

Assume that the utility function $U(z)$ has the form

$$U(z) = \int_0^T \pi(t) \, A \, \exp(-\lambda t) \prod_{\nu=0}^M [z_\nu(t)]^{\beta_\nu} \, dt \qquad (19.8)$$

where

$$\pi(t) A \prod_{\nu=0}^M [\eta_\nu/\Omega(t)]^{\eta_\nu} = V = \text{constant}; \quad \eta_\nu = \beta_\nu \gamma_\nu \quad (\nu = 0, 1, \ldots, M)$$

and $\pi(t)$ is the given "price" attached to the utility. Applying the factor coordination theorem to (19.7) and (19.8), one can determine the optimum strategies for allocating $z_\nu(t)$, assuming $z_0(t)$ to be given.

In the simple case $M = 1$, one obtains from (19.2)

$$\hat{z}_1(t) = \{(\eta_1 \Omega_0(t)/\eta_0 \Omega_1(t)) \, [z_0(t)]^{\gamma_0}\}^{1/\gamma_1}$$

and

$$\hat{U} = U(\hat{z}) = VZ \qquad (19.9)$$

where

$$V = A\pi(t) \, [\eta_0/\Omega_0]^{\eta_0} \, [\eta_1/\Omega_1]^{\eta_1}$$

$$Z = Z_0 + \hat{Z}_1$$

$$\hat{Z}_1 = (\eta_1/\eta_0) Z_0$$

The advantage of the approach used here is that utility (19.9) may be expressed in monetary units (although it is not equal to total per capita consumption Z). It depends on the number V, which depends on prices Ω_0, Ω_1. Because of scale benefits in services ($\gamma_1 < 1$), the value of V in different regions may be different. When total per capita consumption Z and the employment rate $z_0(t)$ are given, maximum utility for access to services is achieved when index $z_1(t)$ equals $\hat{z}_1(t)$.

19.2.2. Regional policies and stability

The regional planning and decision-making process is multiobjective, reflecting the interests of many regional agents or partners (including producers, consumers, and representatives of local and higher-level authorities). Several government commissions or committees evaluate each plan before the final decision is taken. The more members of the committee that support the project, the more chance there is for its approval.

A few concepts from game theory can be used to construct a realistic model which is able to describe the regional planning process and the resulting spatial development. In order to explain how such an approach can be implemented, consider, as a simple example, a committee concerned with evaluating the expected regional budget for a given planning interval; subsidies $S_i (i = 1, 2, \ldots, N)$ are paid from the regional budget to N subregional budgets. The committee consists of $N + 1$ members, representing N subregions or sectors $R_i (i = 1, 2, \ldots, N)$ and the regional authority R_0. Assume that, in addition to S_i, the subregional benefits consist of regional wages $\alpha_{0i} \bar{Y}_i$ (see (19.5)) which are paid by the industries located in the subregion, minus the cost of urbanization (i.e., the population-transfer cost* C_i).

Thus

$$B_i(s_i) = \alpha_{0i} \bar{Y}_i(s_i) + S_i(s_i) - C_i(s_i) \qquad (i = 1, 2, \ldots, N) \tag{19.10}$$

It is also assumed that the policy of the regional authority R_0 is concerned mainly with productive investments $\sum_{i=1}^{N} \alpha_{1i} \bar{Y}_i$, subregional subsidies $S_i(s_i)$, and prices or taxation levels $W_i (i = 1, 2, \ldots, N)$. These are chosen in such a way that the objective function

$$B_0(s) = \sum_{i=1}^{N} \{[W_i - \alpha_{1i} - \alpha_{0i}] \bar{Y}_i(s_i) - S_i(s_i)\} \tag{19.11}$$

is a maximum. The national utility is assumed to be

$$B(s) = \sum_{i=0}^{N} B_i(s) = \sum_{i=1}^{N} \{[W_i - \alpha_{1i}] \bar{Y}_i(s_i) - C_i(s_i)\} \tag{19.12}$$

When the W_i are chosen in such a way that $B_0(s) = 0$, then

$$B(s) = \sum_{i=1}^{N} [\alpha_{0i} \bar{Y}_i(s_i) + S_i(s_i) - C_i(s_i)] \tag{19.13}$$

and the national objective is reduced to maximization of total consumption minus population-transfer costs.

When the function $B(s)$ is concave in s, the unique optimum strategy $s = \hat{s}$ which maximizes the national goal (19.13) can be found. Then, the optimum values of regional employment, investments, and subsidies can be derived. However, the strategy thus derived may be regarded as too distant from the policies of one or more of the participants, the reason being that the model considered is only a simplified version.

---

* The quantities discussed are considered as functions of migration from a particular subregion, represented by migration rate s_i. Migration depends in turn on subregional utility levels; see Kulikowski (1978c, 1980).

For example, subsidy S_i constitutes only a part of the regional budget, the rest being covered by the income from services, taxes, charges, etc.

Real regional fiscal policy is therefore motivated by more objectives than are included in the simplified model (19.10)–(19.13). In addition, a number of random factors (not explicitly specified in the model), such as price changes, may influence the outcome of the game. For these reasons, the participants in the game may wish to bargain in order to realize their goals and policies. They may agree, for example, to depart from $S_i(\hat{s}_i)$, supporting a given subregion i to a greater or lesser extent on the understanding that nobody's interests suffer too much. In other words, the committee may choose to play an $N+1$-person game in order to find, with given taxes $W_i(i = 1, 2, \ldots, N)$, the best or most acceptable scheme of regional subsidies $\hat{S}_i$ $(i = 1, 2, \ldots, N)$.

The game may be played with participants representing all the subregions that constitute a region or all the regions of the whole country. It is also possible to introduce explicitly into the game those industries (or sectors) whose benefits are expressed in terms of productive investments.

In order to determine whether a stable and efficient regional policy exists, one has to check whether a core of the game characterized by (19.10)–(19.13) exists. Using the game-theory concept (Shapley, 1971; Rinaldi et al., 1977; Kulikowski, 1980), it can be shown, given the assumption that profit functions $\alpha_{0i} Y_i(s_i) - C_i(s_i)$ are concave, that a core exists.

19.3. A rural–urban development model

The general methodology outlined in Section 19.2 can be implemented in computerized interactive form. A detailed description of such a system is given in Kulikowski (1980) and Kulikowski and Krus (1980). In this section a brief description of the system will be given.

19.3.1. Production submodel

Using formulae (19.2)–(19.4) for $i = 1$ and 2 (1 = rural subregion and 2 = urban subregion), the production values can be derived as

$$Y_i = W_i \bar{Y}_i = W_i(Y_{0i}/\alpha_{0i}) \qquad (i = 1, 2)$$

where

$$\bar{Y}_i = Y_{0i} + \hat{Y}_{1i}; \qquad \hat{Y}_{1i} = (\alpha_{1i}/\alpha_{0i}) Y_{0i}$$

and

W_i is the index of production efficiency, which can be regarded as the ratio of product price to total production costs

$\hat{Y}_{1i}$ are the capital costs
$\bar{Y}_i$ are the total production costs
α_{1i}, α_{0i} are given parameters

The number of employees can be determined as

$$L_i = \theta_{0i} z_{0i} \bar{P}_i$$

where

θ_{0i} is the share of the population of working age
$\bar{P}_i$ is the subregional population
z_{0i} is the index of access to employment

Labor costs measured in terms of the wages fund can be described as

$$Y_{0i} = \omega_{0i} L_i (1 + s_i (\theta_m / \theta_{0i}))$$

where

$s_i = M_i / \bar{P}_i$ is the migration rate in subregion i
ω_{0i} is the average wage
θ_m is the share of migrants of working age

19.3.2. Consumption submodel

The projected utilities U_i generated at R_i are assumed to take the following form (see 19.6–19.9):

$$U_i = V_i Z_i = V_i (Z_{0i} / \eta_{0i}) \tag{19.14}$$

where

$$Z_i = Z_{0i} + \hat{Z}_{1i}; \qquad \hat{Z}_{1i} = (\eta_{1i} / \eta_{0i}) Z_{0i}$$

and

Z_{0i} is the per capita consumption of personal income
$\hat{Z}_{1i}$ are the expenditures on aggregate (education, health care, etc.) per capita consumption
Z_i is the total per capita consumption
$V_i, \eta_{1i}, \eta_{0i}$ are given parameters

In this section, we are dealing with a linear static production model. The model is used for a given (e.g., 5-year) planning interval. The parameters W_i (production efficiency), α_{1i}/α_{0i} (the ratio of capital to labor costs), and $W_i \omega_{0i}/\alpha_{0i}$ (labor productivity) can be subsequently evaluated using statistical data (instead of using (19.4)). However, the methodology presented in Section 19.2 also enables optimal strategies to be found as a function of time (i.e., in consecutive years of the planning period).

19.3.3. Migration submodel

For the sake of simplicity, the regional system is assumed to be closed with respect to migration. Denoting rural–urban migration by s, then $s = -s_1$ and $s_2 = sl$, where $l = \bar{P}_1/\bar{P}_2$.

It is assumed that migration is a result of the ratio of the consumption levels (measured by utilities) of the rural and urban subregions, according to the econometric relation

$$P_2(s)/\bar{P}_2 = 1 + sl = d(U_2/U_1)^{1/a} \tag{19.15}$$

where

$P_2(s)$ is the urban population dependent on migration $(P_2(s) = P_2(1 + sl))$

d, a are coefficients evaluated ex post facto using the least-squares method

Migration links the production and consumption submodels by the relation

$$Y_{0i}(s) = Z_{0i}(s)P_i(s) = \eta_{0i}Z_i(s)P_i(s) \qquad (i = 1, 2)$$

and introduces costs

$$C(s) = \{[Z_2(s) - Z_1(s)] + Z_h\}\bar{P}_1 s \tag{19.16}$$

where

$Z_2(s) - Z_1(s)$ is the difference in total consumption between urban and rural subregions

Z_h is the per capita urbanization cost

19.3.4. Objective functions and solutions

In the assumed management structure (Figure 19.1), the regional authority R_0 tries to decide on an optimal allocation of resources S between the rural R_1 and urban R_2 parts of the region. At the beginning of each planning period (5 years), it divides the subsidies obtained S into parts S_i, which subsidize the local budgets R_i $(i = 1, 2)$. Subsidies S_i are allocated in turn for capital S_{ki} and aggregate consumption S_{ci} expenditures

$$S_i = S_{ki} + S_{ci}; \qquad S = S_1 + S_2$$

Decision-makers R_0, R_1, R_2 are regarded as participants in the game with the payoff or objective function

$$B_0 = \sum_{i=1}^{2} [\{(W_i/\alpha_{0i}) - 1\} Y_{0i} - S_i]$$

$$B_i = Y_{0i} + S_i \qquad (i = 1, 2)$$

where

B_0 is the value of the products generated in the region minus total consumption, capital, and migration costs $C(s)$

Y_{0i} is the wages fund, when S_i are the subsidies to the subregions obtained from R_0

According to relation (19.15), the number of migrants $\bar{P}_1(s)$ can be controlled by changing the ratio of utilities between urban and rural areas. The utilities depend on rural–urban policies of income distribution (wages) and employment (z_{0i}), as well as on demographic processes and migration. These policies and processes should be correlated to maximize regional benefits.

The effective solution in terms of s can be derived by maximizing the expression

$$B(s) = \sum_{i=0}^{2} B_i(s) = \sum_{i=0}^{2} (W_i/\alpha_{0i}) Y_{0i}(s) - C(s)$$

where, according to eqns. (19.14)–(19.16)

$$C(s) = \bar{P}_1 \hat{Z}_1 [(V_1/V_2)(\{sl + 1\}/d)^a + h] s$$

$$h = (Z_h/\hat{Z}_1) - 1$$

As shown in Kulikowski (1980), the necessary and sufficient (because of the concavity of $B(s)$) condition of optimality $B(s) = 0$ yields the equation

$$(\{sl + 1\}/d)^a [1 + (sla/\{1 + sl\})] = o(s)$$

where

$$o(s) = \frac{V_2}{V_1} \left[\frac{W_2 V_1 \eta_{02} \theta_m d^{-a}}{\alpha_{02} V_2 \theta_{02}} - \frac{W_1 \eta_{01} \theta_m}{\alpha_{01} \theta_{01}} - h \right]$$

Under the optimum $s = \hat{s}$, the necessary subsidies become

$$\hat{S}_{ki} = (\alpha_{1i}/\alpha_{0i}) Y_{0i}(\hat{s})$$

$$\hat{S}_{ci} = (\eta_{1i}/\eta_{0i}) Y_{0i}(\hat{s})$$

while

$$B_i(\hat{s}) = \hat{B}_i = \hat{Y}_{0i} \{1 + (\eta_{1i}/\eta_{0i}) + (\alpha_{1i}/\alpha_{0i})\} \qquad (i = 1, 2)$$

$$B_0(\hat{s}) = \hat{B}_0 = \sum_{i=1}^{2} [Y_{0i} \{(W_i/\alpha_{0i}) - (\eta_{1i}/\eta_{0i}) - (\alpha_{1i}/\alpha_{0i}) - 1\}] - \hat{C}$$

$$B(\hat{s}) = \hat{B} = \sum_{i=1}^{2} [(W_i/\alpha_{0i}) Y_{0i}] - \hat{C}$$

According to (19.6)

$$Y_{0i}(\hat{s}) = \hat{Y}_{0i} = P_i(\hat{s}) Z_i(\hat{s}) \eta_{0i}$$

$$Z_1(\hat{s}) = \hat{Z}_1$$

$$P_1(\hat{s}) = \bar{P}_1(1 - \hat{s})$$

$$P_2(\hat{s}) = \bar{P}_2(\hat{s}) (1 + \hat{s}l)$$

and all the other quantities can be determined from the model equations.

The main problem for the decision-maker is to evaluate losses (in terms of B_i), when the strategies, say $\tilde{S}_{ki}, \tilde{S}_{ci}, \tilde{S}_i$, differ from $\hat{S}_{ki}, \hat{S}_{ci}, \hat{S}_i$ ($i = 1, 2$). Let us assume, for example, that the planner proposes capital expenditures $\tilde{S}_{ki} \neq \hat{S}_{ki}$. As a result of strategy $\tilde{S}_{ki}$, production output decreases because of the decrease in W_i. The coefficients W_i become

$$\tilde{W}_i = \tilde{W}_i (\tilde{\alpha}_{0i}/\alpha_{0i})^{\alpha_{0i}} (\tilde{\alpha}_{1i}/\alpha_{1i})^{\alpha_{1i}} \leqslant W_i \qquad (i = 1, 2)$$

where

$$\tilde{\alpha}_{1i} = \alpha_{1i} \tilde{S}_{ki}/(\alpha_{0i} \hat{S}_{ki} + \alpha_{1i} \tilde{S}_{ki})$$

$$\tilde{\alpha}_{02} = 1 - \tilde{\alpha}_{1i}$$

A similar situation appears when planned expenditures $\tilde{S}_{ci}$ on aggregate consumption are proposed such that $\tilde{S}_{ci} \neq \hat{S}_{ci}$. A corresponding decrease in utilities follows as a result of the decrease in the coefficients V_i, which now become

$$\tilde{V}_i (\tilde{\eta}_{0i}/\eta_{0i})^{\eta_{0i}} (\tilde{\eta}_{1i}/\eta_{1i})^{\eta_{1i}} \leqslant V_i$$

where

$$\tilde{\eta}_{1i} = \eta_{1i} \tilde{S}_{ci}/(\eta_{0i} \hat{S}_{ci} + \eta_{1i} \tilde{S}_{ci}) \qquad (i = 1, 2)$$

$$\tilde{\eta}_{0i} = 1 - \tilde{\eta}_{1i}$$

The value of $\tilde{s}$ resulting from this strategy can be derived as a solution of the equation

$$(\{1 + \tilde{s}l\}/d)^a = \tilde{V}_2 \tilde{Z}_2(\tilde{s})/\tilde{V}_1 \tilde{Z}_1(\tilde{s}) \tag{19.17}$$

where

$$\tilde{Z}_i(\tilde{s}) = (\tilde{S}_{ki} \alpha_{0i} + \tilde{S}_{ci} \alpha_{1i}) [\alpha_{1i} P_i(\tilde{s})]^{-1} \qquad (i = 1, 2)$$

After solving (19.17), it is possible to derive $\tilde{B}_i$ ($i = 0, 1, 2$), $\tilde{B}$, and other endogenous quantities in a similar way, as in the case of optimal $\hat{s}$.

When $\tilde{Z}_i \neq \hat{Z}_i$ we find that $\tilde{B} \leqslant \hat{B}$, while, generally, $\tilde{B}_i \neq \hat{B}_i$ ($i = 0, 1, 2$). The equality signs follow when $\tilde{S}_{ki} = \hat{S}_{ki}, \tilde{S}_{ci} = \hat{S}_{ci}, \tilde{S}_i = \hat{S}_i$, and $i = 1, 2$.

19.3.5. An interactive system

The simplified rural–urban development model was implemented in the form of an Interactive Rural–Urban Development system (IRUD) on a PDP 11/70 computer working under the UNIX operating system. Using IRUD, decision-makers can play a game of subsidies allocation and can evaluate different regional development policies. All input variables and parameters as well as decision variables are introduced into the system in a dialogue manner using screen monitors, which in turn present the results of the calculations. It is not necessary for the users of the system to be familiar with programming techniques: they simply answer the questions appearing on the screen.

The system uses model parameters, exogenous variables, decision variables, and policy variables as input data. The model parameters are coefficients in equations describing production (W_1, W_2, α_{01}, α_{02}, α_{11}, α_{12}), consumption (V_1, V_2, η_{01}, η_{02}, η_{11}, η_{12}), and migration and employment (θ_{01}, θ_{02}, θ_m, a, d). The exogenous variables consist of rural $\bar{P}_1$ and urban $\bar{P}_2$ populations and average wages ω_{01}, ω_{02}. The decision variables are values of subsidies S_1, S_2 (the strategy of the player R_0), and S_{k1}, S_{c1}, S_{k2}, S_{c2} (strategies of players R_1 and R_2). The policy variables are selected from the model parameters and exogenous variables and include minimal rural per capita consumption Z_1, average wages ω_{01}, ω_{02}, and housing costs Z_h.

The model derives the following values:

B_0, B_1, B_2, B – the payoff (goal) functions
s – the migration rate
M – the number of migrants
C – the migration cost

and for the rural and urban sectors the model calculates:

$P_1(s)$, $P_2(s)$ – the populations dependent on migration
Y_1, Y_2 – the production values
y_{01}, y_{02} – the wages fund
L_1, L_2 – the levels of employment
z_{01}, z_{02} – the index of access to employment
Z_1, Z_2 – the total per capita consumption
Z_{01}, Z_{02} – the consumption of personal income
Z_{11}, Z_{12} – the aggregated consumption

All the endogenous variables are derived for the decision strategies introduced at each stage of the game. They are also calculated for the optimum solution maximizing the function B.

To test the model and the operation of the IRUD system, the values of exogenous variables and model parameters were based on statistical data for the Bydgoszcz voivodship (region) of Poland for 1973 (Rocznik Statystyczny, 1977; Rocznik Statystyczny Wojewodztw, 1978). Examples of the calculations are presented in Figure 19.3. Detailed printouts have been published by Kulikowski and Krus (1980).

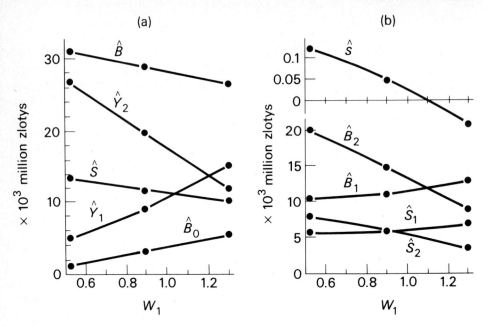

Figure 19.3. Optimal solutions as a function of W_1. The urban economy parameter W_2 was assumed to be 1.5.

It should be noted that the parameters W_1 and W_2 are of great importance. They can be interpreted as the production efficiencies of the rural and urban economies, respectively. Investment in water-system development results in an increase in production efficiencies. The optimum solution as a function of W_1 is shown in Figure 19.3. Strategies are shown for changes in W_1 from 0.52 up to 1.3; W_2 (for the urban economy) was assumed to be constant at 1.5. Figure 19.3(a) shows changes in the objective functions $\hat{B}$ and $\hat{B}_0$, total subsidies $\hat{S}$, and rural and urban production values $\hat{Y}_1$, $\hat{Y}_2$. The total objective function $\hat{B}$, which is the sum of production values minus migration costs, decreases with W_1. However $\hat{B}_0$, which takes into account the value of subsidies, increases with W_1. If W_1 increases, fewer subsidies are needed. In Figure 19.3(b), changes in the optimal migration rate $\hat{s}$ as well as in the values of $\hat{B}_1$ and $\hat{B}_2$ and $\hat{S}_1$ and $\hat{S}_2$ are given.

References

Albegov, M. and Kulikowski, R. (Editors) (1978a). Noteć Regional Development. Proceedings of Task Force Meeting I. RM-78-40. International Institute for Applied Systems Analysis, Laxenburg, Austria.

Albegov, M. and Kulikowski, R. (Editors) (1978b). Noteć Regional Development. Proceedings of Task Force Meeting II. IBS PAN Research Memorandum. Systems Research Institute of the Polish Academy of Sciences, Warsaw.

Gutenbaum, J., Makowski, M., and Owsinski, J. (1981). Noteć Project: System of Models for Regional Planning and Management. Paper presented at the IIASA Conference on Theoretical and Practical Aspects of Regional Development, held at the International Institute for Applied Systems Analysis, Laxenburg, Austria, in March 1980.

Kulikowski, R. (1978a). Optimization of Rural–Urban Development and Migration. Environment and Planning A, 10: 557–591.

Kulikowski, R. (1978b). Regional development: labor, investments and consumption allocation policy impact. In M. Albegov and R. Kulikowski (Editors), Noteć Regional Development. Proceedings of Task Force Meeting I. RM-78-40. International Institute for Applied Systems Analysis, Laxenburg, Austria.

Kulikowski, R. (1978c). Regional utilities, access to services and optimum migration strategy. In M. Albegov and R. Kulikowski (Editors), Noteć Regional Development. Proceedings of Task Force Meeting II. IBS PAN Research Memorandum. Systems Research Institute of the Polish Academy of Sciences, Warsaw.

Kulikowski, R. (1980). Optimization of regional development: an integrated model for studying socio-economic and environmental policies. In M. Albegov, R. Kulikowski, and O. Panov (Editors), Proceedings of the Joint Task Force Meeting on Development Planning for the Noteć (Poland) and Silistra (Bulgaria) Regions. Volume I. CP-80-9. International Institute for Applied Systems Analysis, Laxenburg, Austria.

Kulikowski, R. and Krus, L. (1980). A regional computerized (interactive) planning system. In M. Albegov, R. Kulikowski, and O. Panov (Editors), Proceedings of the Joint Task Force Meeting on Development Planning for the Noteć (Poland) and Silistra (Bulgaria) Regions. Volume I. CP-80-9. International Institute for Applied Systems Analysis, Laxenburg, Austria.

Rinaldi, S., Soncini-Sessa, R., and Whinston, A. B. (1977). Stable Taxation Schemes in Regional Environmental Management. RR-77-10. International Institute for Applied Systems Analysis, Laxenburg, Austria.

Shapley, L. S. (1971). Cores of convex games. International Journal of Game Theory, 1(1): 11–26.

Rocznik Statystyczny (1977) (Statistical Yearbook 1977). Glowny Urzad Statystyczny, Warsaw.

Rocznik Statystyczny Wojewodztw (1978) (Statistical Yearbook of Voivodships 1978). Glowny Urzad Statystyczny, Warsaw.

Zawadzki, S. M. (1973). Polska, Przestrzen, Spoleczenstwo (Poland, Space, Society). Panstwowe Wydawnictwo Ekonomiczne, Warsaw.

Regional Development Modeling: Theory and Practice
M. Albegov, A.E. Andersson and F. Snickars (editors)
North-Holland Publishing Company
© IIASA, 1982

Chapter 20

A DECISION SUPPORT SYSTEM FOR REGIONAL DEVELOPMENT

Boris Mihailov
International Institute for Applied Systems Analysis, Laxenburg (Austria)

Isak Assa*
Institute for Social Management, Sofia (Bulgaria)

20.1. Introduction

In 1979 researchers within the Regional Development Task at IIASA began to construct a set of regional economic models for the Silistra region in Bulgaria (Andersson
and Philipov, 1979). This was done as part of the IIASA case study outlined in Chapter
1, pp. 20–22. Experience has been gained in linking these models to form an integrated system and it is now possible to describe some of the essential relationships
between them (Albegov et al., 1980). The model system developed could be applied
to any region in Bulgaria, and although it has been constructed for the Bulgarian
economy, it could easily be modified to describe the conditions of other planned
economies. The pilot version of the system of models presented here has been tested
using aggregated data. More detailed data will be used in the final version of the
system, which will not be available until the final versions of all of the submodels have
been completed.

20.2. Features of the system of models

20.2.1. The region as part of the national system

Since the national planning system is centralized, the autonomy of each region is
restricted. The regional system of models must therefore take into account the plans
made at the national level (Figure 20.1).

The production volume of a sector within a region should not be optimized locally
but be specified on a national basis (Mihailov, 1979). This is because the sum of the

---

*Currently at the International Institute for Applied Systems Analysis, Laxenburg, Austria.

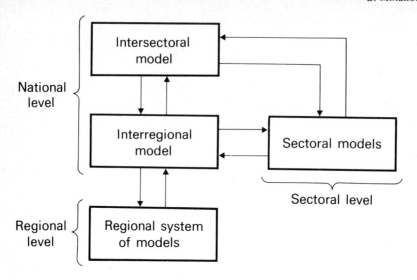

Figure 20.1. Interactions between national, regional, and sectoral models.

optimal solutions for a given sector in each region does not necessarily give the optimal solution for the same sector at the national level. Similarly, the volume of total sectoral production within a given region should not be determined by optimization within the region. This problem must be solved at the national level because the sum of the total sectoral production in each region will not necessarily give the desired national production volume; production in the various sectors of one region cannot be decided without considering production in the other regions.

The level of certain sectoral activities (energy production, etc.) within a given region is determined in an interregional model by the central planning body. The location of these activities and both the level and location of activity in other sectors (pollution control, housing, and social services) are determined by the regional authorities. This is necessary because the region is considered as a single point in the highly aggregated interregional model used at the national level.

20.2.2. Management functions at the regional level

The planning system described above devolves three main functions to the regional managers (Mihailov, 1980):

1. *Local optimization.* The size and output of the most important sectors should be optimized through technological development.
2. *Coordination.* The production volumes and locations of subsystem facilities should be coordinated with the volume and location of available resources.
3. *Strategic planning.* Scenarios for regional development should be generated and analyzed.

The results obtained from the different scenarios generated by strategic planning provide feedback to the national level. These scenarios consider the economic and social problems of the region simultaneously. The regional subsystems are linked at the national level; this can be difficult because the internal links of the system are governed by regional functions.

The relationships between the regional and the national levels take into account the existing planning mechanisms. However, it was not possible to make the system of regional models compatible with a set of national models because no such national models exist.

20.3. The framework of the model

20.3.1. Generation of system forecasts

A regional input–output table is constructed using data gathered from 37 plants within the Silistra region. The basic input–output model, shown in Table 20.1, is referred to as a forecasting model because its results are used as constraints in subsequent time periods. The subsystem models are also constrained by the local resources available and by the results of the population growth model. Thus, the system of models is basically recursive.

Regression-type statistical models are used to analyze the trends in the technological input–output coefficients and to derive consumption and other functions.

The relationship between central and regional levels during the initial stages of the planning process is defined in terms of constraints on regional production volumes, capital investment, size of labor force, and so on.

20.3.2. Linkage procedures

The results of the input–output model are used as the input for a sequence of calculations in which various regional demands are linked to each other. The models are linked in six stages, as shown in Figure 20.2. These six stages are discussed briefly in the following sections.

Stage 1. Generation of constraints for the subsystem models. Results are obtained from the basic input–output forecasting model using an algorithm for the solution of a system of linear equations. A bifactorization method is adopted to deal with the sparseness of the input–output matrix and the special structure of the model.

The regression equations for personal consumption and other exogenous variables that must be forecasted are estimated by least-squares techniques.

Stage 2. Optimization of separate sectors within the region. The nine sectors involved at this stage are first optimized sequentially, using sensitivity analysis.

Table 20.1. Input–output forecasting model.

Producers \ Consumers	x_1	x_2	x_3	x_4	x_5	x_6	Total (1–6)	Personal consumption (C_n)	Exports (E)	Environmental protection (E_v)	Health care (H)	Human settlements (S)	Capital investment (K)	Final consumption (Y) (8–13)	Total (7+14)
x_1															
x_2															
x_3															
x_4															
x_5															
x_6															
Imports															
Wages															
Depreciation															
Total															
Labor															
Labor coefficient															
Wage coefficient															

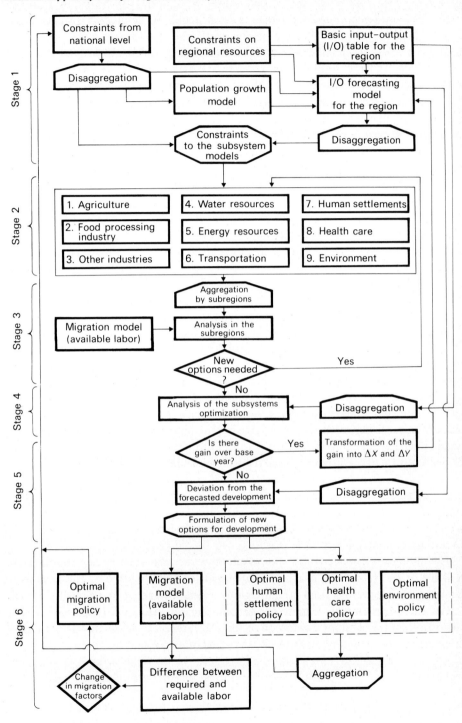

Figure 20.2. The system of regional models.

A mixed integer optimization model is constructed for the production subsystems; the objective function is minimum annual costs (the sum of production and capital investment costs). Technological and locational options are examined, subject to constraints from the input–output forecasting model and some restrictions and limits set at the national level (Christov et al., 1980). A basic assumption here is that economic sectors or subsectors can be divided into those producing goods of national importance and those producing goods of regional importance. Production of goods of national importance is fixed at this level of the analysis.

The solution algorithm is based on an integer-enumeration technique combined with a standard linear programming model.

Transportation is given special treatment at this stage. A transportation model that takes into account various aspects of the structure and use of the transportation system is constructed. Subsystems of the transportation network describing existing and possible modes of transportation are defined. The model considers the total demand for transportation by the production sectors and by the regional subsystems (including traffic traveling to and from other regions). One aspect of the modeling work involves defining the scope for technological development in each mode of transportation. The ultimate aim is to arrive at a distribution of modes of transportation that will satisfy the demands of all sectors at minimum total cost.

The model is formulated as a general flow problem. The solution algorithm is based on a modified Ford–Fulkerson technique combined with a shortest-path algorithm.

Stage 3. Congestion problems. In stage 2, each region is allocated a certain level of activity in each sector. In stage 3, these allocations are evaluated with the help of various indicators. One such indicator, for example, shows whether the labor force available is consistent with the labor force required by the solution of the various subsystem models.

Analysis of these solutions reveals the subregions most active in each sector (Kulikowski, 1980), and the resulting degree of environmental pollution may be measured. It is also possible to find out which activities are complementary or incompatible in terms of improving economic efficiency, achieving social goals, or reducing environmental pollution. New options may then be formulated and fed back to the subsystem optimization models described in stage 2.

The discrepancy between the labor required by the subsystem models and the labor available in the subregions is also estimated and fed back to stage 2. The amount of labor available is derived from a multiregional migration model based on the propensity of the population to migrate (Andersson and La Bella, 1979; Mihailov, 1979). The migration propensities depend on regional differences in economic factors such as employment opportunities, wage levels, and standard of living.

An iterative solution of the models in stage 2 finally yields a result in which the problems discussed above have been reduced to an acceptable level in each region.

Stage 4. Optimization of the development of the region as a whole. The reduced annual costs obtained from stages 2 and 3 are compared with the costs assumed for the initial period, and the differences determined recursively. The total changes expressed by these differences are then transformed into additional intermediate production (ΔX) and consumption (ΔY) volumes, which are introduced into the input–output fore-casting model. Consistency between additional production and consumption volumes is achieved by an interactive procedure.

Stage 5. Coordination of the subsystems within the region. The results produced in stage 4 usually differ from the results of the regional input–output forecasting model. In addition, they may not satisfy all the constraints set at the central level. It is there-fore necessary for the regional planning body to coordinate the development of all sectors (activities) in the region. The inconsistencies described above provide a basis for new patterns of regional development. The consistency analysis is performed by means of a dialogue between regional and national levels, the aim of which is to maxi-mize the level of final consumption in the region. In the approach taken here, this criterion is expressed indirectly in terms of minimizing production costs subject to lower bounds on consumption volumes.

Stage 6. Strategic development of the region. In this stage the model analyzes the effects of different policies on the future development of the region. The issues examined are those of long-term importance to the region: environmental protection, population growth and migration, development of human settlements, health care, and so on.

Regional policies on each of these issues are obtained by balancing the marginal additional value of total income and the marginal total costs resulting from the adoption of each policy option. Thus, a regional environmental protection policy is selected by taking into account, on the one hand, the additional financial and social gain from the production processes and, on the other, the ecological costs and losses that would be incurred as a result.

Regional migration policy is also very important. The optimal migration flows are determined from the balance between additional income (from increased production and services) and additional costs (for transportation, housing, and services). The optimal migration flows may then be compared with the expected migration flows; the difference between them is the volume of migration that should be encouraged by other means. Migration is then regulated to obtain the optimal flow suggested by the tradeoff between gains and costs.

Human settlement and health-care policies are determined by similar methods at this stage of the analysis.

20.4. Mathematical description of selected models

This section provides some examples of the types of model used in the IIASA case study of the Silistra region in Bulgaria. The region is relatively small, having only 190,000 inhabitants, and therefore the existence of indivisible factors cannot be ignored; indivisibility is a central concept, especially in the production sector models.

Two of the submodels will be described in some detail: the capacity–location models of the production sectors, and the models of the transportation sector. Finally, some of the basic socioeconomic indicators used to assess discrepancies between solutions derived hierarchically and solutions based on local analysis will be presented and discussed.

20.4.1. The mixed integer optimization model for the production sectors

The production cost-minimization problem is formulated as

$$\text{minimize } Z = \sum_{i=1}^{m} \sum_{j=1}^{n} [c_{ij} X_{ij}^{R} + \sum_{k=1}^{T_i} (p_{ijk} + s_{ijk}) A_{ijk} z_{ijk}]$$

where

> m is the number of products
> n is the number of subregions
> c_{ij} is the current unit production cost for product i in subregion j
> X_{ij}^{R} is the production volume of product i in subregion j (products of regional importance)
> T_i is the number of technological options for the development of product i
> p_{ijk} is the unit production cost, including transportation costs, for product i in subregion j under technological option k
> s_{ijk} is the social and other cost per unit of production volume for product i in subregion j under option k
> A_{ijk} is the capacity of a plant that produces product i in subregion j under option k
> z_{ijk} is $\begin{cases} 1 \text{ if option } k \text{ is selected} \\ 0 \text{ if option } k \text{ is rejected} \end{cases}$

These costs are minimized subject to the following constraints:

Production volume constraints

$$X_i^{\min} \leqslant \sum_{j=1}^{n} (X_{ij}^{R} + \sum_{k=1}^{T_i} A_{ijk} z_{ijk}) \leqslant X_i^{\max} \qquad (i = 1, 2, \ldots, m)$$

where $X_i^{\min}$ and $X_i^{\max}$ are the lower and upper bounds, respectively, of the production volume of product i.

Capital investment constraints

$$K^{\min} \leqslant \sum_{i=1}^{m} \sum_{j=1}^{n} \sum_{k=1}^{T_i} f_{ijk} z_{ijk} \leqslant K^{\max}$$

where $K^{\min}$ and $K^{\max}$ are the lower and upper bounds, respectively, of capital investment for the region, and f_{ijk} is the capital investment required for technological option k for product i in subregion j.

Labor constraints

$$L_j^{\min} \leqslant \sum_{i=1}^{m} \sum_{k=1}^{T_i} l_{ijk} z_{ijk} \leqslant L_j^{\max} \quad (j = 1, 2, \ldots, n)$$

where $L_j^{\min}$ and $L_j^{\max}$ are the lower and upper bounds, respectively, of labor available in subregion j, and l_{ijk} is the labor required to produce product i in subregion j under option k.

Subregional production constraints

$$X_j^{\min} \leqslant \sum_{i=1}^{m} (X_{ij}^{R} + \sum_{k=1}^{T_i} A_{ijk} z_{ijk}) \leqslant X_j^{\max} \quad (j = 1, 2, \ldots, n)$$

where $X_j^{\min}$ and $X_j^{\max}$ are the lower and upper bounds, respectively, of production in subregion j.

Constraints on resources (water, land, etc.)

$$U_j^{\min} \leqslant \sum_{i=1}^{m} \sum_{k=1}^{T_i} u_{ijk} z_{ijk} \leqslant U_j^{\max} \quad (j = 1, 2, \ldots, n)$$

where $U_j^{\min}$ and $U_j^{\max}$ are the lower and upper bounds, respectively, of the volume of resources available in subregion j, and u_{ijk} is the volume of resources used to produce product i in subregion j under option k.

Zero–one group constraints

$$\sum_{i=1}^{m} \sum_{j=1}^{n} \sum_{k=1}^{T_i} z_{ijk} \geqslant N$$

where N is a positive integer denoting the total number of options taken into consideration.

20.4.2. The models of the transportation sector

Analysis of the transportation sector takes place in two stages. In the first stage, investment in each form of transportation (road, rail, etc.) is optimized using a predefined set of development options. For example, investment in the railway system might involve extending the railway network and/or modernizing the rolling stock.

In the second stage, it is assumed that the investment recommended in the first stage has been made, and the volume of goods is then distributed over the various forms of transportation to minimize overall transportation costs. This is done using a standard network flow model.

Optimization of each mode of transport. A cost-minimization model is first developed for each mode of transport separately. Investment choices are then made on the basis of the results (Mihailov and Nicolov, 1980).

The following notation is used:

M_g is the set of development options for transportation mode g

X_{gkb}^{ij} is the volume of goods carried by component b of transportation mode g, between supply and demand points i and j, under development option k

Y_{gkb} is the volume of goods carried by component b of transportation mode g under development option k

C_{gkb} is the average unit operating cost for component b of mode g under development option k

K_{gkb} is the capital cost for component b of mode g under development option k

T_{gkb} is the average operational lifetime of traffic system component b of mode g under development option k

A_{gkb} is the capacity of component b of mode g under option k

p_g^i is the supply of goods at point i to be transported by mode g

Q_g^j is the demand for goods at point j to be satisfied by mode g

The variables Y_{gkb} are defined as

$$Y_{gkb} = \sum_i \sum_j X_{gkb}^{ij}$$

It is then possible to formulate the cost-minimization model for each mode of transport separately, disregarding the choice of routes:

$$\min_{X_{gkb}^{ij}} \sum_b (C_{gkb} + K_{gkb}/T_{gkb}) Y_{gkb} \tag{20.1}$$

subject to

$$\sum_j \sum_b X_{gkb}^{ij} = p_g^i \tag{20.2}$$

$$\sum_i \sum_b X_{gkb}^{ij} = Q_g^j \tag{20.3}$$

$$Y_{gkb} \leqslant A_{gkb} \tag{20.4}$$

$$X_{gkb}^{ij} \geqslant 0 \tag{20.5}$$

The solution of this problem yields values of the cost function F_{gk} which will depend on the mode considered g and on the development option k. The option k^* leading to the lowest value of F_{gk} for each g, F_{gk}^*, will then be chosen as the optimal transport investment strategy.

Optimization of all modes of transport. The model used in the previous section assumes a certain distribution in the volume of goods carried by the different modes of transport. This is done to keep the problem of investment in facilities and equipment to a reasonable size in terms of data requirements and variables. Optimization is therefore used at this stage to determine which type of product should be carried by each mode of transport.

The model used is similar to that defined by system (20.1)–(20.5), with k replaced by the optimal development option k^* for each mode of transport. In addition, constraints (20.2) and (20.3) are summed over all modes g so that only total supply and total demand are given. This model considers approximately 140 supply and/or demand points, 17 product types, and 7 modes of transport.

The two-step approach used in modeling transportation in the Silistra region may lead to suboptimal solutions after the first stage has been completed. Iterations are therefore performed, testing at the second stage the total cost effect of alternative development options k close to the optimal k^* deduced from the first stage.

20.4.3. The use of socioeconomic indicators

It would be possible to continue this survey with detailed descriptions of the models used for agriculture, water supply, and human settlements and services. However, in a short report of this nature it would be more useful to discuss some of the indicators used in the models, partly to quantify the efficiency of various solutions and partly to measure the gaps that need to be closed between different submodels, as discussed in Section 20.2.

The distribution of production over subregions is given by

$$X_j = \sum_{i=1}^{m} \sum_{k=1}^{T_i} z_{ijk}^* A_{ijk}$$

where z_{ijk}^* reflects the optimal solution to the production cost-minimization problem formulated in Section 20.4.1.

The results of the production subsystem models may also be used to check the results of the initial sectoral production forecast:

$$\Delta X_i = \sum_{j=1}^{n} (X_i^* + \sum_{k=1}^{T_i} z_{ijk}^* A_{ijk}) - (X_i^0 - X_i^N) \tag{20.6}$$

In (20.6), ΔX_i is the difference between the outcome of the subsystem optimization for sector i and the initial forecast made by the input–output model. Note that production in sectors of national importance, X_i^N, is subtracted from the initial forecast X_i^0.

The subregional labor demand is given by

$$L_j^{nec} = \sum_{i=1}^{m} w_i X_{ij}$$

where w_i is the ratio of labor to output in sector i, X_{ij} is the output of sector i in subregion j, and L_j^{nec} is the amount of labor necessary in subregion j.

The labor available in subregion j, L_j^{av}, is determined using standard demographic methods combined with a rather advanced migration model of the linear-regression type. The difference $L_j^{nec} - L_j^{av} = \Delta L_j$ is one of the main indications of inconsistency at later stages in the modeling.

The difference between subsystem optimization of labor for sector i and the initial forecast made by the input–output model is given by an equation similar to eqn. (20.6):

$$\Delta L_i = \sum_{j=1}^{n} \sum_{k=1}^{T_i} l_{ijk} z_{ijk}^* - (L_i^0 - L_i^N)$$

This should also decrease to zero by the last iteration.

The last example involves capital and investment costs. The difference, ΔC, between the reduced annual costs from optimization of the individual subsystems and the costs from the initial input–output calculations is used as the basic criterion for combining the qualitative and quantitative analyses performed in stage 5 of the iterative scheme presented in Section 20.3.

$$\Delta C = \sum_{i=1}^{m} \sum_{j=1}^{n} c_{ij} X_{ij}^R + \sum_{k=1}^{T_i} (p_{ijk} + s_{ijk}) A_{ijk} z_{ijk}^* - (C^0 - C^N) \tag{20.7}$$

The notation used here is defined in Section 20.4.1. Note that X_{ij}^R denotes the production of goods of regional importance in sector i, subregion j.

The cost discrepancy calculated in eqn. (20.7) is related to annual running costs rather than to investment costs. Of course, there may also be discrepancies between the investment costs computed by subsystem optimization and those forecast by the input–output model:

$$\Delta K_i = \sum_{j=1}^{n} \sum_{k=1}^{T_i} f_{ijk} z_{ijk}^* - (K_i^0 - K_i^N)$$

The ΔK_i are important in the adjustment of the final demand deliveries.

20.5. Conclusions

The system of models has been designed to be compatible with the current organizational structure in Bulgaria. The division of production in each sector into goods of national importance (given exogenously) and goods of regional or local importance (obtained endogenously) is an example of the special treatment required to make the models reflect the centralized planning system.

This case study has shown clearly that a system of models for regional development can only be workable if it is compatible with the national planning philosophy. If this were not the case it would be impossible to achieve consistency between national and regional planning.

A system of models for regional development should reflect the existing economic and social mechanisms, especially with respect to the management functions of the regional planning body.

In future work the proposed system of models will be developed further to provide, among other things, a dynamic solution of the system of regional models and a more detailed description of the feedback links between subsystems. More attention will also be given to sensitivity analyses of the optimal solutions.

References

Albegov, M., Kulikowski, R., and Panov, O. (Editors) (1980). Proceedings of the Joint Task Force Meeting on Development Planning for the Notéc (Poland) and Silistra (Bulgaria) Regions. CP-80-9. International Institute for Applied Systems Analysis, Laxenburg, Austria.

Andersson, Å. E. and La Bella, A. (1979). A system of models for integrated regional development: an application to the Silistra case study. In A. E. Andersson and D. Philipov (Editors), Proceedings of Task Force Meeting I on Regional Development Planning for the Silistra Region (Bulgaria). CP-79-7. International Institute for Applied Systems Analysis, Laxenburg, Austria.

Andersson, Å. E. and Philipov, D. (Editors) (1979). Proceedings of Task Force Meeting I on Regional Development Planning for the Silistra Region (Bulgaria). CP-79-7. International Institute for Applied Systems Analysis, Laxenburg, Austria.

Christov, E., Assa, I., and Panov, O. (1980). Modeling the location and development of industry in the Silistra region. In M. Albegov, R. Kulikowski, and O. Panov (Editors), Proceedings of the Joint Task Force Meeting on Development Planning for the Notéc (Poland) and Silistra (Bulgaria) Regions. Volume I. CP-80-9. International Institute for Applied Systems Analysis, Laxenburg, Austria.

Kulikowski, R. (1980). Optimization of regional development: an integrated model for studying socioeconomic and environmental policies. In M. Albegov, R. Kulikowski, and O. Panov (Editors), Proceedings of the Joint Task Force Meeting on Development Planning for the Notéc (Poland) and Silistra (Bulgaria) Regions. Volume I. CP-80-9. International Institute for Applied Systems Analysis, Laxenburg, Austria.

Mihailov, B. (1979). Migrations in the system of models for integrated territorial development. In A. E. Andersson and D. Philipov (Editors), Proceedings of Task Force Meeting I on Regional Development Planning for the Silistra Region (Bulgaria). CP-79-7. International Institute for Applied Systems Analysis, Laxenburg, Austria.

Mihailov, B. (1980). Some Considerations about the Construction of a System of Regional Models. WP-80-120. International Institute for Applied Systems Analysis, Laxenburg, Austria.

Mihailov, B. and Nicolov, N. (1980). Modeling the transportation system as a part of the system of regional development models. In M. Albegov, R. Kulikowski, and O. Panov (Editors), Proceedings of the Joint Task Force Meeting on Development Planning for the Notéc (Poland) and Silistra (Bulgaria) Regions. CP-80-9. International Institute for Applied Systems Analysis, Laxenburg, Austria.

Regional Development Modeling: Theory and Practice
M. Albegov, A.E. Andersson and F. Snickars (editors)
North-Holland Publishing Company
© IIASA, 1982

Chapter 21

AN EVALUATION OF JAPANESE TRANSPORT INVESTMENT PROJECTS

Noboru Sakashita*
Institute of Socio-Economic Planning, University of Twukuba, Sakura, Ibaraki (Japan)

21.1. Introduction

The aim of the present study is to evaluate, using a relatively simple system of regional econometric models, the interregional socioeconomic effects of adopting various national transport investment projects in Japan, such as the extension of the expressway system or the new trunk-line (Shinkansen) railway system to remote areas of the country. For a more detailed assessment, see Sakashita (1974, 1981). The analysis is performed for twelve alternative cases, representing different combinations of transport investment projects, the details of which are given in Table 21.1.

The time horizon of the study extends up to 1995, because a long-range forecast is needed for decision-making on the implementation of the alternative projects. Usable observed data are, however, limited to those for the period 1960–1975; it is therefore almost impossible to adopt an orthodox method of forecasting based on a firmly built econometric model system. It has therefore been necessary to rely on the following, somewhat unorthodox procedure. First, a system of econometric submodels is constructed, based on the available data and using normal methods of estimation. Second, experimental simulations for the future are performed using the estimated model and the simulations are examined to detect any possibly "pathological" movements** of particular variables. If pathological movements are found, the values of a number of relevant parameters are changed systematically in order to stabilize the time paths of important variables. We refer to this procedure as the "stabilization method". Having obtained a well-behaved model using this procedure, simulations corresponding to the twelve alternative cases are performed with the same model system. Alternative simulations differ only in the altered time paths of exogenous policy variables.

---

* The author wishes to express his sincere gratitude to S. Kunihisa and other staff of the Institute of Behavioral Science in Tokyo and to Professor Kaiyama of Saitama University for their cooperation in this simulation analysis.
** Here "pathological" movements implies extraordinary behavior of particular variables over time.

Table 21.1. Alternative transport investment projects in Japan: the 12 cases studied.

Case	New trunk-line	Expressway	Airports
1	No investment after 1982	No investment after 1982	No investment after 1982
2	No investment after 1982	Planned networks completed	Planned local airports completed
3	All planned networks completed	Planned networks completed	Planned local airports completed
4	All planned networks completed	No investment after 1982	No investment after 1982
5	No investment after 1982	Planned networks completed	No investment after 1982
6	No investment after 1982	No investment after 1982	Planned local airports completed
7	Sapporo–Aomori (Hokkaido) route completed	Planned networks completed	Planned local airports completed
8	Hakodate–Morioka (North Tohoku) route completed	Planned networks completed	Planned local airports completed
9	Sapporo–Morioka (Hokkaido and North Tohoku) route completed	Planned networks completed	Planned local airports completed
10	Tokyo–Osaka (Hokuriku) route completed	Planned networks completed	Planned local airports completed
11	Fukuoka–Kagoshima (South Kyushu) route completed	Planned networks completed	Planned local airports completed
12	Fukuoka–Nagasaki (West Kyushu) route completed	Planned networks completed	Planned local airports completed

The method of simulation to some extent resembles that of system dynamics. There is, however, an essential difference in that the method used here depends substantially on actual past data as well as on the estimation method adopted.

21.2. The model

The main part of the model system, which is called the Trunk model, consists of forty-six regional[*] submodels that simultaneously describe aspects of the production, distribution, and expenditure of each regional economy. An outline of the Trunk model for a particular region is shown in Figure 21.1.

The model for a given region is explained by the system of equations presented in Sections 21.2.1–21.2.11 below.

21.2.1. Production capacity function (trend estimation)

$$V_i = f_i\{K_{i,-1}, L_i, NACCC, NACCP\} \tag{21.1}$$

where

V_i is the real net product of industry i ($i = 1, 2, \ldots, 8$, throughout this section)

$K_{i,-1}$ is the real capital stock of industry i at the end of the previous year

L_i is the level of employment in industry i

$NACCC$ is the nonaccessibility[**] in terms of goods transportation

$NACCP$ is the nonaccessibility in terms of passenger transportation

In principle, f_i takes the form of a linear homogeneous Cobb–Douglas function with respect to $K_{i,-1}$ and L_i, and a power function such as $(NACCC)^\delta$ with respect to the nonaccessibility variables. These latter variables will be defined later. Eight industry classifications are recognized in the model; details of the classifications are given in Table 21.2.

---

[*] The terms region and regional are used throughout this chapter instead of the Japanese terms prefecture and prefectural.

[**] The term nonaccessibility is used throughout this chapter to indicate the transformation of an accessibility function into a cost function. Thus, when accessibility increases, nonaccessibility decreases.

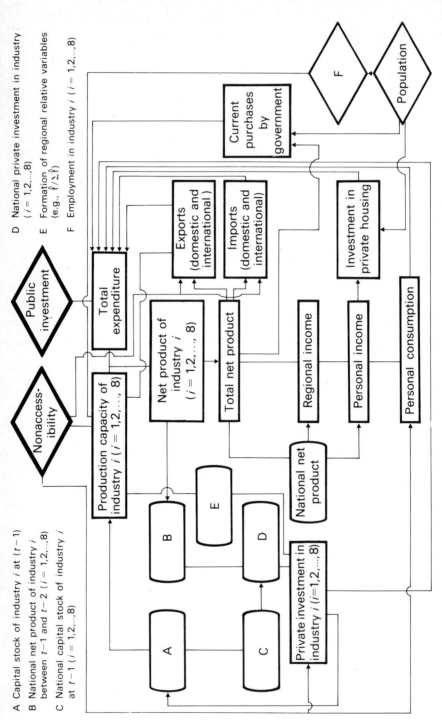

A Capital stock of industry i at $(t-1)$

B National net product of industry i
between $t-1$ and $t-2$ ($i = 1,2,...,8$)

C National capital stock of industry i
at $t-1$ ($i = 1,2,...,8$)

D National private investment in industry i
($i = 1,2,...,8$)

E Formation of regional relative variables
(e.g., $\hat{f}/\Sigma\hat{f}$)

F Employment in industry i ($i = 1,2,...,8$)

Figure 21.1. The regional econometric model. The diamond-shaped boxes represent exogenous variables, those with a bold border being possible policy instruments. The boxes with curved sides represent predetermined endogenous variables while the rectangles (with straight sides) represent current endogenous variables.

Table 21.2. The eight categories of industry recognized
in the model.

No.	Industry
1	Agriculture, forestry, and fisheries
2	Mining
3	Construction
4	Manufacturing
5	Wholesale and retail trading
6	Banking, insurance, real-estate, and private and public services
7	Transportation and communications
8	Electricity, gas, and water supply

21.2.2. Net product-determining function

$$V_i = a_i + b_i \hat{f}_i + c_i \{C + GC + I + HP + (E - M) + GI\} \tag{21.2}$$

where

$\hat{f}_i$ is the value of production capacity calculated from (21.1)
C is real personal consumption
GC is real current purchases by the government
I is real investment by private firms
HP is real private housing investment
E is real (domestic and international) exports
M is real (domestic and international) imports
GI is real gross capital formation (investment) by the government

The second term on the right-hand side of eqn. (21.2) represents the supply factor for net product determination, and the third term represents the demand factor for the same. In this sense, eqn. (21.2) can be considered as a reduced form derived from structural equations.

21.2.3. Import function

$$M = a + b \left(\sum_{i=1}^{8} V_i \right) + c\, NACCC + d\, NACCP$$

where $\sum_{i=1}^{8} V_i$ is the real domestic net product of the region. Several forms were tested as alternative import functions, but finally the simple form given above was adopted.

21.2.4. Regional income function

$$Y = a + b\left(\sum_{i=1}^{8} V_i\right) + c\,NV \tag{21.3}$$

where

Y is the real regional income

NV is the real national domestic net product (total of forty-six regional incomes, excluding that of the Okinawa region because of the lack of continuous data)

The regional income is the sum of domestic net product and net income from outside. The latter may be dependent on $(\Sigma_i V_i)$ and $(NV - \Sigma_i V_i)$, and this consideration leads to eqn. (21.3).

21.2.5. Personal income function

$$W = a + bY + c\,NV$$

where W is the real personal income.

Real personal income is considered to be dependent on both Y and NV, but the variable NV is omitted from the equation in later stages.

21.2.6. Personal consumption function

$$C = a + bW + C\,NACCC + d\,NACCP$$

where C is the real personal consumption.

After several test runs, it was decided that personal consumption should be represented by personal income only.

21.2.7. Government current purchase function

$$GC = a + b\left(\sum_{i=1}^{8} V_i\right)$$

$$GC/POP = a + b\left(\sum_{i=1}^{8} V_i\big/POP\right)$$

where POP is the resident population of the region.

21.2.8. Real private investment

In this case, the national level of real private investment in industry i is determined first, using one of the following equations:

$$NI_i = (a_{i1}D^1 + a_{i2}D^2) + b\{NV_{i,-1} - NV_{i,-2}\} + c\,NK_{i,-1} \tag{21.4}$$

$$NI_i = (a_{i1}D^1 + a_{i2}D^2) + b\,NV_{i,-1} + c\,NK_{i,-1} \tag{21.5}$$

where

$NV_{i,-1}$ is the real net product (national total) of industry i at the end of the previous year

$NK_{i,-1}$ is the real capital stock (national total) of industry i at the end of the previous year

D^1 is a dummy variable for 1965–1973

D^2 is a dummy variable for 1974–present

The first equation (21.4) is of the so-called "acceleration principle" type, and the second (21.5) is of the "profit principle" type. For each industry, the equation giving the better estimation result was selected. The parameter c, which should represent a suppressive effect on investment when negatively valued, in fact turned out to have a positive estimated value. This may imply a destabilizing effect on the whole model system. The two dummy variables were inserted to represent a structural change after the 1974 oil crisis. When the simulations are implemented, the value of b in (21.4) is switched to zero if the acceleration factor takes a negative value.

At the next step, the calculated NI_i is disaggregated into forty-six regional values using one of the following disaggregation functions:

$$\frac{I_{ik}(t)}{NI_i(t)} = (a_1 D_1 + \ldots + a_{46}D_{46}) + b\left[\frac{\sum_{x=\theta}^{t-1} GI_k(x)}{\sum_{x=\theta}^{t-1} NGI(x)}\right] \tag{21.6}$$

$$\frac{I_{ik}}{NI_i} = (a_1 D_1 + \ldots + a_{46}D_{46}) + b\left[\frac{\hat{f}_{ik,-1}}{\sum_{j=1}^{46} \hat{f}_{ij,-1}}\right] \tag{21.7}$$

$$\frac{I_{ik}}{NI_i} = (a_1 D_1 + \ldots + a_{46}D_{46}) + b\left[\frac{I_{ik,-1}}{NI_{i,-1}}\right] \tag{21.8}$$

$$\frac{I_{ik}}{NI_i} = (a_1 D_1 + \ldots + a_{46}D_{46}) + b\left[\frac{\{\hat{f}_{ik,-1} - \hat{f}_{ik,-2}\}}{\left\{\sum_j \hat{f}_{ij,-1} - \sum_j \hat{f}_{ij,-2}\right\}}\right] \tag{21.9}$$

$$\frac{I_{ik}}{NI_i} = (a_1 D_1 + \ldots + a_{46} D_{46}) + b \left[\frac{\left\{ \sum_s \hat{f}_{sk,-1} - \sum_s \hat{f}_{sk,-2} \right\}}{\left\{ \sum_s \sum_j \hat{f}_{sj,-1} - \sum_s \sum_j \hat{f}_{sj,-2} \right\}} \right] \qquad (21.10)$$

where

$I_{ik}(t)$ is the real investment by private firms in industry i in region k in year t

$GI_k(x)$ is the real gross capital formation by the government in region k in year x
(θ is the base year for available data)

$NGI(x)$ is the real gross capital formation (national total) by the government in year x, where $NGI(x) = \sum_{k=1}^{46} GI_k(x)$

D_k is a dummy variable for region k ($k = 1, 2, \ldots, 46$)

$\hat{f}_{ik,-t}$ is the calculated value of the production capacity of industry i in region k, t years previously

The meaning of each functional form is self-explanatory.

Various combinations of the equations for real private investment (21.4 and 21.5) and disaggregation functions (21.6–21.10) are used for each of the categories of industry (1–8), as shown in Table 21.3.

Table 21.3. Combinations of disaggregation functions and private investment equations used for the various categories of industry.

Disaggregation function	Equation for real private investment	
	(21.4)	(21.5)
(21.6)	–	Category 6
(21.7)	–	–
(21.8)	Category 2	Categories 1 and 3
(21.9)	–	Categories 4 and 7
(21.10)	Category 5	Category 8

Thus, for example, eqns. (21.5) and (21.6), respectively, are used for estimating real private investment and for disaggregating the results for industry category 6.

21.2.9. Housing investment function

$$HP = a + bW_{-1} + c\{POP - POP_{-1}\}$$

$$HP = a + bW_{-1}$$

Real private housing investment is represented by the personal income of the previous year and by the change in the resident population of the region. The second equation above was adopted for the Tokyo, Ishikawa, Yamanashi, Tokushima, Ehime, Kochi, and Saga regions, and the first equation was used for the other regions.

21.2.10. Export function

$$E_k = a + b\left(\sum_i \hat{f}_{ik}\right) + c\left(NV - \sum_i V_{ik}\right) + d\,NACCC_k + e\,NACCP_k \qquad (21.11)$$

Real exports from a region are represented by the production capacity of that region and by the net product of the rest of the nation as well as by the nonaccessibility variables. For regions where export and import data are lacking, E in eqn. (21.11) is replaced by a net real exports variable $(E - M)$.

21.2.11. Calculation of nonaccessibility variables

Nonaccessibility variables play an important role in the present model system, since they embody the direct effect of transport investment. There are two sets of non-accessibility variables, which are defined as follows:

1. Nonaccessibility in terms of goods transportation (i.e., physical nonaccessibility) in region k in year t:

$NACCC(k;t)$

$$= \frac{\displaystyle\sum_{l \neq k} \sum_{m=1}^{2} \left[\tfrac{1}{2}\{T_h(m,k,l;t) + T_h(m,l,k;t)\} \cdot \tfrac{1}{2}\{F_{75h}(m,k,l) + F_{75h}(m,l,k)\}\right]}{\displaystyle\sum_{l \neq k} \sum_{m=1}^{2} \left[\tfrac{1}{2}\{F_{75h}(m,k,l) + F_{75h}(m,l,k)\}\right]}$$

where

> $T_h(m,k,l;t)$ is the time/distance for goods transportation by mode m between regions k and l in year t
> $F_{75h}(m,k,l)$ is the flow of goods by mode m between regions k and l in 1975
> l is a running suffix over forty-five regions, excluding region k (and Okinawa)
> m is the mode of transport: $1 = $ railway, $2 = $ road (including expressway if available)

In this definition, parameters T_h may change every year according to the varying levels of transport investment, but the weights F_{75h} are fixed over time. This is a device to prevent the model from being excessively sensitive in simulation practice.

2. Nonaccessibility in terms of passenger transportation (i.e., nonaccessibility to passengers) in region k in year t:

$$NACCP(k;t)$$

$$= \frac{\displaystyle\sum_{l \neq k} \sum_{m=1}^{3} [\tfrac{1}{2}\{T(m,k,l;t) + T(m,l,k;t)\} \cdot \tfrac{1}{2}\{F_{75}(m,k,l) + F_{75}(m,l,k)\}]}{\displaystyle\sum_{l \neq k} \sum_{m=1}^{3} \tfrac{1}{2}\{F_{75}(m,k,l) + F_{75}(m,l,k)\}}$$

where

$T(m,k,l;t)$ is the time/distance for passenger transportation by mode m between
regions k and l in year t
$F_{75}(m,k,l)$ is the flow of passengers by mode m between regions k and l in 1975
m is the mode of transport: 1 = railway, 2 = road (including expressway
when available), 3 = aviation

In this case too, the weights $F_{75}(m,k,l)$ were fixed over time up to 1979; for
1980 a once-only change was made, using external data from a forecast of the future,
in order to account for future increases in numbers of air passengers.

The calculated values of *NACCC* and *NACCP* are obtained only for 1975 because
of the unavailability of data for weighting factors. A special procedure is therefore
needed to estimate the equations containing nonaccessibility variables, and this is
explained in detail in the next section.

21.3. Estimation of equations

All the variables that appear on the left-hand side of the equations in Section 21.2
are of course current endogenous variables, but there are also several current endogenous variables on the right-hand side of some equations. In these latter cases the
direct least-squares method of estimation is therefore not applicable. The simplest
alternative method of estimation is the two-stage least-squares method, in which all
current endogenous variables are regressed on all predetermined variables at the
first stage of estimation. In the present case, however, the sample size (1965–1975)
is too small to implement this first-stage regression. Thus, the variables have to be
"selected" from among the important predetermined variables used in the first stage.
The variables selected are W_{-1}, $\Sigma_i L_i$, *POP*, *NV*, and *NI*.

For equations that contain nonaccessibility variables, as stated before, a special
procedure is needed to estimate the relevant parameters. The relevant method varies
for different equations.

21.3.1. Production capacity function

For each industry, a Cobb–Douglas type equation is estimated using pooled data for
the regions and time series with constant dummy variables, one for each region (D_k,
$k = 1, 2, \ldots, 46$).

In the next stage, the estimates of regional dummy coefficients a_k are regressed first on $NACCC_k$ and then on $NACCP_k$, using cross-sectional data from the year 1975. Thus, two equations that express a_k in terms of accessibility variables are obtained. Then these equations are combined to form one equation that expresses a_k in terms of $NACCC_k$ and $NACCP_k$. Finally, the equation for industry i takes the following form:

$$V_{ik} = \exp\{T_i^c a_i^c + T_i^p a_i^p\} (NACCC_k)^{T_i^c b_i^c} \cdot (NACCP_k)^{T_i^p b_i^p} \cdot \{K_{ik(-1)}\}^{c_i} \{L_{ik}\}^{d_i}$$

(21.12)

Usually t-values for the estimated coefficients of $NACCC_k$ and $NACCP_k$ before consolidation are used for the weighting factors T_i^c and T_i^p in eqn. (21.12); a_i^c and a_i^p are constant coefficients in the estimated equation before consolidation.

21.3.2. Import function

Import functions are estimated for each region separately, then different values are estimated for the constant coefficient a_k of each region. These a_k-values ($k = 1, 2, \ldots, 46$) are regressed on $NACCC_k$ and then on $NACCP_k$ using cross-sectional data from the year 1975. The two equations obtained are then combined to form one equation by so-called t-value weighting. The final form of the equation for region k is

$$M_k = (T^c \alpha^c + T^p \alpha^p) + \{T^c \beta^c (NACCC_k) + T^p \beta^p (NACCP_k)\} + b_k \left(\sum_i V_{ik} \right)$$

21.3.3. Personal consumption function

The consolidation technique used for the import function is also used for the personal consumption equation. The final form of the equation for region k is

$$C_k = (T^c \alpha^c + T^p \alpha^p) + \{T^c \beta^c (NACCC_k) + T^p \beta^p (NACCP_k)\} + b_k W_k$$

21.3.4. Export function

First, the following two equations are estimated using cross-sectional data from the year 1975:

$$E_k = a_k^1 + b_k^1 \left(\sum_i \hat{f}_{ik} \right) + c_k^1 \, NACCC_k$$

(21.13)

$$E_k = a_k^2 + b_k^2 \left(\sum_i \hat{f}_{ik} \right) + c_k^2 \, NACCP_k$$

(21.14)

Equations (21.13) and (21.14) are then combined using multiple correlation coefficients of both equations as weighting factors. Finally, for each region, the constant coefficients of the resulting equation are adjusted according to the data for the year 1975.

21.4. Simulation

21.4.1. Model for simulation and setting of exogenous variables

The estimated Trunk model is then used to simulate the various alternatives up to the year 1995, starting from 1976. The Trunk model consists of forty-six regional models with one additional equation that defines the real national net domestic product NV as the sum of forty-six real regional net products $\Sigma_i V_{ik} (k = 1, 2, \ldots, 46)$.

In addition, a submodel is needed that explains the spatial and temporal behavior of population and employment. This population–employment model and the Trunk model are mutually dependent but we will not describe the former model here. The combination of the Trunk model and the population–employment model makes it possible to describe the future course of events for alternative values of the exogenous variables.

The set of exogenous variables is divided into two subsets. The first subset consists of the future time series of real gross capital formation by the government (GI) and its regional breakdown. For alternative plans of transport investment, there are different spatial and temporal patterns of GI. The second subset consists of the future course of the space–time dependent variables which enter into the definitions of the accessibility variables. Changes in these space–time variables directly reflect changes in transport investment.

All in all, we have twelve different patterns for the future course of the exogenous variables, which correspond to the twelve scenarios of future transport development presented in Table 21.1.

21.4.2. Benefit–cost analysis: concepts

In order to obtain a summarized evaluation of alternative simulations, it is most convenient to rely on the concept of benefit–cost analysis. In this study, the benefit–cost analysis is based on the differences between case 1, which is taken as the reference case, and the other cases. (The 12 cases studied are outlined in Table 21.1.)

There is no difference in either the total amount of transport investment or its regional breakdown among the alternative cases for 1976–1982 and for 1990–1995. Therefore, it seemed reasonable to take the year 1983 as the starting point for discounting purposes.

Let $DGI(x; t)$ be the difference in the total national transport investment between case x and case 1 in year t ($t = 1984$–1989). Then, the present value in 1983 of the flow of costs in case x, with a discount rate r, is

Table 21.4. Benefit–cost analysis of the transport projects proposed.

Case[a]	Average annual growth rate of real national domestic net product, 1995/1976 (%)	National net product 1995 (billion yen in 1970 prices)[b]	Benefit, present value in 1983 (billion yen in 1970 prices)				Cost, present value in 1983 (billion yen in 1970 prices)				Benefit–cost ratio			
			Rank	Discount rate			Rank	Discount rate			Rank	Discount rate		
				4%	6%	8%		4%	6%	8%		4%	6%	8%
1	4.73	226.555	8	11.797	10.295	9.045	9	1.608	1.539	1.479	2	7.337	6.688	6.127
2	4.78	228.634	1	19.491	17.073	15.057	1	3.774	3.613	3.466	9	5.165	4.725	4.344
3	4.81	229.749	10	7.353	6.493	5.772	4	2.166	2.074	1.990	10	3.395	3.130	2.901
4	4.76	227.568	9	11.455	9.993	8.777	10	1.500	1.436	1.377	1	7.634	6.958	6.374
5	4.78	228.584	11	0.286	0.255	0.229	11	0.107	0.103	0.099	11	2.661	2.472	2.306
6	4.73	226.590	6	12.604	11.007	9.677	5	2.047	1.960	1.880	7	6.158	5.617	5.149
7	4.78	228.702	4	13.011	11.369	10.003	7	1.992	1.907	1.829	4	6.532	5.962	5.469
8	4.79	228.760	3	13.519	11.810	10.387	3	2.299	2.201	2.112	8	5.880	5.365	4.919
9	4.79	228.835	2	16.192	14.168	12.483	2	2.539	2.431	2.332	6	6.377	5.828	5.353
10	4.80	229.320	5	12.816	11.195	9.847	6	1.964	1.880	1.803	5	6.526	5.954	5.460
11	4.79	228.770	7	12.448	10.866	9.551	8	1.832	1.754	1.682	3	6.795	6.196	5.678
12	4.79	228.736												

[a] The twelve cases are described in Table 21.1.
[b] For comparison, the 1976 value was 94.101 (billion yen in 1970 prices).

$$C(x) = \sum_{t=1983}^{1989} DGI(x;t)/(1+r)^{t-1983}$$

On the other hand, the overall benefit generated by the transport investment of case x can be expressed by the flow of benefits $DNV(x;t)$ ($t = 1983$–1995) which is the difference in the real national net product between case x and case 1. The present value of the benefits for case x in 1983 is given by

$$B(x) = \sum_{t=1983}^{1995} DNV(x;t)/(1+r)^{t-1983}$$

with a discount rate r. The year 1995 is the final year of our evaluation, and in this sense the benefit–cost analysis presented here is a truncated version.

21.4.3. Benefit–cost analysis: results

The results of the benefit–cost analysis are summarized in Table 21.4. From this table it can be seen that there is a very small difference in the growth rates of the real national net product for the alternative cases and also in the absolute values of the same in 1995.

The absolute values of $B(x)$ and $C(x)$ are both highest in case 3, in which all of the planned transport investments are actually implemented. Case 3, however, has the third smallest value for the benefit–cost ratio, i.e., 4.725 (at a 6% discount).

In case 5, which represents mainly the construction of expressways, the absolute values of $B(x)$ and $C(x)$ are smaller than for most other cases, but the benefit–cost ratio for this case is highest of all, i.e., 6.958 (at a 6% discount).

For cases 7–12 one can compare the relative efficiency of building different new extensions of the trunk-line railway system. The best is case 12, where a new trunk-line is built for the relatively short route from Fukuoka to Nagasaki on Kyushu Island.

Although the results obtained here are still tentative, it is interesting to see the results of a direct comparison between the alternative transport projects proposed.

References

Sakashita, N. (1974). Systems analysis in the evaluation of a nation-wide transport project in Japan. Part I: Framework of the model. Papers of the Regional Science Association, 33: 77–98.

Sakashita, N. (1981). Application of the spatial econometric model (SPAMETRI) to the evaluation of the economic effects of the Shinkansen. In A. Straszak and R. Tuch (Editors), The Shinkansen High-Speed Rail Network of Japan, IIASA Proceedings Series, Vol. 7. Pergamon, Oxford, pp. 79–102.

Regional Development Modeling: Theory and Practice
M. Albegov, A.E. Andersson and F. Snickars (editors)
North-Holland Publishing Company
© IIASA, 1982

Chapter 22

REGIONAL ECONOMIC INTERACTIONS IN THE USSR

Alexander G. Granberg
*Institute for Economics and Industrial Engineering, Siberian Branch of the USSR
Academy of Sciences, Novosibirsk (USSR)*

22.1. National and regional interrelations in the socioeconomic development of the USSR

In the USSR, one of the most important tasks of management is to achieve effective, harmonious development of the national economy and of its individual regions. Spatial factors have a considerable effect on Soviet national economic development because of the vast size of the territory, covering one-sixth of the world's surface, and the very diverse natural, geographical, and cultural conditions which exist.

The uneven territorial distribution of the main factors of production (primarily labor and natural resources, both of which have limited spatial mobility) causes some difficulties in the economic development of the USSR, and it has been found necessary to deploy considerable economic resources to overcome these problems.

The spatial heterogeneity of the Soviet economy affects the solution of the problem of combining in a rational way what might be termed the "intensive" and "extensive" paths toward national economic development. In the existing economic centers, especially in the European zone, the use of resources already available is being intensified: this involves the improvement of operating enterprises and interregional infrastructure systems, increasing the fertility of agricultural areas already under cultivation, etc. Owing to a comparable increase in labor productivity, the number of employees in the productive sector will probably remain constant. On the other hand, accelerated development of eastern Soviet regions, especially northern Siberia, and the deserts and semideserts of Kazakhstan and Central Asia, is carried out in an extensive manner by using new rather than existing resources: this involves setting up new construction projects, utilizing potential natural resources, and recruiting labor from more developed regions.

The most important national economic goal is the gradual raising of living standards. The achievement of such a goal is influenced by regional inequalities in the conditions of essential economic activities and the degree to which the requirements

of the population are satisfied. The present stage of economic growth is also characterized by irregular changes in the marginal efficiency of natural-resource utilization.

Currently, the role of spatial factors in the socioeconomic development of the
USSR at all levels of the territorial division of labor, from macrozones to primary
territorial units, is being radically revised. The importance of certain elements of
scientific and technological development, the territorial distribution of skilled labor,
regional variations in living requirements, and opportunities for extending interregional
economic relations is increasing.

We have reached two main conclusions from our analysis of the role of spatial
factors. First, general economic development trends in the USSR are largely a synthesis of different, and sometimes even contrasting, regional trends and are directly affected by territorial differences in the reproductive patterns of the population, in the
national product, and in national wealth. Second, an economic policy that is efficient
for the state as a whole must be a synthesis of a variety of different regional policies.

22.2. Research methods

22.2.1. Spatial models of the Soviet economy and the statistical data used

One of the most recent methodological trends in Soviet national economic planning
is the construction of multilevel systems of mathematical models. Such systems are
being developed by the Central Institute for Economics and Mathematics (CIEM) of
the USSR Academy of Sciences (Fedorenko, 1972; Baranov and Matlin, 1976), the
Council for Studies of Productive Forces at the USSR State Planning Committee
"Gosplan" (Albegov, 1975), and the Institute for Economics and Industrial Engineering (IEIE) of the Siberian Branch of the USSR Academy of Sciences (Aganbegyan
et al., 1972). The model systems under construction may be categorized in terms of
their degree of generality, their internal structure and information linkages, and the
principles used for deriving coordinated solutions.

In the late fifties and early sixties, the interregional input–output models of North
American researchers (such as W. Leontief, W. Isard, L. Moses, and H. Stevens) became
known in the USSR. The first publications dealing with interregional–intersectoral
models appeared in the USSR in 1963 (work by A. G. Aganbegyan and V. V. Kossov).
Within a short time more than 10 models, mainly of the optimization type (Granberg,
1973), were proposed. In 1967, the IEIE carried out the first runs of their interregional
(10 regions) and intersectoral (16 sectors) models for a 10-year period (Granberg,
1968). Later, runs of another interregional model were initiated by Nikolayev (1971).

Currently in the USSR, three types of interregional–intersectoral model are under
development: input–output models, national optimization models, and local optimization models.

Interregional–intersectoral input–output models. Models of this general type elimin-
ate the choice of variants of interregional and intersectoral linkages by introducing

structural linking parameters: a model becomes a system of equations with a single solution. The first Soviet interregional–intersectoral input–output models were constructed by the IEIE for three republics in the Trans-Caucasian economic region and for the USSR divided into two zones (the Russian Soviet Federal Socialist Republic and the rest of the country). These were based on the models developed by Moses. One of the characteristics of these models is the assumption about the stability of the supply structure of each region with respect to the output of each sector (Granberg, 1975b). A somewhat modified assumption is widely used in constructing regional input–output planning models in the USSR. These models include constant coefficients for imports in relation to regional consumption and balances between complementary and noncomplementary imports. Such models may be used in preplanning studies as tools for checking the consistency of regional projections of economic structure, and for analyzing the consequences of selecting certain demand projections.

Models representing the optimum economic interaction between regions, with local optimization criteria. These types of models are commonly associated with a two-level system: one subsystem represents interactions at the national economic level, while the other deals with the regional level. Typical of this approach is the use of regional intersectoral models with additional trade-balance equations and export–import balances with conditions common to all regions. For each region, the local consumption level is usually maximized. Optimal regional solutions are coordinated by the prices of products exchanged and by adjusted balances of interregional exchange (Granberg, 1975b).

Three main schemes for coordinating regional models of this type are possible. The first scheme coordinates the regional models into a single-region model of the national economy, duplicating the constituent regional intersectoral input–output models. All regional balances and constraints are aggregated, input coefficients are averaged, and interregional linkages are excluded. This scheme represents a synthesis of the regional models in the sense that national final output and resources used are formed from the sum of related regional values, and national economic input coefficients are formed from average weighted regional values.

According to the second scheme, regional resource balances are also integrated, but in the total system some elements of the regional models (the input–output coefficients) are retained, while others (regional final output including exports and imports, and regional inputs) are eliminated. The number of regional output variables usually exceeds the number of products.

In the third scheme, interregional or national–regional economic models are built. They retain all the features included in regional models and, in addition, contain conditions for coordinating regional economies.

For a system of static intersectoral models of the simplest type, this third scheme of integration can be presented in the form of the following simultaneous equations:

$$X^r = A^r X^r + Y^r + D^r \qquad (r = 1, 2, \ldots, m) \qquad (22.1)$$

$$\sum D^r = D \qquad (22.2)$$

where

> A^r is the matrix of input–output coefficients for region r
> X^r, Y^r are vectors of gross output and final products, respectively, for region r
> D^r is the vector of net trade (export – import) for region r
> D is the vector of the net trade balance for the USSR

Equations (22.1) and (22.2) have a block structure: regional intersectoral balances are connected by conditions governing interregional exchange balances. However, this is not a complete model for analyzing the territorial structure of the national economy. It represents only the overall pattern of interrelationships between the regional model blocks. Equations (22.1) and (22.2) have many degrees of freedom, for example in the choice of interregional linkages. Thus, the third scheme for integrating regional intersectoral models can be improved primarily in terms of improving the representation of transport activities (Granberg, 1975b; Granberg and Suslov, 1976; Granberg, 1979).

22.2.2. Interregional–intersectoral optimization models and their uses

Modified versions of the Interregional–Intersectoral Optimization Model (IIOM) have been developed and implemented at the IEIE. The general structure of IIOM is shown in Figure 22.1. A full description of the model is given in Granberg (1973, 1975a, 1976).

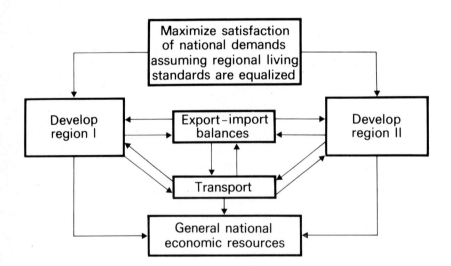

Figure 22.1. General structure of the interregional–intersectoral optimization model (IIOM).

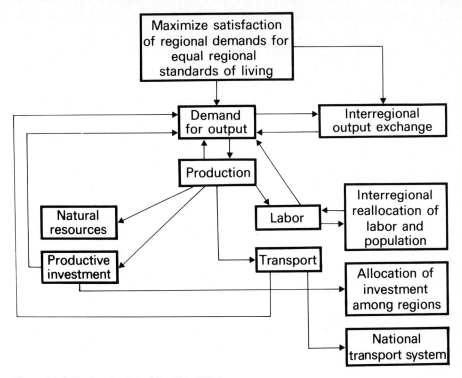

Figure 22.2. Regional submodel within IIOM.

Regional blocks (submodels) in IIOM are linked by regional living-standard correlations, interregional exchange balances and transportation system development, and constraints on national resources. Each regional block is a regional model with open "inputs" and "outputs" (a system of sectoral inputs and outputs and parameters representing employment of labor, investment, natural resources, and transportation). Blocks simulating the development of individual regions are connected by conditions for equalized regional living-standards, interregional exchange balances, main transport activities, and general national economic resources (see Figure 22.2).

In the main version of IIOM, the objective function expresses the maximization of the personal consumption level Z under certain regional conditions ($Z^r \geqslant \lambda^r Z$). In the simplest case, Z and Z^r characterize nonproductive consumption and nonproductive accumulation (in constant prices), respectively.

The solution of the model with a fixed vector $\Lambda = \lambda^r$, $\lambda^r \geqslant D$, $\Sigma_r \lambda^r = 1$, produces an efficient (Pareto-optimal) regional and national development scenario: it cannot be improved for one (or several) regions without deterioration of the consumption level for at least one other region. IIOM simulates the possibilities and effects of implementing the scenarios for equalizing different regional levels of demand. Various ratios of regional consumption and economic development and also varying national economic development rates correspond to different vectors Λ.

After national and regional economic development scenarios have been selected, the possibilities for their implementation under the given economic mechanism (price formation, differentiation of regional wages, investment conditions, principles for allocating social consumption funds, etc.) are analyzed and proposals for state regional policy are elaborated.

The various versions of IIOM differ from each other primarily in the mechanisms that are used to allocate funds between different regions, e.g., by determining regional shares of national consumption funds, by regulating employment and income, by allocating funds for social consumption, by changing interregional exchange prices, or by fixing the interregional exchange balance. For example, in the models of interregional economic relations, in which the objective function is to maximize the living-standard in each region, the most efficient scenario is selected by means of an economic price mechanism built into the model. Equilibrium conditions are given and equivalent solutions to this price-oriented model and to the basic IIOM version are derived (Rubinstein, 1976). Another example is the so-called West–East model, in which an iterative procedure is used to coordinate locally optimal plans for two Soviet regions (West and East). A central reallocation of general national economic resources (including export–import balances) is performed with the aid of optimum prices produced by the models of the individual regions (Granberg and Chernyshov, 1970).

The type of economic planning problem that may be solved using IIOM depends largely on the actual stage of planning. In the initial stages of preparing a national economic plan, IIOM may be used to study the influence of regional factors on national economic development trends, to elaborate a general scheme for territorial distribution of productive forces (in particular, to determine the possibilities for accelerating economic growth in the eastern regions of the USSR), and to evaluate the consequences of equalizing regional development and welfare levels. An aggregated version of the model would be used for this purpose.

Initially, the IIOM test runs were carried out independently of the studies of individual regions using intersectoral models. Currently, the major task is to prepare the model for use within a two-level national–regional system.

22.3. Application of IIOM to economic development analyses of Siberia

Major development data for all the larger regions of the country can be obtained from the IIOM runs. However, the model can also be used for more detailed studies of the problems of regions whose influence on the national economy is increasing. One of these regions is Siberia.

A major factor stimulating Siberia's economic growth is the abundance of natural resources, i.e., minerals, water, forest resources, and rich soil, available in the area. Currently, Siberia produces 25% of the total national output of the mineral extraction industry, about 40% of total fuel, and over 25% of total commercial timber and related products. The importance of the Siberian economy for national economic growth is demonstrated by the fact that Siberia can satisfy the increasing national demand for fuels, nonferrous metals, timber goods, and chemicals.

The main trend for future development of the national and regional economies is to give increasing attention to socioeconomic issues directly related to increasing the welfare level of the population. Considerable attention is also given to changes that increase the efficient use of labor and other resources and thus stimulate economic growth. The characteristics of the future development of the USSR as a whole will influence the choice of development strategy for the Siberian region. These characteristics include an expected reduction in population growth and labor supply, a tendency toward a reduction in capital investment growth, an increased scarcity of natural resources in the European zone, and a rapid extension of international trade.

In the near future, the problem of labor supply in Siberia will be aggravated not only by a sharp reduction in the rate of natural increase of the labor force but also by the growing trend of out-migration from Siberia to other regions where there are labor shortages. It should be remembered that over a recent 15-year period (1961–1975), when the labor supply in the European regions was relatively favorable, Siberia lost over one million inhabitants through out-migration. Obviously, the main way of solving Siberia's labor supply problems is a more rapid (compared to other regions) increase in labor productivity levels and the implementation of a consistent labor-saving policy in all economic sectors. However, to stabilize the indigenous labor force and attract additional manpower to Siberia, priority must be given to raising living standards in Siberia, and job vacancies in the western and southern regions of the country must be regulated more strictly. These socioeconomic policies will have to be enforced at the national level.

Demand for capital investment in the Siberian economy is growing. This is because of the rapid growth in certain capital-intensive sectors (fuel, iron and steel, chemicals, and timber) and the development of production and social infrastructure. However, with decreasing growth rates of capital investment in the national economy as a whole, any appreciable increase in the share of investment allocated to Siberia may lead to an absolute decrease of capital investment in other regions, primarily in the European zone. Although such a policy would obviously stimulate growth in the eastern regions of the country, its effects in the European zone should not be ignored: it might even be of interest to perform national economic calculations to determine whether it would be reasonable to actually reduce Siberia's share of total national capital investment.

Demand for most of the natural resources found in Siberia is practically unlimited on the international market. The economic opportunities for increasing Siberian exports will almost certainly improve in the future as a result of world price rises, and these projected price changes in international markets should be considered in decision-making concerning Siberia's future economic growth.

Several scenarios for Siberia's future development were tested using IIOM, based on a division of the USSR into 11 zones. They were designed so that the structure and development rates of the Siberian region should harmonize with the proposals for the development of other Soviet regions and with prevalent national economic development conditions. They should also correspond to the maximum possible level of fulfillment of national economic goals.

Initially, national economic growth alternatives were generated and analyzed taking into account tendencies observed in the preceding period. Clearly, not all past trends may be extended to the future, but it is still reasonable to consider those situations that are likely to recur. Special attention was paid to scenarios with the following basic assumptions: a trend toward a decrease in capital investment growth in capital-intensive industries, a constant growth rate in labor productivity equivalent to that experienced in the period 1971–1975 (the national economic average), and continuing unchanged migration trends.

Below we present some fairly typical findings, derived from the "central" scenario, that illustrate the similarities and differences between the development rates and economic structure of the Siberian region and those of the USSR as a whole.

Specialization in the Siberian region must be intensified. Those sectors with growth rates greater than the national average will be nonferrous metallurgy, hydroelectric power generation, fuels, chemicals, timber, building materials, and construction. Growth rates will differ from industry to industry, since simultaneous equal growth among all industries would not allow the required scale of sectoral development because of the overall shortage in labor supply. It seems inadvisable to attempt to develop several sectors, for example, machine-building, light engineering, and the food industry, rapidly in Siberia at the same time. However, in some regions favorable conditions exist for developing particular subbranches of such industries, which would serve mainly to meet local requirements and to regulate structural employment.

Siberia's share in national output (in terms of social product, national income, and gross industrial product) is expected to grow by approximately 2% in the next ten years. Output of nonferrous metals, fuels, electricity, chemicals, and timber will increase especially. Siberia's share in national capital investment will grow by 15–16%. The substitution of labor for investment is 25% higher than the national average, which suggests that it will be relatively profitable to apply capital-intensive labor-saving technologies in Siberia.

Runs of a number of variants of the basic IIOM under varying conditions have shown the following dynamic trends. The annual national income growth rate in Siberia should be 1.2–1.4 times higher than the national average (the same is approximately true for the growth rate of gross output in Siberia). This excess is optimal; i.e., it corresponds to the highest attainable level of satisfaction of national demands. In addition, deviations from the optimal economic growth rates would have the following consequences: if the national rate exceeds the optimal level, the Siberian rate will decrease only insignificantly, but should Siberia's economic growth rate be suboptimal, the national growth rate curve will fall more steeply.

In evaluating Siberia's future development, it is of course not necessary to be restricted to one optimum scenario. The Siberian economy is fairly flexible in comparison with the national economy. In the long term, this feature will become more pronounced for two reasons. First, this region will be the source of almost all of the net increase in output of fuels, many nonferrous metals, energy-intensive chemical production, and timber goods. Therefore, it should be possible to compensate for almost any fluctuations in national demand for these goods exclusively by varying Siberian production.

Second, because of labor-supply shortages, there will necessarily be oscillations in sectoral development, leading to sudden changes in the development levels of main and subsidiary production units.

22.4. Concluding remarks

We have analyzed five development scenarios for Siberia in relation to the rest of the national economy. Each of the scenarios concentrates on a specific problem and is derived from generalizations about changes in: the consumption efficiency of fuels, raw materials, and commodities; productivity growth rates in Siberia, taking into account imported labor; national capital investments; Siberia's share in national nonproductive consumption and accumulation; and transport networks between the east and west of the country.

The results of the model runs based on these scenarios have revealed that increases in productivity naturally lead to a growth in national income and national consumption. Since Siberia is a region with one of the most severe labor shortages in the country, productivity increases have a very significant effect on the economy. This is reflected by the fact that the growth rates of income in Siberia are higher than those in the USSR as a whole.

The Siberian economy responds to a rise in investment efficiency with an increase in growth rates (in particular, in terms of national income, with gross product growing only slowly, the sectoral structure undergoing, in this case, substantial revision). The main reason for this is that decreases in unit capital inputs reduce demand for building-and-assembly works as well as for capital goods used to maintain the investment process (building-materials industry, metallurgy, and electricity production). In addition, reductions in demand and sectoral output result in the release of labor, which can be employed in sectors such as ferrous metallurgy and machine-building.

The results of our calculations show that for optimum national development it is necessary to give preference to development of the Siberian economy (which is stable with respect to changes in national economic growth) over that of other regions. This conclusion is fully justified with respect to all economic factors that can be changed within a 15-year period.

References

Aganbegyan, A. G., Bagrinovski, K. A., and Granberg, A. G. (1972). Systema Modelei dlya Planirovaniya Narodnogo Khozyaistva (A System of Models for National Economic Planning). Mysl., Moscow.

Albegov, M. M. (1975). Problemy optimizatsii territorialnogo planirovaniya (Problems of optimizing spatial planning). Ekonomika i Matematicheskie Metody, 11(1): 147–159.

Baranov, E. F. and Matlin, N. S. (1976). Ob eksperimentalnoi realisatsii sistemy modelei optimalnogo perspektivnogo planirovaniya (Testing systems of optimal long-term planning models). Ekonomika i Matematicheskie Metody, 12(4): 627–649.

Fedorenko, N. (Editor) (1972). Problemy Optimalnogo Funktsionirovaniya (Problems of the Optimal Functioning of a Socialist Economy). Nauka, Moscow.

Granberg, A. G. (1968). Eksperimentalnye raschety po mnogootraslevoi modeli optimalnogo razvitiya i razmeshcheniya proizvodstva po ekonomicheskim zonam SSSR (An evaluation of optimization models of multisectoral development and the allocation of industry among economic regions of the USSR). Izvestiya Sibirskogo Otdeleniya Akademii Nauk SSSR, Seriya Obshestvenye Nauki, 3(11).

Granberg, A. G. (1973). Optimizatsiya Territorialnykh Proportsii Narodnogo Khozyaistva (Spatial Optimization of the National Economy). Ekonomika, Moscow.

Granberg, A. G. (1975a). The construction of spatial models of the national economy. In A. Kuklinski (Editor), Regional Development and Planning International Perspectives. Martinus Nijhoff, Leiden.

Granberg, A. G. (Editor) (1975b). Mezhotraslevye Balancy v Analize Territorialnykh Proportsii SSSR (The Role of the Intersectoral Balance in Analyzing Land Allocation in the USSR). Nauka, Novosibirsk.

Granberg, A. G. (1976). Prostranstvennye modeli dlja prognozirovania i planirovaniya narodnogo khozyaistva (Spatial models for predicting and planning the national economy). In A. G. Granberg (Editor), Territorialnye Narodnokhozjaistvennye Modeli (Spatial National Economic Models). Nauka, Novosibirsk.

Granberg, A. G. (1979). On using input–output models in studying the territorial proportions of the USSR. In A. Kuklinski, O. Kultalahti, and B. Koskaho (Editors), Regional Dynamics of Socioeconomic Change. Tampere, Finland.

Granberg, A. G. and Chernyshov, A. A. (1970). Zadacha optimalnogo territorialnogo planirovaniya 'Zapad–Vostok' (The problem of optimal spatial planning 'West–East'). Izvestiya Sibirskogo Otdelenia Akademii Nauk SSSR, Seriya Obshestvennykh Nauk, 2(6): 75–87.

Granberg, A. G. and Suslov, V. I. (1976). Use of republic and regional intersectoral balances in the analysis of the territorial proportions of the national economy of the USSR. In A. G. Granberg (Editor), Spatial National Economic Models. Nauka, Novosibirsk.

Nikolayev, S. A. (1971). Mezhraionnyi i Vnutriraionnyi Analiz Razmeshchenia Proizvoditelnykh Sil (Interregional and Intraregional Analysis of the Location of Industrial Capacity). Nauka, Moscow.

Rubinstein, A. G. (1976). Modeli ekonomicheskogo vzaimodejstvia regionov i vozmoznosti ikh ispolzovaniya (Modeling the economic interaction of regions and problems of distribution effects). In A. G. Granberg (Editor), Territorialnye Narodnokhozjaistvennye Modeli (Spatial National Economic Models). Nauka, Novosibirsk, pp. 7–81.

PART VII

APPLICATION OF REGIONAL SYSTEMS ANALYSIS
TO PLANNING AND POLICY-MAKING

Regional Development Modeling: Theory and Practice
M. Albegov, A.E. Andersson and F. Snickars (editors)
North-Holland Publishing Company
© IIASA, 1982

Chapter 23

TOPICS IN REGIONAL DEVELOPMENT MODELING

Jean-Pierre Ancot, Leo H. Klaassen, Willem T. M. Molle, and Jean H. P. Paelinck
Netherlands Economic Institute, Rotterdam (The Netherlands)

23.1. Introduction

Certain aspects of regional, and particularly multiregional and interregional, model-building will be discussed in this chapter. Based on ten years of research in the field of theoretical and spatial econometrics, three main topics are dealt with: a multiregional growth theory, estimation and testing techniques, and policy formulation. Data collection is not discussed; for details, see Ancot et al. (1981).

23.2. Theory and modeling

A threshold theory seems suitable as a starting point for modeling interregional-growth phenomena. A brief summary is given in this section; for a more detailed presentation, see Paelinck (1978c).

Take an elementary economic cell: for example, a company or a family. It can be assumed that for decision-making the "best" combination is chosen out of a finite set of possibilities. The parties concerned do not always choose the "very best" combination, since on the one hand every economic agent has a choice preference of his own and on the other the degree of information available varies. Only the very worst solutions are rejected (except perhaps when supply is insufficient): this leads to the concept of a threshold value that must be exceeded before a particular solution will be chosen.

At any given moment a region is characterized by a profile composed of a large number of elements. Some elements affect decisions taken by a group of industries or by families, while others affect decisions taken by public bodies (local, regional, national, and international). Examples of such elements include access to product markets, labor markets, and decision centers, and factors relating to the locational behavior of companies.

Labor markets influence the decisions of families in both qualitative and quantitative terms. Public agencies also act on the same basis; they try to correct perceived disequilibria. If they wish to penetrate to the core of the problem, they have to examine and boost a region's local profile. In the short run, subsidies help; they provide financial compensation for threshold values that are too low. However, it is far more important to try to improve a region's relative profile, i.e., relative to "better" profiles in other regions. Thus, the following questions must be answered. What is a particular phenomenon's threshold value, and how does a region's quantitative profile relate to that threshold value? What is the quantitative contribution of each of the elements in a regional profile to its overall value, and which elements can most effectively be changed to achieve the threshold value?

In this chapter, studies that focus on such questions are described. These studies use the notion of threshold value as the basic idea from which to derive the contributions that the profile elements make to the overall value of regional profiles. Ultimately, this can and must lead to a division of regions into those that grow spontaneously and those that suffer from nongrowth. With respect to the latter, some indications are given about how growth may be stimulated.

Unequal regional dispersion of activities is not considered optimal by policy-makers. Thus, efforts to accelerate the growth of per capita income in lagging regions will remain a goal of economic policy in the near future. This means that, in order to solve the problem, regions must be evaluated in accordance with their potential to attract and develop economic activities (Paelinck, 1978b). In addition, there should be an assessment of the regional policy measures that are required to accelerate the evolution of lagging regions, and the intensities of their use. Both theory and methodology are therefore needed to integrate all available information on local conditions (Molle and van Holst, 1976; Molle, 1978).

There are two central ideas here. One is that of a regional profile, the elements of which are to be evaluated as measures of local attractiveness; the other is that of regional thresholds that should be reached by the individual elements of the estimated regional profile (see Figure 23.1).

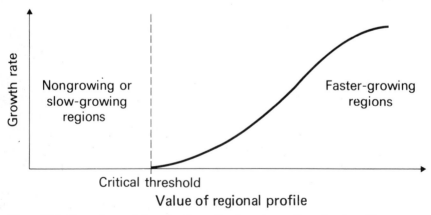

Figure 23.1. Dependence of the rate of growth of a region on its regional profile.

Table 23.1. Factors making up total regional profiles for three regions.

Region	Demand factor	Supply factor	Policy factor	Total value	Critical threshold value	ρ^a
1	20	10	5	35	45	0.78
2	30	20	5	55	45	1.22
3	30	20	5	55	45	1.22

^{a} ρ is the ratio of total value to critical threshold value.

These two ideas have guided the entire development of the FLEUR* model, which is currently being applied to solve this problem. The results obtained allow the regions to be ranked according to the overall values taken by their regional or attraction profiles (the differences between regional and attraction profiles are discussed in Section 23.3), especially with respect to the two growth-rate categories depicted in Figure 23.1.

Moreover, the elements of individual profiles can be evaluated relative to one another. An example of figures showing how the overall regional profiles for three regions are built up is given in Table 23.1. In region 1, the total value falls 22% short of the critical threshold value; assuming the other factors remain constant, the policy factor value must at least triple to make that region cross its threshold value. Regions 2 and 3, in contrast, are spontaneous growers, and policy measures are obviously redundant.

A summary of the formal development of Figure 23.1 and Table 23.1 will now be given, beginning with the discriminant model, which minimizes the expected cost of incorrect classification. Given a classification rule, two kinds of classification error may be made: a vector of measurements from population π_1 may be classified as coming from population π_2, and vice versa. Either error can give rise to a certain cost, $C(2/1)$ and $C(1/2)$, respectively. If the vector of measurements x is a random vector, the probabilities of these errors, $P(2/1, R)$ and $P(1/2, R)$, can be defined, where R represents the classification rule defining classes R_1 and R_2. If one has a priori probabilities of a vector of measurements belonging to one of the two populations, q_1 and q_2, a possible decision criterion is to minimize the expected cost of incorrect classification, namely

$$C(2/1) \cdot P(2/1, R)q_1 + C(1/2) \cdot P(1/2, R)q_2 \tag{23.1}$$

If the densities of vector** x are $p_1(x)$ in π_1 and $p_2(x)$ in π_2, (23.1) can be written as

$$C(2/1)q_1 \int_{R_2} p_1(x)\,dx + C(1/2)q_2 \int_{R_1} p_2(x)\,dx \tag{23.2}$$

---

* Factors of Location in EURope.
** This vector is in fact a regional profile.

Moreover, if a priori probabilities q_1 and q_2 are known, then using the Bayes procedure, which minimizes (23.1) and (23.2), R_1 and R_2 are chosen according to

$$p_1(x)/p_2(x) \geqslant C(1/2)q_2/C(2/1)q_1 \qquad \text{for } R_1$$

and

$$p_1(x)/p_2(x) < C(1/2)q_2/C(2/1)q_1 \qquad \text{for } R_2$$

If a priori probabilities q_1 and q_2 are not known, a unique classification procedure can still be found by applying the minimax principle to the expected loss through misclassification: $C(2/1) \cdot P(2/1, R)$, if the observation is from π_1; and $C(1/2) \cdot P(1/2, R)$, if it is from π_2.

The form of the model also depends on the specification of the distribution of the random vector of measurements x and on the parameters of that distribution. If these parameters are unknown, as they will be in many instances, they must be estimated.

It has been assumed that the elements of vector x follow multivariate normal distributions, with mean vector μ_1 in population π_1 and mean vector μ_2 in population π_2, but with the same variance–covariance matrix Σ. None of these parameters are known a priori; thus, the ratio of densities can be written as

$$\begin{aligned} p_1(x)/p_2(x) &= \exp[-\tfrac{1}{2}(x-\mu_1)'\Sigma^{-1}(x-\mu_1)]/\exp[-\tfrac{1}{2}(x-\mu_2)'\Sigma^{-1}(x-\mu_2)] \\ &= \exp[-\tfrac{1}{2}((x-\mu_1)'\Sigma^{-1}(x-\mu_1)-(x-\mu_2)'\Sigma^{-1}(x-\mu_2))] \end{aligned}$$

$$(23.3)$$

Our treatment of the problem differs from classical discriminant analysis, as summarized above, in that the regions considered are classified into one of two classes, growers and nongrowers, in the absence of any a priori information about the two regimes. The classical framework is, however, useful for constructing a model to estimate the value of the threshold parameter separating the two regimes and to explain the growth process under study in terms of functional relationships.

Starting from

$$P_1(x) = q_1 p_1(x)/[q_1 p_1(x) + q_2 p_2(x)] \qquad (23.4)$$

the a posteriori probability that measurement vector x comes from population π_1 is

$$P_1(x)/[1 \dot- P_1(x)] = p_1(x)q_1/p_2(x)q_2$$

and, using (23.3), one obtains

$$P_1(x)/[1 - P_1(x)] = \exp[x'a_1 + b] \cdot q_1/q_2 \qquad (23.5)$$

where

$$a_1 = \Sigma^{-1}(\mu_1 - \mu_2)$$

and

$$b = -\tfrac{1}{2}(\mu_1 + \mu_2)'\Sigma^{-1}(\mu_1 - \mu_2)$$

Hence

$$P_1(x)/[1 - P_2(x)] = \exp[x'a_1 + a_0] = \exp[g(x)]$$

where

$$a_0 = b + \log(q_1/q_2) \tag{23.6}$$

Thus

$$P_1(x) = \exp[g(x)]/\{1 + \exp[g(x)]\}$$

In the absence of a priori information, one has to set $q_1 = q_2$ in (23.5), so that $a_0 = b$ in (23.6). From (23.4), it is clear that in this case the a posteriori probabilities rest on sample information only. In the present type of application, no a priori information is available; therefore, only this case is treated below.

Moreover, if the growth of the region is represented by Δy with saturation level κ, variable* $(\Delta y + \lambda)/\kappa$ is introduced as a proxy for $P_1(x)$. After some further rearrangements, one obtains the following linear model (which can be estimated) in terms of the elements of the vector of measurements:

$$u = -\log\{[\kappa/(\Delta y + \lambda)] - 1\} = g(x) = x'a_1 + b \tag{23.7}$$

Given a sample of observations of Δy and x, one can thus estimate the parameters of (23.7) using standard techniques.

Finally, it can be shown that threshold parameter c can be computed from

$$c = \tfrac{1}{2}(c_1 + c_2)(a_1' \Sigma a_1)^{1/2}$$

The initial classification is then given by:

$$\hat{u} = x'\hat{a}_1 + \hat{b}$$

for a particular vector of measurements x, and R_1 and R_2 are chosen iteratively according to the rule**

$$\hat{u} \geq c \qquad \text{for } R_1$$

and

$$\hat{u} < c \qquad \text{for } R_2$$

To summarize, starting from a discriminant-analytical approach not only is a threshold value determined to distinguish between two regional regimes, but also a functional explanation of the growth phenomenon is provided in the form of the sigmoid specification (23.7) with saturation level κ.

---

* λ is a suitably small constant, chosen in such a way as to ensure that variable $(\Delta y + \lambda)/\kappa$ remains strictly positive. Apart from a translation effect, the introduction of this constant does not affect the substance of the analysis.
** It is better to approximate Σ.

23.3. Econometric results

The data described above can be used to estimate a workable model (Paelinck, 1978a), as shown by an example based on Spanish data.* These data were collected for a study of the development of value added (at current prices) in seven sectors of the Spanish chemical industry over the period 1967–1971 (Ancot and Paelinck, 1979).** The sectors are given in Table 23.2.

Table 23.2. Sectors of the Spanish chemical industry and their respective code numbers.

Code number	Sector
40	Asbestos and rubber
41	Transformation of plastic material
42	Synthetic materials/artificial fibers
43	Basic chemicals and fertilizers
44	Soap, detergents, and perfumes
45	Other chemical products
46	Oil products

The sample contains observations made in 1967 of the explanatory variables and observations over the period 1967–1971 of the sectoral growth figures to be explained, both sets of data disaggregated by Spanish provinces. The set of potential explanatory variables, from which a regional profile is to be selected by appropriate econometric methods, can be split into five groups or profiles, as follows.

(1) The attraction profile is defined as that part of the regional profile that refers to interindustrial and final supply and demand factors. Since no input–output tables for the study period were available, regional supply factors were quantified on the assumption that a sector's production volume is representative of regional supply. They were, in fact, measured by the production levels (value added in 1967) of the industries that had important input links with the chemical sectors studied. The regional demand factors are represented by the sum of intermediate and final demand. They were calculated from data obtained from national and regional Spanish sources, with the help of various hypotheses that allowed translation of the raw data into material representative of the demand aspect of the attraction profile.

(2) The "classical" profile comprises the locational factors common to all sectors.

---

* Spanish data have been chosen because, with the exception of the explanatory variables, some 35 critically established regional profiles were available.

** The starting point could be a log-probability model, a stochastic utility model with incorporated Weibull distribution, or a discriminant analysis; we chose the last method as our point of departure (Ancot, 1979). Quadratic programming was used. Significance limits for the estimated parameters are being investigated. On these last points, see Ancot and Paelinck (1980, 1981).

These include capital (supply of financial services at the provincial level), labor (represented by variables such as the size of the working-age population, the labor-force participation rate, and the size of the population employed in agriculture), land (a distinction is made between coastal and other provinces), and energy (for which 1967 production values for the large energy sectors were taken as supply indicators).

(3) The sociocultural profile includes variables representing the residential market (proportion of houses with shower or bath, proportion of vacant houses), the level of education (the number of heads of families with a university education, the existence of a university or an institute of technology in the region, the capacity of higher education facilities), the agglomeration advantages (represented by the populations of towns with more than 50,000 inhabitants as percentages of the total regional population), cultural factors (number of cultural facilities per capita and degree of cosmopolitanism*), and certain social characteristics displayed by the population.

(4) The environmental profile comprises only the quality of the local groundwater and surface water and its vulnerability to pollution.

(5) The political profile is represented by three variables: the dynamism of the provincial and local authorities in 1967, the dynamism of the Madrid authorities in 1967*, and per capita public investment in the provinces over the period 1967–1971.

The variables representing the attraction profile and the capital and labor aspects of the classical profile were "potentialized"** in order to incorporate regional interdependence in terms of the accessibility of the other regions relative to the reference region. In total, 40 explanatory variables were retained as a starting set for selecting decisive locational factors.

In short, the model consists of an equation for each sector, and relates the critical threshold value for that sector to a subset of elements of the regional profile. This subset is selected by econometric methods and contains potentialized variables (in the case of variables for which interregional dependencies are relevant, such as access to the labor market or to raw materials) and nonpotentialized variables (whenever the corresponding element of the regional profile is relatively constant, such as land availability or housing quality). To each of the selected variables, potentialized and nonpotentialized, there corresponds a "propensity parameter", the value of which is an indicator of the relative importance of the corresponding element for the "attractiveness" of the regions. Estimated values of the parameters are given in Tables 23.3 (potentialized variables) and 23.4 (nonpotentialized variables). The potentialized variables are entered as space-discounted sums by means of a modified Tanner friction function, containing sector-specific and variable-specific friction parameters. The values of these friction parameters, estimations of which are given in Table 23.5, provide an indication of the spatial relation between the location of the sector and that of the

---

* These variables were quantified in terms of integers in the range 1–5, higher values corresponding to higher intensities.

** These variables were introduced as space-discounted sums, using a Tanner-type function to express friction with respect to interregional distances — see Ancot (1979) and Ancot and Paelinck (1979).

Table 23.3. Estimated values of the coefficients for potentialized variables.

Variable	Sector[a]						
	40	41	42	43	44	45	46
I[b]　Deliveries	—	—	0.25	—	—	9.55	—
II[c]　Agriculture, forestry, and fisheries	—	—	—	—	—	0.007	—
Alcoholic beverages (except wines)	—	—	—	—	—	0.033	—
Paper	—	—	0.003	—	0.002	—	—
Basic chemicals and fertilizers	0.063	—	0.01	—	—	—	—
Electricity	—	—	0.004	0.067	—	—	1.82
River transport	—	—	3.37	—	—	—	—
Commerce	—	—	—	—	—	—	—
Other services	0.085	0.029	0.003	5.63	—	0.062	20.39
III[d]　Credit and insurance institutions	—	0.033	—	—	0.14	—	—
Population of working age	—	—	0.066	0.092	—	0.002	—
Agricultural labor	0.011	0.012	1.01	0.036	—	0.0003	—
Executive staff (%)	—	—	—	—	17.07	—	—
Underemployment	0.88	3.68	2.02	—	0.78	0.62	1.34
Net migration	3.5	—	380.15	7642.8	4320.0	0.96	—

[a] An explanation of the code numbers for sectors 40–46 is given in Table 23.2.
[b] Demand elements: regional (intermediate and final) deliveries.
[c] Supply elements: production levels (value added) of the sectors showing important input links with the chemical-industry sectors.
[d] Set of locational factors.

Table 23.4. Estimated values of the coefficients for nonpotentialized variables.

Variable	Sector[a]						
	40	41	42	43	44	45	46
I[b] Sites in coastal regions	—	—	—	—	—	—	0.01
Mining	—	—	—	—	0.01	—	—
Water, gas, and electricity production	—	—	0.01	0.01	—	—	—
II[c] Dwellings with shower (%)	5.01	5.39	43.25	42.12	4.83	5.78	0.38
Cultural density	1.97	1.57	1.74	0.22	1.66	0.96	0.16
Heads of households with university education	15.69	7.95	45.51	38.01	2.61	7.72	—
Upward social mobility	0.34	0.33	0.01	—	0.36	—	0.71
Cosmopolitanism	32.89	15.88	236.82	154.25	18.50	23.57	—
Population of towns of over 50,000 inhabitants (%)	0.01	0.01	0.97	1.22	0.04	0.06	0.01
Capacity of higher-education facilities (%)	0.13	—	—	—	0.12	—	0.43
Vacant housing	0.92	0.28	—	—	—	0.22	0.41
III[d] Dynamism of provincial and local authorities	2.54	2.08	11.97	7.60	1.24	2.98	—
Dynamism of Madrid authorities	6.80	4.84	45.56	43.14	5.42	5.58	0.35
Public investment (per inhabitant)	0.01	0.01	0.03	0.03	—	—	—

[a] An explanation of the code numbers for sectors 40–46 is given in Table 23.2.
[b] Supply elements: production levels (value added) of the sectors showing important input links with the chemical-industry sectors.
[c] Set of locational factors.
[d] Regional policy factors.

Table 23.5. Estimated values of friction coefficients.

Variables	Sector[a]						
	40	41	42	43	44	45	46
I[b] Deliveries	—	—	-0.34	—	—	5.53	—
II[c]							
Agriculture, forestry, and fisheries	—	—	—	—	—	-1.60	—
Alcoholic beverages (except wines)	—	—	—	—	—	0.80	—
Paper	—	—	-4.86	—	-5.26	—	—
Basic chemicals and fertilizers	-0.47	—	-3.48	—	—	—	—
Electricity	—	—	-3.53	-1.70	—	—	—
River transport	—	—	2.60	—	—	—	-0.64
Commerce	—	—	—	—	—	—	—
Other services	-1.12	-1.42	-3.33	2.60	—	-0.06	7.89
III[d] Credit and insurance institutions	—	-0.68	—	—	0.43	—	—
Population of working age	—	—	-1.90	-1.52	—	-2.57	—
Agricultural labor	-1.06	-0.85	1.12	-1.50	—	-2.51	—
Executive staff (%)	—	—	—	—	-1.29	—	—
Underemployment	-0.58	0.67	2.91	—	-0.40	-0.62	-0.38
Net migration	4.03	—	7.55	18.38	25.00	2.62	—

a An explanation of the code numbers for sectors 40–46 is given in Table 23.2.
b Demand elements: regional (intermediate and final) deliveries.
c Supply elements: production levels (value added) of the sectors showing important input links with the chemical-industry sectors.
d Set of locational factors.

corresponding element of the regional profile. If the value is larger than -1, then a factor of attraction *within* the region is dominant, and increasingly so, the higher the value. If, in contrast, this friction parameter is smaller than -1, the region is attractive because the relevant element is present in *another* region of the system. For example, a particular region may be attractive because a neighboring region contains a market for the sector's products, and/or because another neighboring region has excess labor supply.

The results of parameter estimation based on the data described above are presented in Tables (23.3)–(23.5); for the techniques, see Paelinck and Klaassen (1979). Some comments on these results are in order at this stage.

Among the attraction factors, demand is a determinant factor only for the synthetic-materials/artificial-fibers sector and the chemical-products sector. In both cases, the internal demand of the region rather than the demand of contiguous regions is dominant as a locational factor (friction coefficient positive or negative, but larger than -1). Among the various supply factors selected, the residual sector ("other services") is determinant in all cases except for that of the soap, detergents, and perfumes sector, where the value of the friction coefficient varies from -3.33 (corresponding to a distance of about 200 km between the location of services offered and that of the synthetic-materials/artificial-fibers sector) to 7.89 (indicating the importance of the internal regional supply). The other determinant supply factors are sector-specific, the most important being as follows: for the asbestos-and-rubber sector, the local existence of basic chemical industries; for the synthetic-materials/ artificial-fibers sector, the presence of commercial services; for the basic-chemicals sector, proximity to electrical energy sources; and for the oil-products sector, proximity to river transportation facilities. The link between the alcoholic-beverages industry and "other chemical-products industries" may well come as a surprise.

The variable representing capital (existence of credit and insurance institutions) is determinant only for the plastic-materials-transformation sector and the soaps, detergents, and perfumes sector. Underemployment and the migration balance, expressed in terms of participation rates, are variables representing labor that seem determinant for all sectors. The two other variables selected for this purpose, namely working-age population and proportion of the population employed in agriculture (the selection of the latter variable being rather unexpected), show friction coefficients that are mostly lower than -1 and have a less marked influence. As for the other elements of the classical profile, the availability of sites in coastal regions is decisive for the asbestos-and-rubber and oil-products sectors, while the availability of energy is determinant for the synthetic-materials/artificial-fibers sector, the basic-chemicals sector, and the soap, detergents, and perfumes sector.

Thus, from Table 23.4, it would appear that the sociocultural and political profiles of regions are particularly important as locational factors. The environmental profile was not selected as an explanatory variable in this case study.

23.4. Policy

The central difficulty in analyzing the control aspects of regional problems is that at least three types of policy, national, sectoral, and regional proper, influence the evolution of a system of regions (Paelinck and van Rompuy, 1973). Sectoral and regional policies, for instance, are not independent of each other: industries are necessarily located somewhere, and the stimulation of a given branch of industry inevitably causes regional shifts in income and growth patterns. A scheme of sectoral–regional interdependence in an S-sector, R-region system is presented in Table 23.6. Here the dq_{ij} are regional and sectoral increments, for example in production levels. A model in explicit function form is developed below as an illustration.

Table 23.6. A scheme representing sectoral–regional interdependence.

Sector	Region			Total
	$1 \ldots$	$j \ldots$	$\ldots R$	
1 . . .	$dq_{11} \ldots$		$\ldots dq_{1R}$	$dq_{1.} = \sum_j dq_{1j}$
i . . .		dq_{ij}		$dq_{i.} = \sum_j dq_{ij}$
S	$dq_{S1} \ldots$	dq_{Sj}	$\ldots dq_{SR}$	$dq_{S.} = \sum_j dq_{Sj}$
Total	$dq_{.1} = \sum_i dq_{i1}$	$dq_{.j} = \sum_i dq_{ij}$	$dq_{.R} = \sum_i dq_{iR}$	$dq = \sum_i \sum_j dq_{ij} = \sum_i dq_{i.}$

Take two regions, 1 and 2, with one sectoral factor, s, and one locational factor, f. The sectoral instrument variable is represented by

$$dq_1 + dq_2 = \mu(f_1^\alpha f_2^{1-\alpha})^\beta s^\gamma \tag{23.8}$$

$$dq_1/dq_2 = \nu(f_1/f_2)^\sigma \tag{23.9}$$

The dq_i are defined above; μ and ν are level constants, α and β are elasticities, while σ is more specifically a subsubstitution elasticity for regional locational factors. All parameters are strictly positive.

Relation (23.8) can be interpreted as the reduced form of a production function for the two regions. Relation (23.9) represents the proportional distribution of production as a function of the relative presence of locational advantages.

Solution for dq_1 gives

$$\mathrm{d}\log q_1 = [\beta + (1-\alpha)(\sigma - \beta)]\,\mathrm{d}\log f_1 - [(1-\alpha)(\sigma - \beta)]\,\mathrm{d}\log f_2 + \gamma\mathrm{d}\log s$$

Three effects can be distinguished here: the effect of sectoral policy, $\gamma\mathrm{d}\log s$; an expansion effect, $\beta\mathrm{d}\log f_1$, the other terms in $\mathrm{d}\log f_1$ and $\mathrm{d}\log f_2$ being eliminated if $\mathrm{d}\log f_1 = \mathrm{d}\log f_2$ (i.e., if both regions develop their locational factors equally); and a regional substitution effect, $(1-\alpha)(\sigma - \beta)(\mathrm{d}\log f_1 - \mathrm{d}\log f_2)$.

Intuitively, one would expect the expansion elasticity, β, to be smaller than the

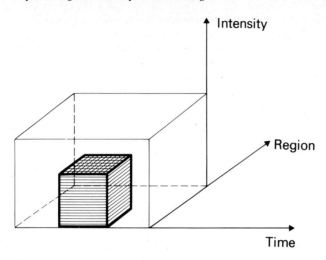

Figure 23.2. A three-dimensional (box) approach to regional policy analysis.

substitution elasticity, σ, in which case the substitution effect would be to the advantage of region 1 if $\mathrm{d}\log f_1 > \mathrm{d}\log f_2$, i.e., if region 1 is relatively favored as far as locational factors are concerned.[*]

The problem must now be translated into econometric terms. Take a linear model including two sectors, 1 and 2, and one region, r.[**] The variable to be explained is y, there is a unique regressor, x, and p is a policy variable. The model could be set up as

$$y_{1r} = a_1 x_r + \epsilon_{1r} + b_{1r}$$

$$y_{2r} = a_2 x_r + \epsilon_{2r} + b_{1r}$$

$$b_{1r} + b_{2r} = bp_r + \eta_r$$

or in vector–matrix terms

$$\begin{bmatrix} 1 & 0 \\ 0 & 1 \\ 0 & 0 \end{bmatrix} \begin{bmatrix} y_{1r} \\ y_{2r} \end{bmatrix} = \begin{bmatrix} x_r & 0 & 0 \\ 0 & x_r & 0 \\ 0 & 0 & p_r \end{bmatrix} \begin{bmatrix} a_1 \\ a_2 \\ b \end{bmatrix} + \begin{bmatrix} \epsilon_{1r} \\ \epsilon_{2r} \\ \eta_r \end{bmatrix} + \begin{bmatrix} 1 & 0 \\ 0 & 1 \\ -1 & -1 \end{bmatrix} \begin{bmatrix} b_{1r} \\ b_{2r} \end{bmatrix}$$

The equations are of the Zellner "seemingly unrelated" type and could probably be handled along these lines (Paelinck and Klaassen, 1979, Chapter 6); this supposes p_r to be measurable. Since regional policy measures have a time dimension (duration of the relevant legislation), an area dimension (only parts of statistical regions are covered by

---

[*] An additional condition is $\alpha < 1$. This is satisfied if the marginal effect of locational advantages slopes downward (compare diminishing marginal product).
[**] The problem can easily be generalized to form an interregional problem similar to that described in Section 23.3.

them), and a relative intensity, a three-dimensional (box) approach could be pursued. Although this approach is only briefly mentioned here, it is currently under investigation (Figure 23.2).

23.5. Conclusions

Even if it has now become possible to design regional models such that they are able to generate alternative scenarios for a system of regions, some issues deserve further investigation. One of these issues concerns conflicting policy measures, such as may arise in regional and industrial policy; another is the integration of phenomena such as the decentralization of functional business units and their dispersion across regions. Spatial econometrics can serve as a valuable aid in the analytical solution of these problems.

References

Ancot, J.-P. (1979). Une approche par analyse discriminante à des problèmes de seuils régionaux et d'analyse de localisation (A discriminant analysis approach to regional threshold problems and an analysis of location). Recherches Economiques de Louvain, 45(3):281–297.

Ancot, J.-P. and Paelinck, J. H. P. (1979). A discriminant analysis approach to regional threshold problems, with an application to Spanish data. Papers of the Regional Science Association, 42: 139–151.

Ancot, J.-P. and Paelinck, J. H. P. (1980). Multiple regression with non-stochastic linear inequality constraints; a simple pre-test and an estimation strategy. Actes du Colloque: Structures Economiques et Econométrie, Lyon, France.

Ancot, J.-P. and Paelinck, J. H. P. (1981). Spatial econometrics: principles and recent results. In D. A. Griffith and R. D. Mackinnon (Editors), Dynamic Spatial Models. Sijthoff and Noordhoff, Alphen aan de Rijn, The Netherlands, pp. 344–364.

Ancot, J.-P., Kemp, P., Paelinck, J. H. P., and Smit, H. (1981). DATONEI, the functional databank of the Netherlands Economic Institute. In A. Kuklinski (Editor), Polarized Development and Regional Policies. Mouton Publishers, The Hague, The Netherlands, pp. 379–391.

Molle, W. T. M. (1978). Input–output in the framework of FLEUR: a model of regional development in the European Community. Foundations of Empirical Economic Research 7. Netherlands Economic Institute, Rotterdam.

Molle, W. T. M. and van Holst, B. (1976). Factors of location in Europe: a progress report. Foundations of Empirical Economic Research 15. Netherlands Economic Institute, Rotterdam.

Paelinck, J. H. P. (1978a). Analysing regional growth. In M. Albegov (Editor), The Strategy of Future Regional Economic Growth. CP-78-1. International Institute for Applied Systems Analysis, Laxenburg, Austria, pp. 67–74.

Paelinck, J. H. P. (1978b). Estimation de systèmes spatiaux complexes (Estimation of complex spatial systems). Actes du Colloque: Structures Economiques et Econométrie, Lyon, France.

Paelinck, J. H. P. (1978c). Une théorie des seuils de croissance régionaux (A theory of regional growth thresholds). In Seuils d'Efficacité de la Planification et de l'Action Régionale (Efficiency Thresholds in Planning and Regional Policy). IDEA, Mons, pp. 101–107.

Paelinck, J. H. P. and Klaassen, L. H. (1979). Spatial Econometrics. Saxon House, Farnborough, Hampshire.

Paelinck, J. H. P. and van Rompuy, P. (1973). Regionaal en sectoraal subsidiebeleid; economische theorie en modellen (Regional and sectoral subsidies: economic theory and modeling). Tijdschrift voor Economie, 18(1): 39–55.

Regional Development Modeling: Theory and Practice
M. Albegov, A.E. Andersson and F. Snickars (editors)
North-Holland Publishing Company
© IIASA, 1982

Chapter 24

TERRITORIAL PRODUCTION COMPLEXES: THE SPATIAL ORGANIZATION OF PRODUCTION*

Mark K. Bandman
Institute for Economics and Industrial Engineering, Siberian Branch of the USSR Academy of Sciences, Novosibirsk (USSR)

24.1. Introduction

In the USSR, steady progress in the theory of the spatial organization of production has led to the implementation of the concept of territorial production complexes (TPCs). The TPC concept is now officially recognized as the most appropriate way of spatially organizing production and as the most efficient means of solving large-scale problems. A new type of complex – the program-and-objective TPC – has recently been created to deal with specific regional problems (Bandman, 1978).

The program-and-objective TPC consists of an aggregate of facilities from interrelated national economic sectors (productive and nonproductive), labor, and natural resources, which is organized in accordance with a specially devised plan. It extends over a relatively small but homogeneous area containing the resources necessary for solving large-scale national economic problems. Within the complex, it is ensured that resources are used efficiently (according to national economic criteria), that environmental protection measures are observed, and that renewable natural resources are replenished. The complex is supplied with an infrastructural system whose composition and mode of development cover the requirements of all the economic sectors within the TPC.

24.2. Reasons for establishing program-and-objective TPCs

The establishment of program-and-objective TPCs was considered essential for solving a number of specific large-scale regional problems: large-scale development of the resources in virgin territory (e.g., the Middle-Ob' River and the Bratsk–Ust'–Ilimsk

---

* This paper is a short summary of a detailed description of TPC methodology presented in Bandman (1979).

TPCs in Siberia, etc.); large-scale use of resources from underdeveloped regions, which results in a fundamental change in the position of these regions within the national economy (e.g., the Sayany, Mangyshlak, and Pavlodar–Ekibastuz TPCs); and rapid development of new, or existing, industries in industrialized regions (e.g., the Kursk Magnetic Anomaly TPC).

Not every spatial combination of national sectoral facilities, labor, and natural resources forms a suitable basis for establishing a TPC. The most important factor is how the sectoral facilities are coordinated and their role in the territorial division of labor.

Only those resources that can be utilized efficiently with respect to the national economy as a whole should be developed: this criterion determines the task to be solved by a particular TPC. It also determines how fully a resource, byproduct, or raw material will be utilized or processed. Both direct and related expenditures, especially those connected with attracting labor and investment, are taken into account.

The complex is assumed to engage extensively in the territorial division of labor. Therefore, development of local resources and specific economic components is governed not by the criterion of individual self-sufficiency but rather by the efficiency of the TPC as a whole.

Program-and-objective TPCs are established as a result of a state decision for rapid development (rather than the usual smooth evolution) of regional economies. Because of the specific nature of the tasks to be solved by the program-and-objective TPCs, it is necessary for both local and central authorities to control their implementation. However, these regulatory authorities do not replace the management of the territorial units in which they are located. The complexes are situated close to the area in which the main resource needed to solve a specific problem is found. Therefore, a program-and-objective TPC may in some cases cover the whole area of a province (or republic).

The life-cycle of a complex conforms to the stages of implementation of the TPC program. The establishment of the complex is divided into four phases: scientific analysis and design, infrastructural preparation, construction, and regular operation. In principle, a TPC may also be completely transformed at any point in its life-cycle if a new large-scale national economic program is launched which gives sufficient impetus to the rapid development of a certain type of production in the area already covered by the TPC.

24.3. Development conditions for TPCs

The number of program-and-objective TPCs created at each stage of national production development depends upon the number of national economic problems to be solved simultaneously. At present, there is a scheme for the development of TPCs over a 20-year period in Siberia. The main indicators used to analyze the development process and to forecast the rate of development are existing structure, creation conditions, construction phases, share in territorial division of labor, and participation in the solution of specific national economic problems (see Table 24.1).

Table 24.1. Main features of the Siberian program-and-objective TPCs.

Feature	Program-and-objective TPC									
	Kuzbass	Irkutsk–Cheremkhovo	Central Krasnoyarsk	Middle-Ob' River	Bratsk–Ust'–Ilimsk	Sayany	Noril'sk	Lower-Angara River	North-Tiumen	Upper-Lena River
Favorable development conditions										
Economic–geographic situation	++	++	++			++				
Energy, industrial raw materials, fuels	++	+	++	++	++	+	++	++	++	++
Agricultural development		+	+			+				
Preliminary phases										
Preparation and design			+	+	+	+		+	+	+
Formation				+	+		+	+		
Operation	+	+		+	+		+	+		
Participation in the solution of national economic problems: creation of bases for										
Fuel and energy	+		+	+	+	+	+	+	+	
Processes with high energy consumption		+	+		+	+	+	+		
Integrated timber processing				+	+			+	+	
Machine building	+	+	+			+				+

A variety of conditions can stimulate the development of Siberian (and other) complexes but the most important factor is always a national economic need to utilize a natural resource found in a particular area. All Siberian TPCs have abundant natural resources and large-scale production facilities. However, significant differences exist in the development conditions in various TPCs.

Certain TPCs were developed as a result of the availability of a single resource in a particular area (e.g., oil in the Middle-Ob' River TPC, coal in the Kuzbass TPC), or a combination of resources (e.g., hydroelectric power, timber, and nonferrous metal ores in the Lower-Angara River TPC). Others were constructed as a result of their economic–geographic situation together with an already established production potential (hydroelectric power and coal in the Central Krasnoyarsk TPC or salt in the Irkutsk–Cheremkhovo TPC). The establishment of one Siberian complex – the Sayany TPC – resulted from the presence of almost all of the above conditions. It covers an area rich in resources (hydroelectric power, water, timber, and nonferrous metal ores) and, in addition, the creation of the complex was greatly helped by a combination of natural conditions favorable to human life and agricultural development, and high transport accessibility.

Siberian TPCs are at different stages of development. Some are in the process of being constructed (e.g., the Middle-Ob' River and North-Tiumen complexes). Others have already been built and are in operation (the Bratsk–Ust'–Ilimsk complex has almost reached this stage). Some TPCs are still at the stage of scientific analysis and design; in the preparation period for the eleventh 5-year plan, certain areas (e.g., the Lower-Angara River) will be reserved for their development.

Preplanning studies and planning of program-and-objective TPCs should be based on the set of documents listed in Figure 24.1. The set includes general TPC and design schemes, together with a TPC construction program and 5-year regional and annual TPC construction plans. These documents do not, of course, replace the socio-economic development plans of the corresponding provinces or territories.

24.4. System of models of TPC development

For investigating the problem of TPC development, a group of optimization models was developed. Territorial production regional mesomodels (TPRM) form the basis of the group, which includes territorial models, production models, regional models, and partial territorial models.

The territorial production models are: optimization models (in which the main criterion is minimum total adjusted costs); spatial models (in which the territory is represented not by a point but by a set of territorial taxonomic units of relevant rank); multicomponent models (in which many TPC elements are considered simultaneously); multicommodity models (in which products, linkages, services, and resources are represented in disaggregated form); and multistage models (in which not only output but also coordination and particular stages of production are analyzed).

The model scheme has a block structure corresponding to three main areas of

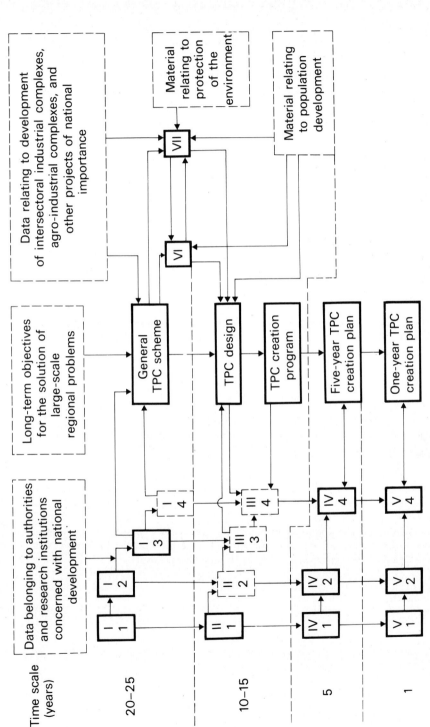

Figure 24.1. The information required to develop territorial production complexes (TPCs) for solving large-scale regional problems. The arabic numbers in the boxes denote the territorial units covered by the information: 1 – USSR; 2 – Soviet Socialist Republics; 3 – economic regions; 4 – provinces or territories. The roman numerals describe the type of information required: I – general schemes for the development and territorial distribution of production capacity; II – main long-term directions in social and economic policy; III – programs for solving particular economic problems or for coordinating industrial development in the regions; IV, V – data on national economic growth; VI – regional planning schemes; VII – plans for environmental protection, land use, or the development of particular production sectors.

research: a functional block, a production–transportation linkage block, and several territorial blocks. All blocks are connected by elements of their objective functions and constraints and include alternative combinations of links to all TPC components. The constraint sets include obligatory targets for sectoral development and obligatory destinations for deliveries of their output, development targets for infrastructural elements whose importance extends beyond the boundaries of the region under consideration, constraints on imports of limited resources or products required outside the TPC, constraints on the capacity of particular facilities, and constraints related to employment and use of local resources. This group of models was used for solving practical problems for Siberian TPCs.

Experience in establishing program-and-objective TPCs has revealed the need to create authorities to monitor their construction. Such authorities should be attached to the Council of Ministers or the State Planning Committee. In addition, it is advisable to have a local unit actually in the area where the program is being implemented. The main organization responsible for the creation of program-and-objective TPCs is the territorial (or provincial) council of People's Deputies and its executive body, i.e., the planning commission of each TPC.

Preplanning preparation, construction, and operation of the Bratsk–Ust'–Ilimsk TPC was discussed in detail during the second IIASA conference on case studies of large-scale planning projects (Knop, 1977). As has been indicated in this chapter, the TPC represents an important example of the practical use of systems of mathematical models in current Soviet regional development planning.

References

Bandman, M. K. (Editor) (1978). Projection of the Formation of Territorial Production Complexes. Institute of Economics and Organization of Industrial Production (IEOPP), Siberian Branch of the USSR Academy of Sciences, Novosibirsk.

Bandman, M. K. (1979). Territorial production complexes as a form of the spatial organization of production: a subject for planning and management. Institute of Economics and Organization of Industrial Production (IEOPP), Siberian Branch of the USSR Academy of Sciences, Novosibirsk.

Knop, H. (Editor) (1977). The Bratsk–Ilimsk Territorial Production Complex. In Proceedings of the Second IIASA Conference on Case Studies of Large-Scale Planning Projects. CP-77-3. International Institute for Applied Systems Analysis, Laxenburg, Austria.

Regional Development Modeling: Theory and Practice
M. Albegov, A.E. Andersson and F. Snickars (editors)
North-Holland Publishing Company
© IIASA, 1982

Chapter 25

THE REGIONAL DEVELOPMENT CONSEQUENCES OF A SHUTDOWN OF NUCLEAR POWER PLANTS IN SWEDEN

Folke Snickars*
Department of Mathematics, Royal Institute of Technology, Stockholm (Sweden)

25.1. The problem summarized**

25.1.1. Background

In March 1980, a referendum was held on a major political issue of the 1970s in Sweden: whether or not Sweden should dispense with nuclear power as soon as possible (specifically, within 10 years). If the development plans current at the time of the referendum were to be fulfilled unchanged, the number of nuclear power plants in operation would be doubled over the same 10-year period, and these new plants would function for at least 25 years.

In 1979, a commission was established by the government of Sweden to assess the consequences of dispensing with the nuclear power sector. This "Commission on Consequences" studied the potential effects on the Swedish economy with respect to employment, regional development, private consumption, and environmental quality. It did not explore the risks attached to the operation of nuclear reactors but addressed the problem of evaluating in quantitative terms the costs to Sweden and its inhabitants of changing the energy system significantly.

This chapter contains a summary of the methods and models used by the commission, especially for assessing regional effects. The aim is not to discuss the socioeconomic and political implications of the abolition issue but to present the techniques used for measuring impacts in various dimensions and to explain why they were selected. The regional effects, which are central to the presentation, are evaluated not only by quantitative models but also by general economic analysis. The major result that will emerge from the study is that an elaborate systems analysis of a

---

* Currently at the International Institute for Applied Systems Analysis, Laxenburg, Austria.
** Parts of this background description draw heavily on the official English-language summary of the work of the Commission on Consequences (Guteland, 1980).

regional problem, or any other problem of considerable complexity, should include both qualitative and quantitative aspects. The qualitative part consists of a micro-analysis of the regional development of production and employment in energy-intensive industries, and the quantitative section of a macroscale modeling exercise aimed at measuring the total direct and indirect regional effects of different energy scenarios.*

25.1.2. The Commission on Consequences – alternatives investigated

The basic methodological problem faced by the commission was to produce a set of alternative scenarios for providing the Swedish economy with energy and to compare their positive and negative effects. The approach chosen was to elaborate a reference scenario, in line with the current economic development trends and including the current nuclear power program, and then to specify alternative energy-system scenarios in which nuclear power is abolished before 1990. These energy-system scenarios cover a period of twenty years, from 1980 to 2000.

Since the nuclear power plants primarily produce electricity (waste heat is not yet used in remote-heating systems), the alternative energy systems may be compared by describing the electricity-production subsystems (Table 25.1) and by assessing the structure of electricity demand, at equilibrium, in various cases (Table 25.2).

The reference case assumes that existing plans will be put into effect. Twelve nuclear power plants will operate for their planned service lifetimes, but no longer. In 1990 nuclear power will produce 58 TWh of electricity, which, as mentioned above, is more than double the 1980 level. The structure of electricity consumption will change toward a more intensive use of electricity in the "miscellaneous" sector. A considerable portion of this increase relates to the increased use of electric heating in dwellings and workplaces.

The abolition alternative has two variants. The total level of electricity consumption is lower in both variants in 1990 and 2000 than in the reference case. This stems from the difficulty of completely replacing nuclear power over a period of only ten years.

The 105 TWh level may be attainable by 1990. However, this presupposes a swift expansion of coal-fired electricity generation, and there may be difficulties in establishing and commissioning the necessary plants with sufficient speed. In this alternative no expansion of electric heating of dwelling places is allowed.

The 95 TWh level by 1990 represents a situation in which the growth in electricity consumption is heavily restricted. It presupposes a particularly strong emphasis on saving in the miscellaneous sector. This alternative has been advocated by those desiring the rapid closure of the nuclear power sector. Since it may be difficult to

---

* The work reported here was performed by Bo Erixon, Lars Lundqvist, and Mats Reidius, together with the present author.

Table 25.1. Comparison of electricity-supply systems in the reference case and in the two alternatives involving abolition of the nuclear sector, for 1990 and 2000.

Supply system	1980	1990				2000			
	Observed	Reference	Abolition			Reference	Abolition		
	91 TWh	125 TWh	105 TWh	95 TWh		140 TWh	120 TWh	105 TWh	
Hydroelectric power	62	65	65	65		65	65	65	
Nuclear power	23	58	2	1		58	–	–	
Power generated by industry and returned to grid[a]	5	7	8	7		9	10	9	
Geothermal energy[a]	6	6	16	15		15	18	18	
Conventional oil-fired power generation[b]	4	1	8	8		–	1	1	
Conventional coal-fired power generation[b]	–	–	16	8		3	34	19	
Wind power	–	1	1	1		4	4	4	
Total generation	100	138	116	105		154	132	116	
Transmission losses	– 9	– 13	– 11	– 10		– 14	– 12	– 11	
Final electricity consumption	91	125	105	95		140	120	105	

[a] Both heat and power are produced, but the figures refer to electricity output only.
[b] Producing electricity only.

Table 25.2. Structure of final electricity consumption in 1978 and in 1990.

Sector	1978	1990		
	Observed	Reference	Abolition	
	82 TWh	125 TWh	105 TWh	95 TWh
Industry	39	57	53	50
Transport and communications	2	3	3	3
Miscellaneous	41	65	49	42
Total	82	125	105	95

keep the miscellaneous sector at the low level indicated in Table 25.2, a further alternative has been tested where a larger portion of the necessary saving is taken from the industrial sector by means of the price mechanism.

The abolition of nuclear power will mean that resources will have to be used to expand other electrical energy production systems and to conserve energy in homes, in industry, and elsewhere. In addition, more coal and oil will have to be imported. Sooner or later this will have to be paid for in terms of living standards for the population lower than those of the reference case. The commission assumed that the cost of abolition would have to be met from private consumption because, even in the reference case, the growth of the public sector has been kept very low. Another conceivable recourse would be to reduce other investments or to increase international borrowing, thus deferring some of the costs to future generations. This was considered to be unacceptable.

The central result of the Commission on Consequences is that the total cost to Swedish society, in terms of private consumption, would correspond to some 2–3% less private consumption in 1990 than if the nuclear power sector were retained. This means a capital loss of 20,000 Swedish krona per worker for the period 1980–2000. Another important result is that if labor-market policies are implemented such that full employment is attained, no drastic effects are found for the development of different production sectors. The case of electricity price increases for industry at the 95 TWh level is an exception, with strongly negative effects in the pulp and paper industry. Price increases of the order of 50% for households and 30% for industry were deemed necessary to keep aggregate electricity demand at the required supply level during the phase of replacement of nuclear power (i.e., the 1980s).

25.2. The system of models employed

25.2.1. Organizational structure

The Commission on Consequences worked under a severe time constraint. Therefore, no major model development work was attempted. Instead, the work was organized

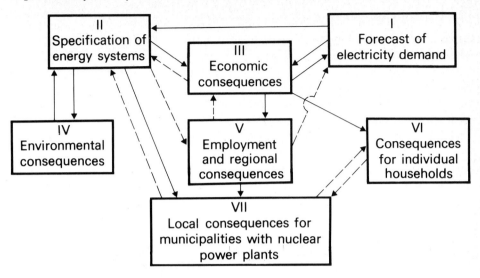

Figure 25.1. The organization of the working groups making up the Commission on Consequences.

among various groups, which used suitable models and methods available at the outset of the investigation. These tools were used as a basis for drawing conclusions about the magnitudes of the consequences resulting from the closure of nuclear power plants. A serious problem occurred because of this organizational framework: how were the analyses of the separate working groups to be integrated?

In Figure 25.1 an outline of the links between the working groups is given. The division into working groups may be seen as analogous to a division of the systems-analytic problem of developing a set of models and techniques to cope with the complex problem of assessing effects in an interdependent system. The Commission on Consequences provides a good example of a solution to an organizational as well as a factual problem.

In the same way that a decomposition approach to mathematical programming consists of isolating subsystems, the internal workings of which need not be fully considered at the higher level, so it was not necessary for each subgroup of the commission to deliver all its information to the other subgroups; only those items of central importance were necessary.

The idea was that a number of reiterations of this information would lead to a fully consistent impact analysis. However, the time constraint meant that such overall consistency could not quite be achieved.

The iteration in the analysis began with a forecast of the development of electricity demand (I). This initial forecast was consistent with earlier Swedish energy consumption forecasts although the total demand level of 125 TWh in 1990 was somewhat lower than in the earlier forecasts. This forecast, disaggregated as shown in Table 25.2, was passed on to subgroups II and III, which dealt with specifying energy systems and assessing economic consequences, respectively.

The next step (II) in the iteration procedure was the specification of energy systems capable of producing the energy demanded at minimum production costs, under environmental, regional, local, and other external constraints. The results of the analysis in subgroup II were then reported to subgroups III (investment costs, composition of primary energy supply), IV* (size and location of plants and primary energy sources), V (size and location of energy production facilities), and VI (location of new energy production plants at former nuclear power plant sites).

An important stage in the whole investigation was the pooling of information from subgroups I and II into an overall analysis of economic consequences in subgroup III. There the total costs of transition to a new energy system were evaluated for short-term, medium-term, and long-term perspectives. The costs were measured in terms of the level of private consumption that could be attained for alternative electricity demand levels and different energy production systems. The results of these analyses were reported back to subgroups I (scarcity prices, production structure, etc.), V (production structure), VI (level of public and private consumption), and VII (local impacts).

The next step in the iteration scheme was an analysis of the results of other subgroups in the environmental, regional, and household groups. Strong feedback was exerted from the environmental to the energy systems subgroup. Various criteria were applied to see whether the energy production scenarios were permissible from an environmental point of view. Information about the degree of fulfillment of regional efficiency and welfare goals was also fed back from the employment and regional group to subgroups II and III. An example of such feedback was the cost–benefit analysis of an alternative location in the northernmost part of Sweden for a coal-fired plant.

As mentioned earlier, this organizational framework was actually used in the investigation, with explicit deadlines for the reports of external information between the subgroups. It is fair to say that some integration was actually achieved in this way, although complete consistency could not be attained. One reason for this was the time constraint, but institutional frictions also played a role.

25.2.2. The models used

The description of the work organization given above does not reveal the extent to which the subgroups used quantitative models or other quantitative analytical techniques in their work.

Quantitative analyses were attempted in all of the working groups. However, these analyses were conducted at very different levels of sophistication. The methods used in groups IV, VI, and VII were quite rudimentary from a mathematical point of view,

---

* In effect, there was no formal subgroup IV in the commission, but some members of the existing working groups formed an informal subgroup.

amounting to a more or less systematic application of various types of multipliers and ratios. They were not necessarily internally or externally consistent.

In subgroup II, models were employed, for example, to find an efficient way of using renewable, domestic primary energy resources in industrial processes. In these models, for instance, local biofuel sources such as peat were considered as alternatives to coal and oil in combined heat and power-production facilities.

By far the most sophisticated set of models was used in the economic analyses. The model exercises included the use of both the medium-term and the long-term economic forecasting models of the Swedish Ministry of Economic Affairs. The medium-term model is of an input–output Keynesian type, with consumption and import functions but with exogenous investment variables. The long-term model is a variant of multisectoral growth models with linear energy demand and export and import functions. The core model used in the investigation, however, was the general equilibrium model of Bergman and Por (1980). One reason for choosing this as the central model was the fact that it simulates more effectively than other models the substitution possibilities over the long term in an open economy as the production of energy becomes more expensive in real terms. However, it is not necessary to elaborate on its construction at this point. In Section 25.3 further details of its properties are given where necessary.

The economic models used in the investigation are almost an order of magnitude more complex than the regional models, at least with respect to their applicability to the problem of assessing the impacts of different energy scenarios. Nevertheless, the rest of this chapter is devoted to a presentation of the regional models, and to a discussion of how they might be developed for integration with regional–national impact models.

25.3. The regional model analyses

25.3.1. The break-down model

The prerequisite for using any general equilibrium model is a balanced situation in the national economy. Production factors are adjusted such that no excess supply or demand exists. This situation is of course attainable only in the long term, especially if the current situation is characterized by significant economic imbalances.

The idea behind the use of the economic model at the national level is simply to compare the equilibrium states in the economy in terms of the scope for private consumption and the equilibrium economic structure for different energy scenarios. The impact of the abolition of the nuclear power sector is simulated by a higher depreciation rate for capital in the nuclear energy production sector. The output of the model contains both factor inputs, production levels, capital stock, foreign-trade data, and employment. Employment is measured in terms of the input of work-hours needed in the various sectors.

The purpose of the break-down model is to transform these employment results into forecasts of the total number of persons needed to perform the necessary work-hours, and to disaggregate these figures to the regional level. Thus, the aim of the break-down model is to outline the consequences of alternative national energy scenarios with respect to total employment — or rather total demand for labor — at the regional level. This involves two major questions: (A) How will the total labor demand in the regions be affected by changing equilibrium patterns of production and employment by sector at the national level? (B) What are the direct effects — and the consequent indirect ones — of varying regional developments in individual sectors, for example in the energy production sector?

In these disaggregation exercises, the break-down model was used in conjunction with the medium-term and long-term economic forecasts for Sweden (Snickars, 1979). In the economic forecasts a judgment was also made about the regional development of the labor supply. Since this is assumed to be independent of the energy system, the current application amounts to a comparison of different labor demand scenarios.

The break-down model is very simple and rests strongly on the historical development of employment by sector and a basic–nonbasic hypothesis. It may be summarized in mathematical form by two sets of equations.

Let the following definitions hold:

$\alpha_{ik}(t)$ is the proportion of total basic employment in sector i located in region k at time t (the sum over k of $\alpha_{ik}(t)$ is equal to 1)

$\sigma_{ik}(t)$ is the number of service jobs in sector i in region k, per job in region k, at time t, in relation to the relative size of sector i at the national level

$\bar{S}_i(t)$ is the total number of work-hours in sector i as given by the equilibrium model

$\mu_i(t)$ is the number of hours worked per year per person in sector i at time t

$x_i(t)$ is the number of basic employees in sector i at time t (to be determined)

$y_k(t)$ is the total employment in region k at time t (to be determined)

Given these definitions the break-down model may be summarized by the formulas presented below.

$$\sum_i \alpha_{ik}(t)x_i(t) + \sum_i \sigma_{ik}(t)\cdot y_k(t) = y_k(t) \tag{25.1}$$

$$\sum_k \alpha_{ik}(t)x_i(t) + \sum_k \sigma_{ik}(t)y_k(t) = \bar{S}_i(t)/\mu_i(t) \tag{25.2}$$

$$0 \leqslant x_i(t) \leqslant \bar{S}_i(t)/\mu_i(t) \tag{25.3}$$

The nonnegativity condition for $y_k(t)$ is automatically fulfilled under (25.3) and appropriate choices of $\sigma_{ik}(t)$.

Forecasting using model (25.1)–(25.3) amounts primarily to making good projections of the model parameters. The model contains a fundamental basic to nonbasic multiplier $1/[1 - \Sigma_i \sigma_{ik}(t)]$ that relates total employment to total basic employment, reflecting the combined effect of all the parameter forecasts.

The data set for the parameter forecasts contains consistent time-series data for a five-year period. The projections are made subject to existing detailed sectoral data or by some submodel, for example, of the Salter analytic type. Two empirical observations may be made in relation to these forecasts. The first is that forecasts for the whole time series seem to be necessary to get a clear view of the regional development in, for instance, the pulp and paper industry. The second is that the service ratios are very high in the stagnating regions in the north of Sweden. This is true for the construction industry and also for several of the public sectors in Sweden as a whole (education, health care, etc.). The fact that the service ratios differ significantly between regions and that service sectors account for an expanding proportion of total employment in Sweden make it important to disaggregate the model considerably for the service sector. Thus, the Swedish break-down model for the public sector is one of the most disaggregated of national–regional models.

25.3.2. Microsectoral studies

An aggregated break-down model of the type described above is too coarse to identify all the regionalized labor-demand consequences, even though it is theoretically possible to build in a submodel for each sector, in each case including detailed knowledge about the sector in question. In the nuclear power application, a significant result was that scarcity prices had to be used to keep down electricity demand for households and industry. This implies that a closer investigation should be made concerning the electricity-dependence of energy-intensive production units at the regional level.

Such an analysis is especially necessary in a country such as Sweden, with its large surface area and low population density. Many small towns and villages in Sweden, particularly in the northern and central parts of the country, are dependent on only one dominant industrial enterprise. A further complicating factor is that these units are most common in energy-intensive sectors. Economies of scale in, for example, the pulp and paper industry, the metal industry, and to some extent the chemical industry, have led to a concentration of production in a few large plants.

To show the regional concentration of industry, the geographical structure of the Swedish steel industry is given in Figure 25.2. There are only twelve commercial steelworks in the country. According to earlier Swedish investigations into the future of the steel sector, some 2,000 out of 46,000 jobs (in 1978) will have to disappear by 1985, if the sector is to remain viable.

In the Commission on Consequences, a special study was performed to assess the effects of a 50% price increase in electricity on energy-intensive industries. The results showed that the scrap-iron steelworks would be the most seriously affected branch of the steel sector. Some 1,200 jobs would have to disappear within a five-year period as a result of this price shift. However, some of these jobs are also likely to disappear even in the reference case. This illustrates the basic methodological problem in

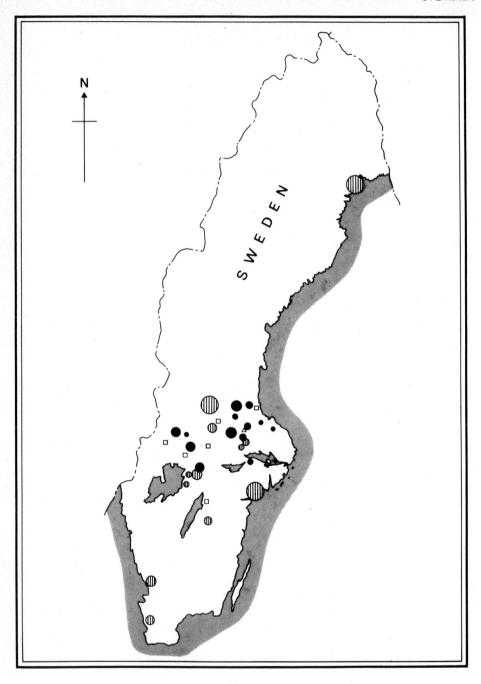

Figure 25.2. Regional distribution of the Swedish steel industry. The type of steel produced is indicated as follows: ● – special steel; ◍ – commercial steel; □ – secondary steel and plates; △ – crude iron only.

performing an impact analysis without a comprehensive modeling framework within which to evaluate, in a consistent way, various direct and indirect effects.

The purpose of this chapter is not to present and analyze the results of the studies of energy-intensive sectors. However, it should be strongly emphasized that such an analysis was regarded as a useful complement to the macro break-down model exercises of the commission. Since one central result of its economic analyses was that only small effects on the sectoral structure of the whole of Sweden could be expected (assuming a situation of full employment), the regional effects were also rather small as shown by the break-down model. At least, this was true for large regions and for the total demand for labor. The microsectoral studies indicated a considerable negative effect at the local level in certain industrial sectors.

Since the predicted effects operate at a number of different levels of aggregation, it is not possible to ascertain offhand whether or not they are mutually consistent. In addition, the fact that the microanalyses are rather short-term makes them somewhat less comparable with the medium-term macroanalyses. In an attempt to reconcile the two approaches, interregional input–output analysis was applied to estimate the total indirect employment effects in the regional production system of the disappearance of jobs in industries with a high dependence on electricity. These computations are outlined in Section 25.3.3.

25.3.3. Interregional employment multipliers

Interregional input–output analysis is a useful method for estimating indirect effects in a multiregional framework. By means of this technique interactions throughout an interrelated system of regions may be taken into account when assessing both total production effects of investment programs or other final-demand changes and the indirect production effects of initial production changes.

An approach based on interregional input–output would be an alternative to the basic–nonbasic impact model described above. It is a theoretically much more elaborate approach but its usefulness is severely restricted by its massive data requirements. A possible way of removing part of this disadvantage, although with some loss of realism, was described by Snickars (1978), where a generalized entropy method was used to derive a full interregional input–output matrix from incomplete data. The method employed by Snickars has been used to derive an interregional input–output matrix for Sweden, adapted to regional and sectoral production data for 1975. This matrix could have been used in a number of ways for the regional-effect studies of the Commission on Consequences: (A) to obtain a regional break-down of national production forecasts and a comparison of alternatives in production rather than employment terms; (B) to apply regionally varying energy input coefficients to the regional production forecasts in order to estimate energy use by region — this might also be used as a consistency check on the national energy demand forecasts; (C) to derive the production necessary to fulfill the regionally specified investment program to replace the closed nuclear power plants and/or the investment program for the energy savings in the building stock; (D) to determine the indirect effects on the national economy of rationing the energy supply; and (E) to estimate the indirect effects of a decline in production in electricity-intensive sectors (see Section 25.3.2 for a description of the initial production decreases).

Attempts were in fact made to perform analyses of types (A), (B), and (E) in the work of the commission. The type (A) analysis was not successful because of the difficulties in calibrating the reference regional structure of employment by this method compared to the basic–nonbasic method. The type (B) analysis was abandoned because of problems in the consistency checks with the national energy demand forecasts. In both examples, the difficulties in fulfilling the initial aims were partly a result of the classical conflict between national sector forecasting and regional forecasts. Inconsistencies at the national level, as indicated by regional analysis, are likely to be regarded as reflections of bad regional data rather than as correct results of more detailed analysis.

The only analysis made by the commission in which the interregional input–output method was finally used was an analysis of type (E). To describe this application, the following notation will be used:

a_{ij}^{kl} is the input of intermediate products from sector i, region k, required to produce one unit of output in sector j, region l

f_i^k is the final demand of deliveries from sector i, region k

x_i^k is the production level in sector i, region k

λ_i^k is the labor/output ratio in sector i, region k

Let $\mathbf{A} = \{a_{ij}^{kl}\}$ be an $n \times n$ matrix, and let $f = \{f_i^k\}$ and $x = \{x_i^k\}$ be n-vectors. Then, the fundamental matrix equations of input–output analysis are

$$x = \mathbf{A}x + f$$

$$x = (\mathbf{I} - \mathbf{A})^{-1} f = \mathbf{B}f \tag{25.4}$$

where $\mathbf{B} = \{b_{ij}^{kl}\}$ is defined through relation (25.4).

A unit reduction in final-demand deliveries from energy-intensive sector j in region l leads to a direct reduction in production by the same amount. Thus, on average, employment is directly decreased by λ_j^l. From (25.4) the total interregional effects may be outlined. Let Δs_j^l be the employment decreases in energy-intensive sector j, region l that emerge from the analysis in Section 25.3.2. Then, the total employment effect of that set of decreases should be Δl^k, given by

$$\Delta l^k = \sum_i \lambda_i^k \sum_j \sum_l b_{ij}^{kl} \Delta s_j^l / \lambda_j^l \tag{25.5}$$

Of course, formula (25.5) is only valid in an average sense, since the labor/output ratios, for instance, are gathered from employment and production data aggregated by sector and region.

Table 25.3 contains an example of the results of such an employment multiplier

Table 25.3. Direct and indirect employment effects (in numbers of jobs) within the production system in 1990 if electricity-intensive branches of the steel sector are closed.

Region	Effect on employment		Reference development for the whole sector
	Direct	Indirect	
Stockholm	0	− 2,900	+ 300
East central	0	− 2,500	− 4,200
Southeast	0	− 1,500	+ 800
Southern	− 200	− 2,400	− 1,400
Western	− 2,100	− 4,000	− 300
Northwest central	− 700	− 2,100	− 5,300
Northern	− 1,200	− 3,700	− 100
Far north	0	− 1,200	− 700
Sweden	− 4,200	− 20,300	− 10,900

exercise, based on possible production and productivity conditions in the year 1990. It is clear from Table 25.3 that the indirect employment multiplier is approximately five for the steel industry. This is a considerable multiplier effect, which is due to the fact that the steel industry produces primarily intermediate products and is highly capital-intensive.

It is also evident from Table 25.3 that the reference development level for the steel sector in the eight Swedish regions is not quite in line with the direct employment decreases resulting from energy price increases. The negative indirect employment effects stated in Table 25.3 are of course compensated by employment increases in other sectors. Among industrial sectors, the major employment increase is foreseen in the machinery and equipment sector. Outside industry, the major increase is forecast for the public sector.

25.3.4. Regional effects: main findings of the commission

A summary of the basic results of the Commission on Consequences is given below, together with some observations on the possible methodological causes for these results.

Three conclusions may be drawn: the break-down studies indicate that the total regional effects of the national energy scenarios are minor; the special studies of electricity-intensive sectors indicate serious problems for small communities with a heavy reliance on only one or a few energy-intensive sectors; and the input–output studies indicate a spreading of the negative effects of the problems in the electricity-intensive sectors to all parts of Sweden, but with local concentration in the regions where electricity-intensive sectors are dominant.

One disadvantage of the break-down method is that it is basically a reflection of national development at a rather coarse sectoral level. This level is not detailed enough to single out the energy-intensive parts of the economy, and positive and negative effects may be smoothed out against each other. The basic regional paradigm behind this approach is that the regional adjustments are just as smooth as the national ones. If there is reason to believe that this is not true, a more detailed and microscale approach is necessary. An approach of this type has also been used by the commission.

25.4. Toward an integrated regional analysis of energy scenarios

The organization of the work and the quantitative methods used in an investigation of the effects on the Swedish economy of dispensing with the nuclear power sector have been described. Several types of quantitative models of varying complexity were used by the commission. An integration of the model results was attempted by a procedure of iteration between the working groups.

Several fundamental questions of considerable methodological and factual interest may be raised in relation to such a study. Are the results of the studies subject to uncertainty and, if so, for what principal reasons? To what extent could a possible degree of uncertainty in the results be removed by using a more integrated set of model approaches? Would the economic consequences at the national level be different if a regionally specified general equilibrium model had been used?

Without going into a detailed argument concerning these questions, it should be emphasized that an assessment of the consequences in the medium- and long-term of different energy scenarios is of course subject to a considerable degree of both static and dynamic uncertainty (international economic development, energy prices, and stabilization problems). It is also self-evident that an ideal set of models could sharpen the results further. Such models would have to be specially designed as a set and specifically constructed to evaluate energy systems; the major feature of the set should be that it produces internally consistent results. Using such a set of models, the number of sensitivity tests performed on exogenous data could be increased, which would in fact tend to reduce the degree of uncertainty in the results. The success of an integrated model system when used by a body such as the Commission on Consequences would of course still depend on whether the approach were fully accepted by the working-group members.

It might well be found that such a set of models should operate at the regional level. This should definitely be the case in a country where there are large regional differences in economic conditions and structures. In a country where large regional differences in factor endowments and demand conditions, for example, for the energy sector, do not exist, such a regional specification may not be necessary.

Returning to the Swedish study, it is quite clear that a more integrated regional analysis of national energy scenarios is desirable. Sweden has considerable resources of renewable energy with a nonuniform regional distribution. It has a regionally varying production structure. This means that the regional impacts of different national energy scenarios would vary, at least if the effects are evaluated in dimensions other than that of employment. It also means that regional and local variants of energy supply systems are quite conceivable and should be analyzed.

A coordinated set of regional models for energy systems analysis could possibly be integrated into a comprehensive regional–national policy evaluation model of the type developed, for example, in France (the REGINA model of Courbis and Cornilleau, 1978) and in several other European countries. Promising modeling work has also been initiated in Sweden; for examples, see especially the work of Lundqvist (1980) and also the national–regional forecasting model of Granholm and Snickars (1979).

References

Bergman, L. and Por, A. (1980). A Quantitative General Equilibrium Model of the Swedish Economy. WP-80-04. International Institute for Applied Systems Analysis, Laxenburg, Austria.

Courbis, R. and Cornilleau, G. (1978). The REGIS Model. A Simplified Version of the Regional–National REGINA Model. Paper presented at the Regional Science Association's European Congress, held in Fribourg, Switzerland, June.

Granholm, A. and Snickars, F. (1979). An Interregional Planning Model of Private and Public Investment Allocation. Paper presented at the Regional Science Association Congress, British Section, held in London, August.

Guteland, G. (1980). Suppose We Go Non-Nuclear? Summary of the Work of the Government Commission on the Consequences for Sweden of Abolishing Nuclear Power. Ministry of Industry, Stockholm.

Lundqvist, L. (1980). A Dynamic Multiregional Input–Output Model for Analyzing Regional Development, Employment, and Energy Use. Royal Institute of Technology, Stockholm.

Snickars, F. (1978). Estimation of interregional input–output tables by efficient information adding. In C. Bartels and R. Kettellapper (Editors), Exploratory and Explanatory Analysis of Spatial Data. Martinus Nijhoff, Leiden.

Snickars, F. (1979). Regional Break-Down of Forecasted National Demand for Labor (in Swedish). Ministry of Industry, Stockholm.

Regional Development Modeling: Theory and Practice
M. Albegov, A.E. Andersson and F. Snickars (editors)
North-Holland Publishing Company
© IIASA, 1982

Chapter 26

POSTWAR INFLUENCES ON REGIONAL CHANGE IN THE UNITED STATES: MARKET FORCES VERSUS PUBLIC POLICY

Bernard L. Weinstein
Southern Growth Policies Board, Washington, D.C. and University of Texas at Dallas, Richardson, Texas (USA)

26.1. Introduction

At the present time, the United States has no national policies dealing with population growth, population distribution, or regional economic development. While a wide range of federal programs, from military procurement to Federal Housing Administration (FHA) mortgage insurance, has undoubtedly had differential regional impacts, these programs are not specifically designed to influence the regional distribution of economic activity.

Though there are several federal government agencies concerned with the spatial aspects of economic development, each has limited interventionist powers and a miniscule budget. For example, the Economic Development Adminstration of the U.S. Department of Commerce, which makes grants and loans to low-income communities for public works and infrastructure projects, is authorized to spend only $1.1 billion in fiscal year 1980. This amounts to less than one-fifth of one percent of total federal outlays.

In the mid-1960s, a group of regional commissions was established to examine developmental problems in selected areas. As a result of political pressures, however, they have now grown to encompass at least part of every state in the continental part of the United States (see Figure 26.1). In areal terms, only northern California, central Texas, the southern tier of the upper Great Lakes states, and central Virginia and North Carolina lie outside an economic development region. The regional commissions prepare five-year development plans and also award grants for infrastructure and public works. Their combined budgets for fiscal year 1980 amounted to $768 million, about one-tenth of one percent of total federal outlays.

Despite the absence of regional development programs per se, the lagging regions of the United States have grown steadily since World War II. Since the 1960s, this growth has been quite dramatic — especially in the South, which historically has been the least-developed and most poverty-ridden region of the nation.

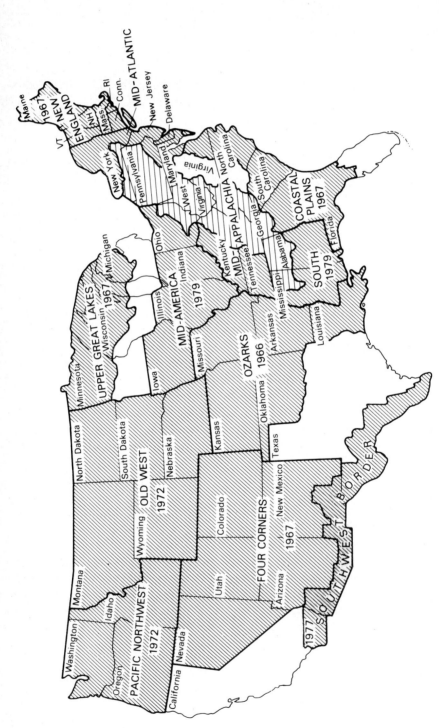

Figure 26.1. Economic development regions of the USA. The shading represents the regions in which regional development commissions have been set up; the year in which each commission was first organized is also given. The heavily shaded region in the Southwest Border withdrew from the Four Corners Economic Development Region in October 1979.

Since 1970, population growth in the South has been nearly double that of the nation as a whole. A large portion of the population gain has stemmed from net in-migration from other regions. This population trend is universal across the South — every southern state grew at a rate faster than the national average between 1970 and 1977. By contrast, the states of the northern industrial group all grew at rates below the national average (Weinstein and Firestine, 1978).

Along with the rapid population growth in the South has come rapid employment growth. Total nonagricultural employment grew 27% between 1970 and 1977 compared with a 16% gain nationally. A significant proportion of southern employment growth has occurred in the manufacturing sector, while industrial employment has remained flat or dropped slightly in other regions of the United States.

Personal income has also grown faster in the South than in the nation as a whole. The region recorded a 111% gain between 1970 and 1977 compared to a 92% gain nationally. In per capita terms, every southern state is showing faster income growth than the nation as a whole. In contrast, many northern states, in the old industrial heartland of the nation, are exhibiting per capita income growth less than the national average. This slow income growth has led some politicians and journalists to refer to the North as the lagging and most distressed region in the United States. In fact, most northern states still have incomes above the national average, while every southern state, despite rapid economic growth over the past two decades, remains below the national average — in some cases by 25–30%.

Of course, the South is not the only region showing above-average income gains. As Table 26.1 indicates, all major regions of the United States with per capita income below the national average in 1965 recorded above-average income gains during the following decade. This attests to the acceleration of a long-term income convergence trend among regions, which goes back to at least 1929.

26.2. The impact of federal spending on regional development

Before World War II, economic growth in the various regions of the United States depended almost entirely on private-sector activities. But the postwar era has seen a tremendous increase in the size and scope of federal government. Thus, federal spending itself has become a major force in shaping regional economic activity. It is not surprising, therefore, that much of the ongoing public debate about regional growth and decline has focused on the differential impacts of federal tax and spending policies. In particular, many northern politicians have alleged that favorable federal spending policies have been instrumental in bringing about the rapid growth of the South while accelerating the decline of the industrial North — especially New York state.

Such assertions are often backed up by statistical computations showing that the Northeast and Midwestern states are running "balance of payments" deficits with the federal government; i.e., they are sending more to Washington D.C. in the form of taxes than they are receiving back in federal outlays. For example, a recent study by the National Journal (1977) calculated that the five Great Lakes states ran a $20.1 billion

Table 26.1. Per capita income by region, 1965 and 1976.[a]

Region	Per capita income ($)		Percentage change, 1965–1976
	1965	1976	
New England	2,992	6,590	120
Mid-Atlantic	3,158	6,932	120
Great Lakes	2,992	6,793	127
Great Plains	2,632	6,130	133
South Atlantic	2,244	6,007	168
South Central	2,133	5,544	160
Mountain	2,493	5,990	140
Pacific	3,185	7,081	122
U.S. Average	2,770	6,441	133

[a] Source: U.S. Bureau of the Census.

Table 26.2. Federal government spending and taxes per capita by region, fiscal year 1976.[a]

Region	Federal spending per capita ($)	Federal taxes per capita ($)	Spending/tax ratio
New England	1,599	1,676	0.95
Mid-Atlantic	1,405	1,718	0.82
Great Lakes	1,142	1,633	0.70
Great Plains	1,438	1,401	1.03
South Atlantic	1,577	1,427	1.11
South Central	1,447	1,227	1.18
Mountain	1,701	1,372	1.24
Pacific	1,904	1,650	1.15
U.S. Average	1,524	1,524	1.00

[a] Source: National Journal (1977).

deficit with Washington D.C. in fiscal year 1976 while the Mid-Atlantic states ran a $12.6 billion deficit. By contrast, according to the study, the South ran a "surplus" with Washington D.C. of some $12.6 billion while the West gained $10.4 billion.

As indicated in Table 26.2, the New England, Mid-Atlantic, and Great Lakes regions all show spending/tax ratios less than one on a per capita basis, while the other regions show ratios greater than one. But such ratios do not, by themselves, support northern claims of regional discrimination. Indeed, any a priori expectation of regional balances between spending and taxes would totally ignore the significant distributional aspects of federal fiscal policy. For example, progressive income taxes take more from wealthy persons (and states) than they do from poor persons (and states). Similarly, many federal expenditure and transfer programs have been specifically designed to raise the incomes of the nation's lowest socioeconomic groups.

As shown in Table 26.1, per capita incomes in the South Atlantic, South Central, and Mountain regions are well below the national average while the New England, Mid-Atlantic, and Great Lakes regions show per capita incomes above the national average. Thus, it is not surprising that the average tax payment per person is lower in the South and Mountain states than it is in the North. The tax differentials simply reflect the relative income positions of the regions.

Though growing rapidly, the South Central region is still the poorest in the nation. Yet per capita federal spending in this region in 1976 was considerably below the national average of $1,524.

State-by-state or region-by-region comparisons of federal taxes and expenditures are also misleading on other grounds. For example, a number of federal spending programs are channeled to persons rather than local governments or private contractors. Social security payments, federal retirement and military pensions, and welfare payments are received by highly mobile individuals. The fact that many retirees are moving to the South, and bringing their social security, military and federal retirement pensions with them, is not a matter of discretionary fiscal policy.

In any case, the importance of federal spending to the economic development of the South has been much overstated (see, for instance, Rees, 1978). The North has been losing people, jobs, and investment to the Southeast and Southwest for over two decades. To a large extent, the migration of population and employment opportunities is occurring in response to economic forces affecting the cost and efficiency of production. Lower living costs, taxes, energy costs, and land costs in many parts of the South have facilitated the economic development of the region and, at the same time, have improved the overall efficiency and productivity of the national economy. Economic growth in the region has become self-sustaining as a result of growing markets and a broadening industrial base. Given the present diversity and dynamism of the region's economy, it is unlikely that a redirection of federal funds to the North would stem the flow of people, investment, and jobs to the South.

26.3. Contemporary regional change in the United States: a response to market forces

Although interregional shifts in population and industry have only recently emerged as a major public-policy issue, such movements have been occurring since the early days of the republic. Most historians and demographers cite three major phases of internal migration prior to the southern surge in the late 1960s. The first and longest phase began after the Revolutionary War and ended with the closing of the western frontier around 1880, a period during which the United States remained a predominantly agricultural society. The second phase of internal migration coincided with the industrialization and urbanization of the U.S. economy, covering the period from around 1890 to 1920. The third significant phase of internal migration occurred between 1920 and 1960 when net black out-migration from the 16 southern states exceeded three and a half million persons.

A common thread to all of the mass migrations in the history of the United States has been the search for economic opportunity. People left the areas where they were living because they perceived better opportunities for themselves and their families elsewhere. This was true of the nineteenth-century pioneers and the southern blacks of the 1920s and was also true of the educated, middle-income whites who migrated in large numbers to the South during the 1970s. Indeed, migration is perhaps best defined as one form of human response to the uneven spatial distribution of opportunities and resources.

Obviously, there is a strong relationship between migration and regional economic development. Migration affects personal income, employment, investment, and the demand for public services in both growing and declining areas. Typically, regions receiving large numbers of migrants show per capita income growth above the national average, while those areas losing population show slower than average income growth (see Table 26.1). Rapidly growing regions often require heavy public expenditure for roads, schools, and utilities, while, in contrast, declining regions find themselves faced with heavy public outlays for unemployment, welfare, and other social services (Keeble, 1967).

While federal fiscal flows have obviously had some impact on regional development, the growth of the South and the relative decline of the North can be best understood in the context of convergence toward regional equality in a major economic system — that system being the United States as a whole. Indeed, neoclassical economic growth theory would predict such a convergence. In its simplest form, neoclassical growth theory posits that factors of production — labor, capital, entrepreneurship, etc. — are free to move within an economic space to seek their "opportunity costs" or highest returns. Eventually, an equilibrium is reached where returns to factors (income) are equalized among regions. In short, the theory suggests that any differences in income levels among regions are temporary and will disappear over time.

Differential growth rates in per capita income over the past half century indicate that factor price equalization is indeed occurring among the regions of the United States. In 1929, per capita income in the Southeast was only 53% of the national average and 38% of the Mid-Atlantic average. Per capita income in South Carolina, the state with the lowest income in 1929, was only 23% that of New York, the state with the highest income. By 1976, however, per capita income in the Southeast had reached 84% of the national average and 74% of the Mid-Atlantic average. Per capita income in Mississippi, the state with the lowest income in 1976, was 62% of that in Illinois, the state with the highest income (excluding Alaska).

With the exceptions of Indiana and Wisconsin, all of the states of the Northeast–Midwest manufacturing belt showed relative declines in per capita income between 1929 and 1976, and in some states, such as New York, Connecticut, and Delaware, the declines were dramatic. In contrast, all of the southeastern and southwestern states recorded substantial gains in relative per capita income. The Southeast and Southwest have also been the largest recipients of migrants in recent years, suggesting that individuals are, in fact, seeking their "opportunity costs" and thereby bringing about income convergence among the regions.

There is also growing evidence that rapid economic development is helping to reduce the long-standing poverty levels so endemic to the South (see Table 26.3).

Table 26.3. Families below poverty level in the South, the Northeast, and the United States as a whole, 1969 and 1975.[a]

Region	Percentage below poverty level		Percentage change, 1969–1975
	1969	1975	
South	16.2	12.1	− 14.7
South Atlantic	13.9	10.9	− 9.1
East South Central	20.9	14.5	− 22.2
West South Central	16.8	12.4	− 15.9
Northeast	7.6	7.2	− 1.6
New England	6.7	6.5	+ 4.6
Mid-Atlantic	7.8	7.4	− 3.4
United States	10.7	9.0	− 7.5

[a] Source: U.S. Bureau of the Census and Department of Health, Education, and Welfare (1978).

In 1969, 16.2% of all families in the South were classified as poor, compared to 10.7% for the nation as a whole. By 1975, however, poverty status in the South had dropped to 12.1% of all families, compared to 9.0% for the United States as a whole. This was equivalent to a 14.7% decline in the number of poor families living in the South compared to a decline of only 7.5% in the number of poor families nationwide. The region's poorest states recorded the sharpest reductions in poverty. For example, Mississippi saw its poverty-level population drop from 28.9% of all families in 1969 to 20.4% in 1975, while Arkansas' poor dropped from 22.8% to 14.1% of all families over the same period.

At the same time the Northeast, which showed little economic growth during the period 1969–1975, recorded only a modest decline in the number of poor families. In fact, in New England, the number of poor families actually increased.

Another explanation of differential regional economic performance may perhaps be found in Schumpeter's theory of capitalist development (Schumpeter, 1947). In Schumpeter's view, the process of economic development emerges from the fiercely competitive environment of the capitalist system. This competitive struggle he called "creative destruction". Capitalism grows by destroying old economic structures and creating new ones. Old firms and products are driven out of business by more efficient and innovative producers.

Schumpeter predicted that the very success of capitalism would sow the seeds of its eventual demise for three reasons. First, investment opportunities, so critical to the process of creative destruction, would vanish as human wants were increasingly satisfied. Second, the entrepreneurial function would become obsolete. Third, the capitalist system would generate growing social hostility among intellectuals and laborers that would lead to a decomposition of the political framework on which capitalism rests.

While Schumpeter did not have a spatial context in mind when discussing the process of creative destruction, it is easy to visualize the South and the Northeast of the United States as rising and declining systems of entrepreneurial capitalism, respectively.

Because of industrial obsolescence and slow population growth, investment opportunities have diminished in the Northeast. The South, by contrast, offers a wide range
of new investment opportunities as import substitution proceeds.

The politically conservative environment of most southern states reinforces the
institutional setting for competitive capitalism. Taxes are low, and labor unions are
generally weak. This near-absence of union influence reflects both a deep cultural
bias against unions and the existence of restrictive labor legislation in many southern
states. In the Northeast, by contrast, Schumpeter's predictions of social hostility to
capitalism may have been realized. High state and local taxes on people and businesses,
liberal social welfare legislation, strong union pressures, and the coalescence of organized interest groups have stifled the atmosphere in which competitive capitalism can
thrive.

Several other factors can be cited as stimulating the growth of the southern region.
In general, it is still cheaper to live in the South than the North. For example, in 1976
a family of four at a middle-income budget level required 36% more income in the
Boston metropolitan area than in Austin, Texas to maintain the same standard of
living. Recent census data on migration to and from metropolitan areas show that all
of the high-cost areas experienced net out-migration during the period 1970–1974
while all of the low-cost metropolitan areas, located primarily in the South, had considerable net in-migration (Liner and Lynch, 1977). While it is likely that cost-of-
living differentials among regions will narrow significantly over the next decade, for
the immediate future lower living costs will continue to be an inducement to south-
ward migration.

Another factor contributing to the rapid growth of the South has been the promise
of a better quality of life than can be found in other regions of the United States.
Though concepts such as "quality of life" and "environmental quality" are quite
subjective and difficult to quantify, they clearly exert strong influence on migratory
decisions, both intrametropolitan and interregional. The perceived amenities (quality
of life) of the rapidly growing southern and western regions — less environmental
degradation, lower population densities, more moderate climates, ease of transportation,
access to recreational activities, lower crime rates, etc. — are apparently striking a
responsive chord with many residents of the older, congested areas of the Northeast
and Midwest.

But the most cogent explanation of southern prosperity stems from the fact that
structural differences between the South and the rest of the nation are disappearing
rapidly. The South was late in moving into sustained and diversified industrialization
and the modernization of its agriculture; but once the process took hold, the region
moved ahead faster than the older industrial areas because it could adopt the most
modern and efficient technologies. The rapid rise of income and employment in
the South is partly a manifestation of this transition to a modern, urbanized, industrialized society.

26.4. Implications of U.S. experience for regional development strategy

It must be remembered that the United States has an open economy and that there are no artificial barriers to the movement of capital, technology, or labor. To a large extent, the migration of people and jobs is occurring in response to real economic forces affecting the costs of production and service delivery. The result has been an increase in the overall efficiency and productivity of the national economy. Over time, market forces appear to be creating an optimal spatial distribution of economic activity.

Are there lessons to be drawn from the experience of the United States with its "lagging" regions that might be applicable to other nations? For the nations of eastern Europe and the USSR, with little or no private economic activity and a commitment to central planning, the U.S. experience is probably irrelevant. The institutional and political differences are simply too great to allow any transfer of experience or policy.

In the case of those countries in western Europe and southeast Asia that retain a viable private sector – i.e., one producing more than 50% of GNP – the U.S. experience would seem to argue for little or no governmental interference with investment, migration, and industrial relocation decisions. Factors of production should be permitted, and perhaps encouraged, to move freely. But direct or indirect governmental subsidies to stimulate growth in a particular region or subregion should be avoided. Such activities distort market signals and invariably bring about a serious misallocation of resources.

References

Keeble, D. E. (1967). Models of economic development. In R. J. Chorley and P. Haggett (Editors), Socio-Economic Models in Geography. Methuen, London.

Liner, E. B. and Lynch, L. K. (Editors) (1977). The Economics of Southern Growth. Southern Growth Policies Board, Research Triangle Park, North Carolina.

National Journal (1977). Federal Government Spending and Taxes Per Capita by Region in Fiscal Year 1976. July 2. Government Research Corporation, Washington, D.C.

Rees, J. (1978). Manufacturing change, internal control, and government spending in a growth region of the U.S. In F. E. I. Hamilton (Editor), Industrial Change: Challenge to Public Policy. Longman, London.

Schumpeter, J. A. (1950). Capitalism, Socialism, and Democracy. Harper and Row, New York.

U.S. Bureau of the Census (1975). Historical Statistics of the United States, Ser. F, pp. 297–348.

U.S. Bureau of the Census and Department of Health, Education, and Welfare (1978). Statistical Abstract of the United States, 99:470.

U.S. Bureau of Economic Analysis (1977). Survey of Current Business, April.

Weinstein, B. L. and Firestine, R. E. (1978). Regional Growth and Decline in the United States. Praeger, New York.

Regional Development Modeling: Theory and Practice
M. Albegov, A.E. Andersson and F. Snickars (editors)
North-Holland Publishing Company
© IIASA, 1982

Chapter 27

THE REGIONAL PROBLEM IN THE NETHERLANDS

Paul Drewe
*Department of Architecture and Urban Planning, Delft University of Technology
(The Netherlands)*

27.1. Introduction

There is of course a spatial aspect to regional socioeconomic policy as the word "regional" might suggest. Remarkably, however, in The Netherlands the formulation of (regional) socioeconomic and spatial policies is not synchronized. There is in preparation a new Note on Regional Socioeconomic Policy setting the course for the eighties. However, the direction of spatial policy for the period 1980–1990 was already defined as early as 1977 in the Third Note on Urban Development (1977).* How will regional socioeconomic policy respond to aspects of spatial policy? This question should be considered by policy-makers at the State Planning Office and especially by persons with first-hand experience of the "regional problem", for example, the inhabitants of East Groningen and South Limburg. It is all the more urgent because spatial planning objectives have changed substantially since the publication of the Second Note on Urban Development (1966).

27.2. New-style spatial distribution policy: reemphasizing the center

There has been a change of direction in spatial dispersion policy in The Netherlands (with respect to population, employment, welfare provision, etc.). The old-style policy was designed to encourage a deceleration in population growth in the West** and an acceleration in the North,** with a view to lessening the population pressure in the West and to bringing more prosperity to the North. The target was to have three million inhabitants in the North by the year 2000 (starting from a population of 1.3 million in 1965), but so far the actual population dispersion achieved has remained

---

* The idea is to revise this structural sketch every five years.
** The West consists of the provinces of Utrecht, Noord-Holland, and Zuid-Holland; the North comprises Friesland, Groningen, and Drenthe (see Figure 27.1).

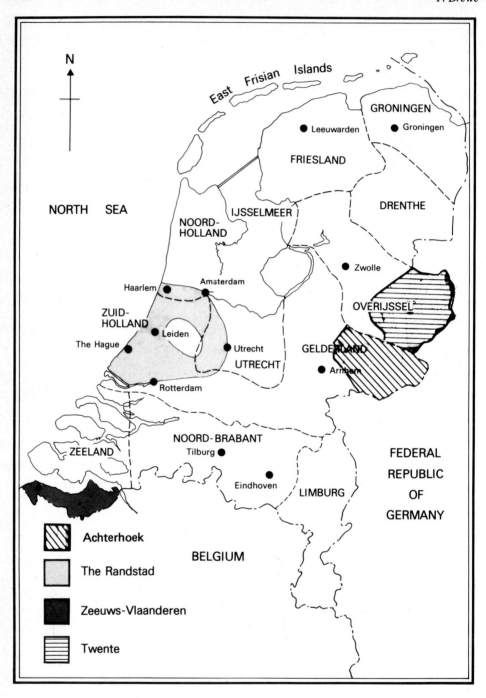

Figure 27.1. Provinces, regions, and major towns of The Netherlands.

far below that target. The reason for this, according to a government statement in 1975, was that the government was unwilling to make a firm commitment on the issue, judging the objective to be too ambitious. A second reason given is that the policy efforts required to achieve the objective had been underestimated.

On the other hand, national and provincial population data also show remarkable changes. On the regional level the most striking change is the reversal of net migration in the North: from an average annual loss of 1,700 persons during the period 1960–1969,* there was an average annual gain of 4,800 during the years 1970–1976. A new policy concept has gradually been developed through a series of intermediate steps.** The ambitious target of 1966 has ultimately given way to development following existing trends, no policy intervention being envisaged up until 1990.

So much for the North; however, ideas about the West have also changed. The average annual net out-migration from the western part of the country increased from 5,600 in the period 1960–1969 to 24,000 in the period 1970–1976,*** which is a favorable development in terms of the criteria of the Second Note on Urban Development (1966). But by the time the Third Note on Urban Development (1977) was produced, other ideas had been developed, leading to a new style of spatial dispersion policy: a policy that aimed at considerably decreasing the existing trend of migration flow from the Randstad to North Brabant and Gelderland (see Figure 27.1). Thus, attention shifted from the peripheral areas (the North, Limburg, Twente, Achterhoek) to the Randstad and its hinterland zones (also called the intermediate zones, at least as far as North Brabant and Gelderland are concerned).

27.3. Gloomy prospects for the periphery

To achieve a proper relation between regional-economic and labor-market policies on the one hand and spatial policy on the other, it will be necessary (particularly in the new Note on Regional Socioeconomic Policy) to focus attention on the Randstad. However, it is feared that peripheral areas, and specifically the North of the country, may suffer as a consequence. But is this anxiety well-founded? Can the unflagging efforts of the Ministry of Economic Affairs be relied upon to aid traditionally weak regions? The rather pessimistic expectations of the present author are based on: the approach to the problems of the North displayed in the Third Note on Urban Development (1977); the experience gained with the Integrated Structural Plan for the North of the Country; and the national economic recession and its consequences for regional economic policy.

---

* In the period 1950–1959, migration from this region was even greater, with an average annual loss of 8,100 persons.

** One of the main intermediate steps was taken in the Integrated Structural Plan for the North of the Country: 8 (1975); see also Steering Group for the Integrated Structural Plan for the North of the Country (1974).

*** There was a positive migration surplus of 7,500 in the years between 1950 and 1959.

In the Third Note on Urban Development (1977), the North was considered mainly from a demographic point of view. Areas with chronic net out-migration tended to receive more attention than areas with positive or balanced net migration.* More recently, the North has developed a positive net migration; policy corrections to the trend-wise demographic development are therefore considered to be less urgent than before. However, the important question is whether the transition from negative to positive net migration has brought at least some relief with respect to the "regional problem". Indeed, it is still the inequitable regional distribution of job opportunities and incomes that causes concern. The plea for three million inhabitants in the North by 2000 was implicitly based on the assumption of a simple positive relation between population size and regional prosperity. The same was true of later pleas for reinforced population growth in the North, and of the idea that no policy intervention is needed as long as the population tends to develop in a "favorable" way. But is there any proof that, for instance, relative unemployment has really decreased? Or is it plausible that it will decrease in the near future?

Over the period 1952–1975, the employment position of the three northern provinces did not improve (Bartels, 1977),** which is more remarkable if compared to the situation in the three western provinces. The average unemployment percentage in the North was relatively high, employment being relatively low and unstable, and unemployment showing a relatively strong growth trend. Prospects for the future are also not bright. A comparison between the future developments of active population and employment gives an idea of the regional problems to be expected (Table 27.1). If the present policy is continued, unemployment in the three northern provinces will have grown to no less than 12.3% by 1985; in the period 1975–1985 employment will shrink by about 1% while the labor force will grow by about 7%; the latter development is a result of the birth wave of the sixties.

Table 27.2 presents comparative data on the potential labor force, number of employed persons, and unemployment in a number of western European countries; the table includes both 1974 data and projections to 1990. Unemployment in The Netherlands may rise to as much as 14.4% in 1990 if the potential labor force increases by 16.1% between 1974 and 1990 and the number of employed persons (jobs) by only 1.7%.*** A figure of 14.4% for The Netherlands represents the huge total of 818,000 unemployed persons.

There is a real danger that by 1990 the North will be worse off than The Netherlands as a whole; indeed, unemployment in the North will probably reach a level of 12.3% as early as 1985. But, a critical reader may protest, policy will be

---

* "Political pressures do much to explain why chronic out-migration areas, mostly economically distressed rural or old industrial areas, are usually regarded as the problem areas for purposes of migration policy" (Klaassen and Drewe, 1973, p. 56). For the relative diagnostic value of migration surpluses (because of the selectiveness of migratory flows), see Koch (1979).

** Comparison with the unemployment analysis on the COROP level, or the income analysis on the same level, both of which paint an equally unfavorable picture of the situation in the North, gives some insight into the intraregional distribution.

*** In the period 1970–1975, there was a 2% growth in employment.

Table 27.1. Projected labor force, employment, and employment deficit in the three northern provinces[a] of The Netherlands, 1975, 1980, and 1985.[b]

Year	Labor force[c]		Employment		Employment deficit	
	(thousands)	(%)[d]	(thousands)	(%)[d]	(thousands)	(%)[e]
1975	501.9	100	474.9[f]	100	27.0	5.4
1980	513.9	102	464.7	98	49.2	9.6
1985	534.8	107	469.2	99	65.6	12.3

[a] The provinces of Drenthe, Friesland, and Groningen.
[b] Data from the Integrated Structural Plan for the North of the Country: 9a (1976a, p. 119a).
[c] Assuming an average annual net migration of zero from 1976 onward.
[d] Based on a 1975 value of 100.
[e] As a percentage of the labor force.
[f] Including existing known vacancies.

changed in the next few years. Future governments will take up the fight against national unemployment in earnest. And the Ministry of Economic Affairs, always supporting the weaker regions, is sure to create additional employment there.

That may be so, but it is also true that a policy such as the Integrated Structural Plan for the North of the Country (ISP) calls for some critical comments. Appreciative remarks made elsewhere about its policy relevance hardly touch the core of the matter (van Hamel and Janssen, 1978). For what exactly are the concrete gains of the ISP? For the period 1979–1981, ". . . an additional policy effort has been agreed upon referring to 10,000 job opportunities". And in the period 1982–1985, ". . . in order to decrease unemployment an additional policy effort will be required representing the creation or additional maintenance of 10,000 to 12,000 job opportunities" (Steering Group for the Integrated Structural Plan for the North of the Country, 1979). This figure seems insignificant when one considers that the employment deficit in the three northern provinces for 1985 is estimated at about 65,000 (see Table 27.1). Even if there were no unemployment at present, that number of jobs would have to be created just to maintain the status quo. The real matter at stake is what unemployment proportion is felt to be acceptable, or, expressed in another way, what is to be understood by an "equitable" regional distribution of employment.

While the ISP refers in particular to "further reduction or elimination of the regional component of unemployment", the national unemployment percentage (or the total volume of unemployment) in The Netherlands certainly also plays a role, as is evident from a "discussion" held in connection with the Integrated Structural Plan for the North of the Country (1976c, pp. 29–30, annexes). Table 27.3 shows that the forecasts of northern unemployment made by the Working Group on Economic Development and by the Central Planning Office are greatly at variance; this difference is "smoothed away" in a remarkable manner. The Working Group reduces unemployment in the North to just over 40,000, on the assumption that half the unemployment growth of the period 1975–1980 should be regarded as "disguised". Furthermore, it is assumed that *national* unemployment will be reduced to a level of

Table 27.2. Projected labor force, employment, and employment deficit in various western European countries, 1974 and 1990.[a,b]

Country or region	Labor force[c] (thousands)		Employment (thousands)		Employment deficit[d] (thousands)		(%)	
	1974	1990	1974	1990	1974	1990	1974	1990
Belgium	3,979	4,213	3,829	3,762	112	410	2.8	9.7
FRG	26,959	24,296	26,164	25,190	584	1,895	2.2	6.9
Denmark	2,482	2,582	2,399	2,363	83	219	3.3	8.5
France	21,679	24,033	20,944	21,542	735	2,491	3.4	10.4
Ireland	1,121	1,372	1,066	1,171	55	201	4.9	14.6
Italy	19,391	21,173	18,898	19,194	493	1,979	2.5	9.3
Luxemburg	143	143	150	161	–	–	–	–
The Netherlands	4,888	5,676	4,765	4,848	115	818	2.3	14.4
United Kingdom	25,262	26,235	24,715	23,439	547	2,797	2.2	10.7
EEC Total	105,904	112,723	102,930	101,670	2,724	10,810	2.6	9.6
Austria	3,102	3,372	3,043	2,964	41	391	1.3	11.6
Switzerland	2,968	2,844	3,069	2,707	1	240	0.1	8.4
EEC + Austria and Switzerland	111,974	118,939	109,042	107,341	2,766	11,441	2.4	9.6

[a] Data are from Koch (1979, p. 20).
[b] The various countries were disaggregated into regions for analysis and projections; the 1990 figures quoted here for each country are sums of the respective regional totals, assuming zero net interregional migration. The Netherlands was treated as a system of five regions: West, North, East, North Brabant + Zeeland, and Limburg.
[c] Based on projected labor force participation rates for 1990.
[d] The figures for unemployment in 1990 take commuting into account.

Table 27.3. Projected labor force, employment, and employment deficit in the North[a] of The Netherlands, 1975 and 1980.[b]

Year	Labor force[c] (thousands)	Employment (thousands)	Employment deficit (thousands)	(%)[d]	National unemployment (thousands)	Jobs to be created in North (thousands)[e] Excluding disguised unemployment[f]	Including disguised unemployment[f]
1975	538.3	508.5	29.8	5.5%	~ 205.0	–	–
1980 (WEO[g] estimate)	551.5	498.2	53.3	9.7%	> 250.0	36.5	~ 73.0
1980 (CPB[h] estimate)	–	–	~ 25.0	~ 5.0%	~ 150.0	~ 10.0	~ 20.0

[a] The three provinces, Drenthe, Friesland, and Groningen, considered in Table 27.1, together with the Northeast and Northwest Overijssel.
[b] Data from the Integrated Structural Plan for the North of the Country (1976a, p. 119a; 1976c, pp. 29, 30, annexes).
[c] Assuming an average annual net migration of zero from 1976 onward.
[d] As a percentage of the labor force.
[e] Assuming 3% unemployment.
[f] Disguised unemployment manifests itself as additional supply when the labor market improves; here it is assumed that registered and disguised unemployment have a one-to-one relationship.
[g] The Working Group on Economic Development.
[h] The Central Planning Office.

some 150,000 persons by 1980; that is, about 3% of the Dutch working population (the reasoning being that 100,000 fewer unemployed in The Netherlands as a whole would proportionally correspond to 15,000 fewer in the North).

The second correction is particularly noteworthy. The argument goes as follows: "Given the national policy measures announced, however, we are justified in assuming a national level of 150,000 unemployed". The conclusion is however unwarranted, for such a level of unemployment may not be assumed without reservations. Indeed, it is not at all certain that the announced national measures will have the desired effect, which suggests that a set of alternative plans will be required for the North. Just as confidently, and almost casually, it is assumed that the 20,000 new jobs that will then be needed (or the 10,000 additional jobs needed at present) will certainly be created.

The ISP gives the impression of consciously or unconsciously avoiding the most fundamental question of regional economic policy: how effective are current policy tools? From the Note on Regional Socioeconomic Policy 1977–1980, it appears that the evaluation of present policy instruments in most cases is no more than an ad hoc inspection (Ministry of Economic Affairs, 1977, pp. 29–37, 125, 126). The Note mentions the amounts spent on improving the infrastructure and on employment-creation programs, the number of migrations affected by the Migration Regulations of 1971 (or, as the case may be, the Migration Regulations for the North), and finally the number of registrations and applications for employment according to the Selective-Investment Regulation (SIR). With regard to the results achieved by the "regional-development companies", the Note simply refers to the relevant annual reports. Only in relation to the Investment-Premium Regulation (IPR) has an attempt been made to estimate the "effect". The number of jobs created by each project, especially in the province of Groningen, is modest. Admittedly, a more thorough investigation of policy effectiveness would be a difficult proposition.* On the other hand, the functioning of the most important policy instruments is hardly receiving the attention it merits, and the doubts or uncertainties about the effectiveness of these instruments are not heeded during policy selection.

27.4. National economic crisis: interregional equity, a politically unacceptable diversion?

The field of regional economic policy is certainly one in which tensions between efficiency and equity occur, that is to say tensions between national economic efficiency and interregional equity. Such tensions are apt to increase in times of national economic recession, when efficiency tends to be measured in terms of national economic performance (employment, price stability, economic growth, etc.). Thus,

---

* Although difficult, it is not impossible; see, for instance, Bartels and Roosma (1979). For general aspects, see Richardson (1978, pp. 253–256); for concrete data problems, see Bartels and Folmer (1979).

Monod and de Castelbajac (1978) are justified in stating, with respect to France, that a policy aiming at diminishing regional disparity can succeed only in an expanding economy, while in times of recession the position of the Paris region tends to be reinforced; such a situation has international relevance (see Goze, 1976). If The Netherlands follows the French example, there will be little need for those responsible for spatial policy to claim more attention for the Randstad from a regional economic and labor-market point of view, and even less so when the change in course, including the regional effects of nonregion-specific measures of support and stimulation, has already been effected; in other words, when factual policy has actually preceded "memorandum policy". Assuming again that the French model will be followed in The Netherlands, the new Note on Regional Socioeconomic Policy will serve only to legitimize factual policy. Is it not true that the (pilot) study of the urban labor market commissioned by the Ministry of Social Affairs* already suggests a changed course? There are, overall, very good reasons for pessimism as far as the future of peripheral regions in The Netherlands is concerned.

But is it not a fact that there is more in total to distribute when things go well (or better) in the West? And should we not stop being obsessed by purely material things such as jobs and incomes? Is it really necessary to focus regional economic policy on the very region that is already the most developed in The Netherlands? In fact, none of the arguments suggested in these questions stand up under severe criticism. The first should be placed against the background of the figures presented in Tables 27.1–27.3. How could the unemployment problem in the North be solved by creating more employment in the Randstad? Migration of job-seekers from the North to the West could offer a solution if, in a purely quantitative sense, there were sufficient jobs left, after the Randstad's own demand for employment had been fulfilled, to reduce unemployment in the North at least to an acceptable level, and if job-seekers from the North were mobile and would agree to move to the Randstad. Whether both conditions can be met is uncertain, to say the least. There are those who believe that, with the present macroeconomic policies, we should not count on there being sufficient jobs (we shall consider this idea later). This destroys the second illusion of a positive emanation of more employment in the Randstad toward the peripheral areas or of a subsequent overspill of employment in the West into the periphery, i.e., the three northern provinces, Twente, the Achterhoek, the whole of Limburg, and Zeeuws–Vlaanderen (cf. Molle and Vianen, 1979).

But even without such macroeconomic doubts there is no reason to expect an improvement in the situation to result from overspill, especially as far as total employment is concerned. In the recent past the Randstad's share in total Dutch employment has remained constant (in fact, it has increased slightly); it is first and foremost the hinterland areas** that have achieved a slight gain at the expense of the periphery

---

* Indeed, the Note on Regional Socioeconomic Policy is also submitted on behalf of the Minister of Social Affairs.
** These consist of the IJssel region, large parts of Gelderland and Zeeland, and the whole of North Brabant.

Table 27.4. Development of the percentage shares of zones and areas in total Dutch employment, 1960–1975.[a]

Zone/area	Year			
	1960	1965	1970	1975
I. Randstad				
Large towns	25.0	24.9	24.4	23.2
Urban areas	15.7	15.9	16.5	17.2
Rural areas	4.8	5.0	5.1	5.5
Total	(45.5)	(45.8)	(46.0)	(45.9)
II. Hinterland zone				
Urban areas	13.4	13.7	14.1	14.4
Rural areas	15.1	15.0	15.4	15.8
Total	(28.5)	(28.7)	(29.5)	(30.2)
III. Peripheral zone				
Urban areas	8.5	8.2	7.6	7.3
Rural areas	17.6	17.3	16.9	16.7
Total	(26.1)	(25.5)	(24.5)	(24.0)
The Netherlands	(100.0)	(100.0)	(100.0)	(100.0)

[a] Data from Molle and Vianen (1979, p. 18).

(Table 27.4). If the trend continues, job-seekers from peripheral areas have at best a choice of two alternatives: they may either migrate to the Randstad or become long-distance commuters to hinterland areas. The distribution pattern of activity classes between 1960 and 1975 does not indicate much likelihood of a considerable overspill (for details, see Molle and Vianen, 1979).

27.5. Not by bread alone?

Suppose one were to stop thinking in material (that is, purely economic) terms. From the point of view of nonmaterial (noneconomic) attraction, is it not the periphery that scores highest, with relatively attractive surroundings, smaller-scale average living and working environments, sparse population, absence of congestion, and a reasonable level of sociocultural provision (in terms of accommodation)? Are these not the positive features of the North that should be emphasized in regional policy? And should not the nonmaterial bottlenecks – if any – be the ones to be removed with high priority (Integrated Structural Plan for the North of the Country, 1975, 1976a)? That would certainly be more in line with alternative visions of the future on the macrolevel.

Take for instance the Scientific Board for Government Policy (WRR) and its reconnaissance into the future of The Netherlands.* A movement toward a gradually decreasing economic growth rate, reaching zero by the end of the century, will in the long run have a beneficial effect on energy consumption, land development, and certain forms of environmental pollution. Therefore, such a trend is relatively favorable in terms of nonmaterial considerations (this is designated variant B to distinguish it from variant A, which is based on a continuing annual growth in production by 3%). However, from a material point of view it is hard to see how it can produce an adequate number of jobs (generating an income), even on the assumption that the supply of workers, and in particular male workers, will grow less rapidly in future (more part-time employment).

The question is: will there be many people who feel that the nonmaterial advantages of living in the periphery outweigh the disadvantage of being unemployed and having a lower income than Randstad inhabitants? This question concerns the inhabitants of the weaker regions.** That material losses can be compensated for on an adequate scale is doubtful. This seems unlikely, at any rate in the early eighties, unless there are sudden discrete changes (the WRR's reconnaissance is, indeed, "surprise free"). Rather it is to be expected that inhabitants (and administrators) of weak regions will compare their situation with that of the better-endowed West. Moreover, the level of income greatly influences the appreciation of matters material and non-material, as an example from the USA illustrates (Los Angeles Goals Council, 1969). In 1969 the average income of the total population of Los Angeles was significantly higher than that of the black population. It was found that, on average, the total population considered air-pollution (smog) abatement, and the extension of police protection and fire prevention to be more important than the creation of additional employment, the construction of additional cheap housing, and urban renovation; the black population clearly had a reverse order of priorities.

27.6. The "new" regional problem

What would be the prospects if there were to be a decision against changing the course of regional economic policy or, if already changed, to switch it back? That is to say, if the West were not, once again, favored by economic policy?

The matter is complicated by the fact that, on close examination, the situation in the Randstad is far from ideal. It is true that the unemployment figures of the three western provinces look good in comparison with those of the other provinces, but in Amsterdam, Rotterdam, and The Hague the official unemployment figures are above the national average.

---

* Nothing more can really be expected from the reconnaissance, which the Scientific Board for Government Policy (1977) has clothed in rather relative terms.
** On this subject compare, for example, an elaboration of variants A and B for Twente and its employment situation. See Provincial Planning Service, Overijssel (1979a) and also Provincial Planning Service, Overijssel (1979b).

It is the distribution within the towns, however, that is most revealing: in the four largest towns the unemployed are clearly concentrated in certain quarters, notably the old ones (Valkenburg, 1979).* Moreover, unemployment is only one aspect of the urban problem; another is the unequal distribution of collective consumption goods. Autochthonous Dutchmen of the lowest income classes and immigrants from Mediterranean countries, Surinam, and the Antilles find themselves segregated into prewar residential areas, where they are relatively deprived of collective consumption goods. The housing situation is especially depressing (see Drewe, 1979b). Indeed, the urban problem in the Randstad stems in part from intraregional (personal) equity problems. Intraregional efficiency is also affected; for example, as mobility levels increase, residential, work, and leisure areas grow further apart. Bottlenecks of this type of development on the regional level are taken to demonstrate the necessity for an urban planning policy that pays special attention to the West (Third Note on Urban Development, 1977). The types of bottlenecks referred to here are quite familiar (see, for example, de Boer and Heinemeyer, 1978).

It is striking how ideas of efficiency have altered, as is evident from the current concept of "costs of urban growth" (Drewett, 1979), which has outgrown the limited notion of economic costs and benefits of urban land use (see, for example, International Institute for Urban Studies, 1974). However, this change has only partly been carried through into actual urban-planning policy. Attempts to carry out council-housing programs in and around the central urban area make it very clear that conflicts between efficiency and equity are not limited to regional economic policy and that a conflict between purely market-economic intraregional efficiency and intraregional (personal) equity has to be reckoned with.**

Summarizing the arguments, we conclude that: the ancient problem of *inter*regional equity will continue to exist and even intensify (owing to, among other things, the national economic recession) in the near future; and that the new problem of *intra*regional efficiency and equity, especially in the Randstad, will require much attention.

27.7. Alternatives, yes — appeasement, no

What is to be done? Are there any alternatives? Much will depend on the course of macroeconomic policy and its impact on employment levels. The alternatives, most recently listed by Salverda and van den Doel (1979), are: (1) to boost production in the private sector; (2) to decrease worktime; and (3) to expand the quaternary sector, the public sector, and subsidized institutions.

---

* Quite rightly, the Ministry of Social Affairs gives great attention to the urban labor markets in the West.
** For instance, a conflict of council versus private housing in the old residential area and in the central area (see Drewe, 1979b); also, of course, a conflict between the use of land for housing and for other purposes.

It is important to know what such solutions of the unemployment problem are going to cost in collective provisions (alternative 1) or in wages and incomes (alternatives 2 and 3). But it is equally important to know how they will affect various social groups and various regions. Distribution among income groups is already a weighty point in daily political discussions on macroeconomic policy, but regional distribution is mostly treated only summarily. In current practice, regional effects give rise to corrections only after the goal of national economic efficiency has been satisfied, distribution being adjusted on the macroeconomic level. Macroeconomic quantities are no more than abstract notions, however, and the "economy" is really vested in towns and regions.

The alternatives given above should be judged not only by their contribution to equity, but also by their contribution to intraregional efficiency. However, there is no reason to limit efforts toward achieving intraregional efficiency and equity to the Randstad alone.

What are the prognoses for the future? It is impossible to give clear verdicts until the regional effects have been examined more closely. At any rate the prognoses cannot be divorced from an appreciation of the present regional economic policy. As Richardson (1978, p. 264) has found for the UK, the evaluation of regional economic policy is a controversial matter. The majority of experts think that current practice, dominated as it is by indirect intervention by the government, is just right. Without the present policy, the (old) regional problem would certainly be more serious than it is now. There are some, however, who maintain that the problem has remained unsolved for lack of direct state intervention. Another minority, finally, holds that the problems are caused by a surfeit of government interference.

27.8. ". . . With entire national economies, such as Holland and Great Britain's, at the mercy of transnational concerns . . ." (Friedman and Weaver, 1979)

As far as The Netherlands is concerned, I tend to agree with Holland (1976), in view of the present and future problems of the North (outlined above) and taking into account my opinion of the ISP. According to Holland, indirect state intervention on the level of regional policy is bound to fail in the face of the rising mesoeconomic sector, especially the multinationals. Multinational establishments in Third-World countries have proved far more profitable for leading companies than multiregional establishments in the less-developed regions of developed countries such as The Netherlands. The therapy recommended by Holland is twofold: a purposeful discriminating sectoral policy in favor of the less-developed, peripheral regions such as the North, the mesoeconomic sector being kept in check by direct state intervention, for instance, by the foundation of new state companies; combined with control of the establishment of leading and multinational industries in the more-developed metropolitan regions such as the Randstad.

However, I doubt that this therapy will be effective because: (i) it does not take proper account of the worsening economic recession, which seems to make establishment control in the Randstad hardly feasible; and (ii) the sectoral policy advocated is confined to the industrial sector. Apart from these points, Holland's point of view seems to be in line with the concepts underlying the three macroeconomic alternatives, none of which can be implemented without increased direct state intervention. For it is the government's responsibility to ensure that savings in the collective sector or shorter working hours are in fact translated into more jobs. Expansion of the quaternary sector can be achieved not only by more government orders to private companies, but also by the state acting more frequently as an employer in this sector.

Given the present standpoint, how can the various forms of state intervention help to solve "old" as well as "new" regional problems? Solutions that fail to increase the periphery's share of employment in The Netherlands cannot contribute to enduring equity (see Table 27.4 together with Table 27.3). The results will be negative if newly created jobs, financed from savings in the collective sector and from shortened worktime, are allocated proportionally to existing locations; positive discrimination in favor of the peripheral regions is essential. The same is true of government orders for private companies in the context of an expanding quaternary sector.

The state, as an employer in the quaternary sector, will have a better chance of controlling interregional equity if it determines only the total wage sum and the total volume of investments. Solutions that do not imply a greater production of collective goods (such as urban renovation, improved education and health-care facilities, a cleaner environment, etc.) will stand in the way of intraregional efficiency and equity. This argument shows a bias toward alternative (3) and against alternatives (1) and (2),* but in particular against alternative (1), since it is difficult to understand how more collective and more private consumption could be achieved simultaneously. On the other hand, solutions involving a cutback in private consumption are still liable to meet with opposition, as would a future according to variant B (see above). This would make both shortened worktime and extension of the quaternary sector difficult to achieve.

27.9. Toward a more decentralized approach

The relation between macroeconomic and regional economic policy has procedural as well as conceptual aspects. The crucial question with respect to the procedure is whether to use "top-down" or "bottom-up" regional planning. In present practice, macroeconomic policy is broadly regionalized with the help of current regional models;** regional disaggregation stops at certain "parts of the country" or provinces, and is therefore inadequate, in particular for dealing with problems of intraregional efficiency and equity.

---

* Possibly with the exception of worktime contraction in the government sector, provided that on balance, owing to increased productivity, more collective goods can be produced.
** The regional economic model (REM) and the regional labor-market model (RAM) operated by the Central Planning Office.

Incidental cooperation between state and provinces as achieved in the ISP cannot replace a socioeconomic policy plan covering several years that is drawn up by the province. The advantages of such a plan are that: in principle the planners have the regional expertise they require at hand, including data on subregions, sectors, and individual industries;* there is more opportunity to involve the local population and to initiate a meaningful discussion process; the province is mobilized politically, which promises a better basis than the present one for negotiations with various ministries and other provinces (with the northern part of the country, for example); and on the regional level spatial and socioeconomic policies can be more easily integrated.**

If eleven or more provincial socioeconomic policy plans were submitted to the government, and negotiations started, the government would be compelled to develop efficiency and equity criteria to coordinate these provincial plans and fit them into the prevailing government policy. The centralization of public funds enables the government to implement its policy. The approach envisaged implies that interests are weighted interregionally; the policy ensuing may be one of discrimination for or against a given province. Close scrutiny reveals the procedure proposed to be a compromise between bottom-up and top-down planning.

Research, too, has to be adapted to the new system of regional socioeconomic policy-making, and should be understood to include the models, methods, and techniques available for policy preparation and process control. There is a clear difference between research on the regional and on the national level. Research on the regional level is marked by subregional and sectoral differentiation, while on the national level multiregional consistency and interregional weighting are the main features.*** In the National Program of Labor-Market Research (NPAO), it is necessary to solve simultaneously national and regional labor-market models,† which have different functions in policy preparation and in process control.

When work on the ISP was first undertaken, it was hoped that, given a sounder scientific foundation, a "better policy" could be produced. That hope has not been fulfilled because, rather than performing selective research, those responsible were aiming at integrated development analysis, i.e., analysis of the interdependencies of development. However, the fact that the ISP has failed in this respect is not an

---

* Which is not to say that there is no need for coordinating the collection of regional statistical material.

** To mention an example (in translation): "Little attention has so far been given to the way developments in the service sector can best be monitored in the framework of (regional) physical planning" (Dekker, 1979).

*** A model like REM can in that case be operated as a simulation instrument. A hypothetical example of a policy-relevant model simulation is given by van Delft et al. (1977). Given a constant easy national labor market, an investment impulse of 10% of the capital stock into the industrial sector in the north would lead, among other things, to less unemployment in that part of the country with increasing unemployment in other parts.

† Compare the discussion in the Working Group on Labor-Market Models at the Conference on the NPAO. See National Program of Labor Market Research (1979) and also Heijke and Maas (1978).

argument against having a better scientific foundation, but rather in favor of taking a different approach to the investigation, by tuning the research to the plans that are being developed (Drewe, 1977, pp. 4–37; 1979a).

The argument for a more decentralized approach to regional economic policy is not novel; the Socioeconomic Council has advocated such an approach (SER, 1978) and so have spatial planners "in the field" (Dekker, 1979) as well as "alternative" regional economists and planners from the academic world (Friedman and Weaver, 1979; Stoehr and Toedtling, 1979; see also Scientific Board for Government Policy, 1977). Even in the Note on Regional Socioeconomic Policy 1977–1980, tentative suggestions in that direction were put forward (Ministry of Economic Affairs, 1977). But as van Voorden (1979) has rightly pointed out, so far a concrete formulation has not been produced.*

27.10. Conclusion

It is to be hoped that the arguments presented in this chapter have convincingly proved why a new approach to regional economic policy in The Netherlands is currently very urgent. Given that the formulation of (regional) socioeconomic and spatial policy in The Netherlands is not synchronized, it is difficult to see how the complicated regional problems faced can be relieved, let alone solved, by separate piecemeal measures. Spatial policy such as that planned for the eighties might well turn out in future to be an "unpaid bill", and that applies to policies of dispersion as well as to those of physical planning.

References

Bartels, C. P. A. (1977). Economic Aspects of Regional Welfare. Martinus Nijhoff, Leiden.
Bartels, C. P. A. and Folmer, H. (1979). Hoe controleerbaar is het overheidsbeleid? (How check-able is government policy?) Economisch Statistische Berichten, 64:1197.
Bartels, C. P. A. and Roosma, S. Y. (1979). De dienstensector in het regionale beleid I (II) (The service sector in regional policy I (II)). Economisch Statistische Berichten, 64: 311–316.
de Boer, N. A. and Heinemeyer, W. F. (Editors) (1978). Het Grootstedelijk Milieu, Kansen en Bedreigingen (The Metropolitan Environment, Chances and Menaces). van Gorcum, Assen.
Dekker, A. (1979). Het Streekplan Nieuwe Stijl, Ervaringen in Overijssel (The New-Style Regional Plan, Experiences in Overijssel). Introduction to the seminar on Regional Development and Planning, held at the Technical University, Eindhoven on 14 and 15 March, 1979.

---

* He thinks that the new policy can only begin to be developed when five crucial choice problems have been solved (in translation): "a. the relation between the government and social organizations; b. the relation between centralized and decentralized decision-making; c. the coordination of regional and sectoral economic policy; d. the extent of government influence or compulsion in regional socioeconomic policy; e. institutional formation on the regional level".

van Delft, A. et al. (1977). Een Multi-regionaal Model voor Nederland (A Multiregional Model for The Netherlands). Paper presented at the seminar of the Dutch branch of the Regional Science Association, held in Rotterdam on 5 April, 1977.

Drewe, P. (1977). Planning methods and techniques – state-of-the-art. In Recent Developments in Planning Methodology: Netherlands Colloquium, October. Ministry of Housing and Physical Planning, The Hague, pp. 4–37.

Drewe, P. (1979a). Integrated regional planning as applied to the northern Netherlands. In H. Folmer and J. Oosterhaven (Editors), Spatial Inequalities and Regional Development. Martinus Nijhoff, Boston, Mass., pp. 219–254.

Drewe, P. (1979b). Segregatiebeleid Rotterdam (Segregation Policy in Rotterdam). Paper presented at the Netherlands Economic Institute (NEI) Anniversary Congress on the Dynamics of Urban Development, held in Rotterdam on 4 and 5 September, 1979.

Drewett, R. (1979). Changing Urban Structures in Europe. Paper presented at the NEI Anniversary Congress on the Dynamics of Urban Development, held in Rotterdam on 4 and 5 September, 1979.

Friedman, J. and Weaver, C. (1979). Territory and Function, the Evolution of Regional Planning. Edward Arnold, London.

Goze, M. (1976). Analyse sommaire de la politique française depuis la Seconde Guerre Mondiale (Brief analysis of French politics since the Second World War). Revue Economique du Sud-Ouest, 1 : 3–39.

van Hamel, B. A. and Janssen, A. M. C. (1978). The integrated regional plan for the North of The Netherlands. Planning and Development in The Netherlands, 10:42–96.

Heijke, J. A. M. and Maas, R. J. M. (1978). De arbeidsmarkt in model (Modeling the labor market). In Proceedings of the Conference on the National Programme of Labour Market Research. NPAO, The Hague.

Holland, S. (1976). Capital versus the Regions. Macmillan, London.

Integrated Structural Plan for the North of the Country: 7 (1975). Interim Rapport van de Werkgroep Sociaal-culturele Ontwikkeling (Interim Report of the Working Group on Socio-cultural Development). Ministry of Economic Affairs, Assen.

Integrated Structural Plan for the North of the Country: 8 (1975). Eindrapport van de Werkgroep Belvolking (Final Report of the Working Group on Population Aspects). Ministry of Economic Affairs, The Hague.

Integrated Structural Plan for the North of the Country: 9a (1976a). Eindrapport van de Werkgroep Economische Ontwikkeling I: Bedrijfstakvooruitzichten (Final Report of the Working Group on Economic Development I: Sector-Wise Prospects). Ministry of Economic Affairs, The Hague.

Integrated Structural Plan for the North of the Country: 10 (1976b). Eindverslag van de Werkgroep Sociaal-culturele Ontwikkeling (Final Report of the Working Group on Sociocultural Development). Ministry of Economic Affairs, Assen.

Integrated Structural Plan for the North of the Country (1976c). Het Sociaal-economisch Beleid voor het Noorden des Lands (The Socio-economic Policy for the North of the Country). Staatsuitgeverij, The Hague.

International Institute for Urban Studies (1974). Rotterdam–Rijnmond Study. Final Report. Volumes 1 and 2. Ramat-Gan, Israel.

Klaassen, L. H. and Drewe, P. (1973). Migration Policy in Europe. Saxon House–Lexington Books, Farnborough, Hampshire.

Koch, R. (1979). Demographic Changes in Regions and Cities and Their Implications. Council of Europe Proposed Seminar on the Impact of Current Population Trends in Europe's Cities and Regions. Strasbourg, September, 1979.

Los Angeles Goals Council (1969). Summary Report. Los Angeles, California.

Ministry of Economic Affairs (1977). Nota Regionaal Sociaal-economisch Beleid 1977–1980 (Note on Regional Socioeconomic Policy 1977–1980). Staatsuitgeverij, The Hague.

Molle, W. T. M. and Vianen, J. G. (1979). Werkgelegenheid, Spreiding en Verstedelijking (Employment, Spatial Dispersion, and Urbanization). Paper presented at the NEI Anniversary Congress
on the Dynamics of Urban Development, held in Rotterdam on 4 and 5 September, 1979.

Monod, J. and de Castelbajac, P. (1978). L'Aménagement du Territoire (Regional Planning). Second
edition. Presses Universitaires de France, Paris.

National Program of Labor Market Research (1979). Verslag van de Conferentie over het Nationaal
Programma Arbeidsmarktonderzoek (Proceedings of the Conference on the National Programme of Labour Market Research), held in Delft on 25 April, 1979. NPAO, The Hague.

Provincial Planning Service, Overijssel (1979a). Herziening Streekplan Twente, Onderzoeknota
t.b.v. de Hoofdlijnen 03, Werkgelegenheid (Revision of Regional Plan for Twente, Research
Note Regarding the Main Lines 03, Employment), Zwolle.

Provincial Planning Service, Overijssel (1979b). Note on the Formation of Plans P5, Policy Space,
Zwolle.

Richardson, H. W. (1978). Regional and Urban Economics. Penguin Books, Harmondsworth,
Middlesex.

Salverda, F. and van den Doel, H. (1979). Den Uyl loses all his old friends and makes no new ones.
Vrij Nederland, 40(3): 5–6.

Scientific Board for Government Policy (1977). Wetenschappelijke Raad voor het Regeringsbeleid,
De Komende Vijfentwintig Jaar, een Toekomstverkenning voor Nederland (The Next Twenty-
five Years, A Reconnaissance into the Future of The Netherlands). Staatsuitgeverij, The Hague.

Second Note on Urban Development (1966). Ministry of Housing and Physical Planning,
Staatsuitgeverij, The Hague.

SER (1978). Advies Inzake de Hoofdlijnen van het Regionale Sociaal-Economische Beleid (Advisory
Note on the Main Lines of Regional Socioeconomic Policy) (1978). SER, The Hague.

Steering Group for the Integrated Structural Plan for the North of the Country (1974). Het
Noorden, een Versterkte Bevolkingsgroei of Juist-Niet (Population Growth in the North,
to Stimulate or Not to Stimulate? That is the Question). Staatsuitgeverij, The Hague.

Steering Group for the Integrated Structural Plan for the North of the Country (1979). Samen-
valtting van het Beleidsrapport Integraal Structuurplan Noorden des Lands (Summary of the
Policy Report on the Integrated Structural Plan for the North of the Country). Staatsuitgeverij,
The Hague.

Stoehr, W. and Toedtling, F. (1979). Spatial equity: some antitheses to current regional development doctrine. In H. Folmer and J. Oosterhaven (Editors), Spatial Inequalities and Regional
Development. Martinus Nijhoff, Boston, Mass.

Third Note on Urban Development (1977). Part 2d, Versteddijkingsnota, Ministry of Housing and
Physical Planning, Staatsuitgeverij, The Hague.

Valkenburg, F. C. (1979). Segmentering van de Stedelijke Arbeidsmarkt (Segmentation of the
Urban Labor Market). Paper presented at the NEI Anniversary Congress on the Dynamics of
Urban Development, held in Rotterdam on 4 and 5 September, 1979.

van Voorden, W. (1979). De regionaals sociaal-economisch beleidskader (The region in a socioeconomic policy framework). Economisch Statistische Berichten, 64: 200–203.

AUTHOR INDEX

SUBJECT INDEX

of market, in continuous flow modeling,
257–258
under partial employment, 230
in service-provision systems, 243, 245, 247
Equilibrium service area radius, 240
Equity
criteria of, 248
versus efficiency, 394, 398
interregional, 394–396, 399, 400
Ergodicity, of transition matrices, in vintage
approach, 220
Estimation, and testing, techniques of, 341
Estonia, IMFB model for, 262–263
Euler equation, 253, 256
Evaluation criteria, for land allocations,
174–178
Expansion effect, in sectoral–regional
interdependence scheme, 352–353
Experimental design, 37–38
Export function, in Trunk model, 323
estimation of, 325–326

Facet (environmental) policies, 160–161
Factor coordination, in production and
consumption, 287–290
Factor supply considerations, 100
Federal agencies, in USA, 377
Federal taxes and spending, impact of on
regional development in USA, 379–381
Feedback, 34, 69, 97
interregional, 88, 91
regional–national, 91
Fermat's principle, 253
Financial aspects, in IMFB model, 260, 261,
265, 268
Financial flows, in IMFB model, 259, 260, 261,
262, 265
Firms, individual, behavior of, 97
First-stage decision, in multistage planning
situation, 190, 191
Fiscal policy
Federal, in USA, distributional aspects of, 380
regional, 292
Fixed costs, 218
FLEUR (Factors of Location in Europe)
model, 343
Flexibility, 186, 188, 190
Flow modeling, 251–258
Flows, of commodities, 257–258
Flows, in discrete and continuous models, 252
Following regions, 72
Forecasts, generation of in Silistra regional
model system, 303
France, INSEE regional projections for, 67
Freedom of action, 192, 193
expressions of desire for, 191
interpretations of in Swedish energy policy,
188, 190

limits on, 202
in Stockholm land-use–network-design model,
198, 200
in Swedish energy program, 186–187
Free-located activities, regional distribution of,
80
Friction coefficients, 347, 350, 351
Friction function, Tanner-type, 347
"Frostbelt" versus "Sunbelt" debate, 85
Fukuchi model, for Japan, 68, 76, 77n
Funck–Rembold model, for FRG, 66

Game, in subsidy allocation, 297
Game theory, in modeling of regional planning
process, 287, 291–292
Gaming models, 62
Gauss' theorem of divergence, 254–255, 258
General equilibrium analysis, 110
solutions in, in a static situation, 18
General equilibrium model, used by Committee
on Consequences, 366–367
General equilibrium models, spatial, 5–6
General interregional input–output model, 5
Generalized entropy method, 371
Generalized Regional Agriculture Model
(GRAM), see GRAM
Goal functions, 17, 182
Gödel's theorem, 30
Government current purchase function, in
Trunk model, 320
Government spending, effects of spatial shifts
in, 90, 92
Gradient, of potential function, 254
Gradient law, in continuous flow modeling, 256
GRAM, 211
overview of, 42–43
Graph, oriented, 271, 272
Gravity models, 33, 93
Gravity-type interregional input–output
models, 77
Gross migration, and decision-making, 116–122
Gross migration structure, 123
Gross national product, 107
of USA, 94
Gross profits, 218, 224
Gross profit share, 223
vintage classification of plants based on, 226
Growth
economic, multisectoral models of, 193
functional explanation of, 345
long-term, regional, 92
national and regional patterns of, 93
in postwar USA, 377, 379
Growth models, 9
Growth poles, 7
Growth rate, as function of regional profile,
342, 343